A History of the Native Woodlands of Scotland, 1500–1920

T. C. Smout, Alan R. MacDonald
and
Fiona Watson

EDINBURGH UNIVERSITY PRESS

© T. C. Smout, Alan R. MacDonald and Fiona Watson, 2005, 2007

First published in hardback in 2005 by
Edinburgh University Press Ltd
22, George Square, Edinburgh

This paperback edition 2007

Typeset in Minion and Gill Sans
by Pioneer Associates, Perthshire,
with corrections by Servis Filmsetting Ltd, Manchester, 2007 and
printed and bound in Spain by
GraphyCems, Spain

A CIP record for this book is available from the British Library

ISBN 978 0 7486 3294 7 (paperback)

The right of T. C. Smout, Alan R. MacDonald and Fiona Watson
to be identified as authors of this work
has been asserted in accordance with
the Copyright, Designs and Patents Act 1988.

Published with the support of the
Edinburgh University Scholarly Publishing Initiatives Fund.

Contents

	Acknowledgements	vii
	List of black and white maps	ix
	List of black and white figures	x
	List of colour plates	xii
	List of tables	xiii
1	Introduction	1
2	The extent and character of the woods before 1500	20
3	The extent and character of the woods, 1500–1920	45
4	Woodland produce	77
5	Woodland as pasture and shelter	102
6	Trading and taking wood before 1800	124
7	Managing the woods before 1770	157
8	Outsiders and the woods I: the pinewoods	192
9	Outsiders and the woods II: charcoal and tanbark	225
10	Woodland management in an industrial economy, 1830–1920 and beyond	258
11	Rothiemurchus, 1650–1900	290
12	The Navy, Holyrood and Strathcarron in the seventeenth century	319
13	The Irish and Glenorchy, 1721–40	340
14	The MacDonald woods on Skye, 1720–1920	364
15	Conclusion	388
	Bibliography	404
	Index	422

Acknowledgements

The research was made possible by the Economic and Social Research Council, assisted by supplementary grants from the British Academy and the Russell Trust, and an illustration grant from the Carnegie Trust. To all four we express our grateful thanks, as we do also to the librarians and archivists of the great national institutions, the National Archives of Scotland (NAS), National Library of Scotland (NLS) in Edinburgh, and the British Library and the National Archives (formerly the Public Record Office: PRO) in London, which held many of the manuscripts that we consulted. We also used papers and rare books at the University of Edinburgh, the University of Guelph, Canada, the University of Glasgow Business Records Centre, the University of St Andrews, Inverness Museum, Tain and District Museum and Clan Donald Lands Trust Visitor Centre, Skye, and owe a debt to them all. To the Campbell Preston family and their housekeeper Mary MacDonald we are grateful for access to the Campbell of Lochnell archives, to John Grant of Rothiemurchus for access to the papers at the Doune (and for much kind hospitality) and to the Duke of Argyll for access to papers at Inveraray Castle.

We should explain that the research was done under the auspices of the Institute for Environmental History at the University of St Andrews, equally by all three of us; Chris Smout was primarily responsible for writing most of the text, but Alan MacDonald for Chapters 12 and 14 and Fiona Watson for Chapter 13. Nevertheless it was all a joint effort of mutual criticism and encouragement. In the course of the work we were helped by a great many people, but we owe a particular debt to John Ballantyne who transcribed for us several hundred woodland sale deeds dating from the sixteenth century to the 1770s, from the Register of Deeds and the Gifts and Deposits in NAS, and the archival holdings of NLS. These related to woods from all over Scotland, a source hitherto unused and to which he drew our attention.

Those who helped us with various aspects of our research, large and small, are not too numerous to mention, but so numerous we fear we may have forgotten some. To all the following, and to any accidentally omitted, we express our thanks: Jane Begg, of the Woodland Trust; Robin Callander, of the Birse Community Trust; Hugh Cheape, of the National Museums of Scotland; Anthony Cooke, of Dundee University; Sandy Coppins, of the Lichen Mapping Scheme; Barbara Crawford, of St Andrews University; Professor Robert Crawford, of St Andrews University; Anne Crone, of AOC Scotland; Roy Dennis, crofter and naturalist; Pat Dennison, of Edinburgh University; Professor Robert Dodgshon, of the University of Wales; Angela Douglas, of the Woodland Trust; Alex Eaton, of Ullapool Museum; David Foot, formerly of the Forestry Commission; Diana Gilbert, of Highland Birchwoods; Bill Gilmour, Ullapool; Jeanette Hall, of Scottish Natural Heritage; Kate Holl, of Scottish Natural Heritage; Jonathan Hughes, of the Woodland Trust; James Kirby, Sunart; Keith Kirby, of English Nature; Professor Fred Last, of Edinburgh University; Margaret MacDonald, of Clan Donald Visitor Centre; Professor Allan Macinnes, of Aberdeen University; Neil MacKenzie, ecological and forestry consultant; Robin Maclean, Brig O'Turk; Donald McNeill, of the Forestry Commission; Professor Alan Mather, of Aberdeen University; John Mitchell, Drymen; Ross Noble, formerly of the Highland Folk Museum, Kingussie; Peter Quelch, of the Forestry Commission; Estelle Quick, of Tain and District Museum; Oliver Rackham, of Cambridge University; Alistair Ross, of Stirling University; Richard Saville, formerly of St Andrews University; Sue Scoggins, of Scottish Natural Heritage; Jim Skelton, Garmouth; Mike Smith, of Borders Forest Trust; Richard Smithers, of the Woodland Trust; Mairi Stewart, of Stirling University; Professor David Stevenson, of St Andrews University; Jack Stevenson, of the Royal Commission on the Ancient and Historical Monuments of Scotland; Domhnall Uilleam Stiùbhart, of Edinburgh University; Ron Summers, of the RSPB; Simon Taylor, of St Andrews University; Richard Tipping, of Stirling University; David Warren, Rothiemurchus estate; Professor Chris Whatley, of Dundee University.

The illustrations are acknowledged individually, on the captions, and we are especially appreciative of the skill of the St Andrews technicians who prepared them for publication. John Davey and James Dale of Edinburgh University Press guided the book through production with much patience. Finally, we could not possibly have done without the secretarial expertise and extreme care of Margaret Richards, who saw the entire project through from typing the first research grant application ten years ago to nurturing and polishing the final version. We know of people who do write books without her help, but it cannot be easy.

CHRIS SMOUT, University of St Andrews
ALAN MACDONALD, University of Dundee
FIONA WATSON, University of Stirling

List of black and white maps

1.1	The ancient semi-natural pinewoods of Scotland.	4
2.1	The major woodland types in Scotland 6,000 years ago.	27
3.1	Woodland in Glenesk, 1590–1946.	57
3.2	Woodland in Muckairn parish, Argyll, 1750 and 1876.	62
5.1	Distribution of shielings in the Forest of Mar.	112
6.1	Destination of pine timber sold in Rannoch, 1779–81.	150
8.1	Main native pinewoods known to have been subjected to commercial exploitation, 1600–1830.	194
8.2	Outlines of Abernethy forest, showing changes in woodland cover since 1750.	215
9.1	Charcoal ironworks with outsider interests.	228
9.2	Supplementary sources of charcoal for the Lorn furnace at Bonawe, 1786–1810.	245
11.1	Rothiemurchus Forest and Glenmore.	291
12.1	Strathcarron and neighbouring glens.	320
13.1	Woodlands purchased by the Irish partnership in Argyll, from 1721.	344
13.2	Glenorchy and Loch Etive pinewoods.	348
14.1	Woodlands on Skye.	365
15.1	Distribution of wood ants in Scotland.	396

List of black and white figures

1.1	Vera's four-stage cycle.	11
2.1	The Wood of Caledon from Ptolemy's description.	21
2.2	The oaken roof of Darnaway Castle.	38
3.1	Strathnaver according to Pont and Blaeu.	50
3.2	Ben Alder and its environs according to Gordon and Blaeu.	54
3.3	The woods of Glencoe according to Pont.	56
3.4	Plan of the battlefield of Glenshiel.	58
4.1	A birlinn on the Macleod tomb at Rodel, Harris.	82
4.2	The pine roof of Castle Grant.	85
4.3	Plan of a cruck-framed building at Corrimony, Inverness-shire.	87
4.4	An apparent creel house in the central Highlands.	94
4.5	Interior of a byre, Isle of Jura.	96
5.1	Cadzow's medieval hunting park, Lanarkshire.	107
5.2	Summer grazing of cattle on Loch Lomondside.	116
5.3	Strathmore in Sutherland, with a creel house.	121
6.1	Inveraray Castle, Argyll.	130
6.2	Timber-fronted buildings in Hyndford's Close, Edinburgh.	132
6.3	The ceiling of Gladstone's Land, Edinburgh.	133
6.4	The Forest of Mar and Glen Quoich.	145
7.1	Types of wood dyke in Sunart.	167
7.2	Ancient dyke and oak, Firbush, Lochtayside.	168
7.3	Heraldic seal of the Earls of Angus.	169
7.4	Plan for dyking an oak plantation in Argyll, 1754.	169
8.1	Glen Affric in 1929.	208
8.2	Floating wood down Strathglass, 1795.	210
8.3	Plaque from Glenmore presented to the Duke of Gordon, 1805.	219
9.1	Sir George Hay, first Earl of Kinnoull.	230
9.2	A view of Bonawe, Loch Etive, in 1836.	240
9.3	The buildings of the Lorn Furnace at Bonawe.	242
10.1	Turning bobbins at Gateside Mill, Fife.	265
10.2	The bucket mill at Finzean, Aberdeenshire.	267
10.3	The pyroligneous acid works at Balmaha.	268
10.4	The scaffolding of the Clyde shipyards.	273

11.1	John Peter Grant of Rothiemurchus.	303
12.1	Plan of the Cromwellian fortress at Inverness.	328
14.1	Plan of unenclosed woods on Sleat, Skye, 1763.	368
15.1	Niche preferences of wood ants.	397

List of colour plates

To be found between pages 178 and 179.

1. Map of the ancient semi-natural woodlands of Scotland.
2. A pinewood and pinewood flora.
3. The birches of Rannoch.
4. A hazel wood and its lichen.
5. Birch, aspen and holly in Assynt.
6. The forest at Rothiemurchus and montane scrub in Glen Doll.
7. Wood pasture in Glen Finglas and the Borders.
8. Regeneration at Inchnadamph and a bird tree at Glen Finglas.
9. Woodland cattle pasture and dykes.
10. Charcoal and bloomery sites in Sunart.
11. Standard and coppice oak in Sunart.
12. Planted oak and a woodland field in Sunart.
13. Veteran trees: sycamore, rowan and oak.
14. Healthy regeneration at Glen Affric and Creag Meagaidh.
15. Pine Marten and Red Squirrel.
16. Wood ants and their nest.

List of tables

1.1	Trees and shrubs native to Scotland.	2
3.1	Sinclair's estimates of natural woodland cover, 1814.	65
6.1	Import of deals into Scotland, 1765.	128
7.1	Management practice indicated from woodland sale deeds, 1600–1769.	161
7.2	'Good management practice' by region as indicated by woodland sale deeds, 1600–1769.	163
7.3	Examples of oak cutting dates.	175
7.4a	Pinewood sales: whole wood examples.	183
7.4b	Pinewood sales: specified quantity examples.	185
9.1	Bloomery sites in Scotland.	226
9.2	Price of oak tanbark delivered at Glasgow, ca. 1690 – ca. 1825.	249
9.3	Sales of coppice advertised in the *Edinburgh Advertiser*, 1790–1814.	253
10.1	'Official' statistics of the percentage of land under wood in Scotland, 1812–1908.	258
10.2	Price of oak bark per ton, 1830–1903.	260
10.3	Imports of timber into the UK in the nineteenth century.	275
13.1	Irish partnership's purchases in the West Highlands, 1721–3.	343
13.2	Values of Glenorchy and Loch Etive pinewoods, ca. 1721.	347

CHAPTER 1

Introduction

We need to start with a few explanations and definitions. 'Native woods' are those which are composed predominantly of tree species that have arrived in the Scottish landscape unassisted by human hand. A full list of native trees is given in Table 1.1. That list does not include a few species of introduced broadleaf that have been here for many centuries, like the sycamore and beech, though these have sometimes now naturalised themselves in the native woods. Nor does it include the exotic conifers that were planted across the countryside from the eighteenth century onwards, but especially in the twentieth century, at first larch and Norway spruce, then Douglas fir and a range of other North Americans, but above all in our time the great, dark, thick swathes, squares and cuboids of the Sitka spruce. The woods of which we write here were also for the most part both 'semi-natural' and 'ancient'. The first term means that they are self-sown, regenerating naturally on or close to the same spot with never a planted generation. The second means that they are directly descended on the same site from the original 'wildwood' of early prehistory, that developed while humans in Scotland were still hunter-gatherers. They are therefore, movingly, a living link to our oldest past. They are also very beautiful, full of light and flowers in summer, rich in the biodiversity of birds, mammals and insects, the older trees encrusted with moss, lichens and ferns. They are both an historical and a natural heritage of the greatest importance, and they deserve more protection than they currently receive.

Since the 1980s, inventories have been made which found that about 1 per cent of Scotland was still covered by ancient semi-natural woodland (Plate 1), and that another 1 per cent was covered by ancient woodland that had been converted into conifer plantations since the middle of the nineteenth century. The latter are sometimes referred to as 'Plantations on ancient woodland sites' (PAWS) and many are still capable of restoration to ancient semi-natural woods if the will and the money to do so can be found, since the seed bank of the native flora can survive for about fifty years. Such restoration is the only way of adding to the stock of these precious woods, and should be a conservation priority.

The native ancient semi-natural woods are by no means of the same

Table 1.1 Trees and shrubs native to Scotland

Common Name	Scientific Name
Common Alder	*Alnus glutinosa*
Ash	*Fraxinus excelsior*
Aspen	*Populus tremula*
Downy Birch	*Betula pubescens*
Silver Birch	*Betula pendula*
Dwarf Birch	*Betula nana*
Blackthorn	*Prunus spinosa*
Bird Cherry	*Prunus padus*
Wild Cherry (Gean)	*Prunus avium*
Elder	*Sambucus nigra*
Wych Elm	*Ulmus glabra*
Hawthorn	*Crataegus monogyna*
Hazel	*Corylus avellana*
Holly	*Ilex aquifolium*
Juniper	*Juniperis communis*
Pedunculate Oak	*Quercus robur*
Sessile Oak	*Quercus petraea*
Scots Pine	*Pinus sylvestris*
Dog Rose	*Rosa canina*
Guelder Rose	*Viburnum opulus*
Rowan	*Sorbus aucuparia*
Rock Whitebeam	*Sorbus rupicola*
Whitebeam	*Sorbus pseudofennica*
Whitebeam	*Sorbus arranensis*
Goat Willow	*Salix caprea*
Grey Willow	*Salix cinerea*
Eared Willow	*Salix aurita*
Woolly Willow	*Salix lanata*
Downy Willow	*Salix lapponum*
Tea-leaved Willow	*Salix phylicifolia*
Mountain Willow	*Salix arbuscula*
Whortle-leaved Willow	*Salix myrsinites*
Dark-leaved Willow	*Salix myrsinifolia*
Net-leaved Willow	*Salix reticulata*

Note: It is questionable whether or not Yew (*Taxus baccata*) may be native in the west.

character in every part of Scotland. The Forestry Commission classifies them into six types, to provide management advice for their owners. First are the Lowland mixed broadleaf woods, a mixture of oak, hazel, wych elm, alder and birch, characteristic of the rich soils of the east. Quite scarce, by contrast, are the upland mixed ashwoods, dominated by ash and hazel, and often with wych elm, oak, birch and rowan. The upland oakwoods are much commoner, though seldom reproducing themselves today, dominated by sessile oak mixed with birch: these are characteristic of Argyll and the acid soils of the west coast, through in a belt to the Trossachs and Perthshire and down to Galloway. The upland birchwoods are ubiquitous on acid soils at any elevation, mixed sometimes with other broadleaves like rowan, hazel, and goat willow. The native pinewoods are restricted to some 18,000 hectares of the Highlands, dominated by Scots pine but often with birch and sometimes other broadleaves (Map 1.1). The term Caledonian pine refers to the genetic strains of Scots pine characteristic of Scotland, but it is not one we have used in this book, preferring to keep to the specific name. Then there are wet woodlands, dominated by alder, willow, downy birch (or a mixture) and found in flushed, riverine soils.

These six types are only an indicative classification, for there is much local variation and intergrading. There may, however, very fairly be added another, the Atlantic hazelwoods that stretch along the exposed west coasts from Knapdale to Sutherland, and form a habitat unique to Scotland and a few other sites on the west coast of Ireland. In some ways these are the least changed of all our woods, the nearest we can hope to come to seeing the original wildwood.

Woodland communities are long-lasting but not unchanging, varying according to climate change, soils and management. Even in historic time there has, for instance, been far more birch and oak in the pinewoods, and far more other broadleaves such as holly, hazel and thorn in the oakwoods. The woods have also varied very greatly in their openness, from densities where the canopies of the trees touch, to 'wood pastures' where they stand well apart, the grassy sward beneath making good grazing for stock. This likewise can be a function of nature, for example trees thinning out at a higher elevation, or of management, where trees are kept this way for the benefit of the animals, and often pollarded in a distinctive manner.

Much of this book is about management. It is a book about the human use of woods rather than about the natural history or ecology of woods, though use interacted with the ecology and influenced the natural history.

It originated from a grant made by the Economic and Social Research Council (under its Global Environmental Change programme) to investigate sustainability in the use of semi-natural woodlands in the Scottish Highlands, 1600–1900. As is the way of things, when we set about our investigations the boundaries of our interests changed and widened. We became intrigued by woods outside the Highlands and by periods somewhat earlier than 1600 and a little later than 1900. The eventual starting date of 1500 was determined by the availability of documentary evidence (thin in the sixteenth century, but sufficient to draw certain conclusions), and the end date of 1920 by the commencement of operations by the Forestry Commission. After that, a chain of events was set in motion that so altered woodland history in Scotland that it will demand another set of authors to do it justice.

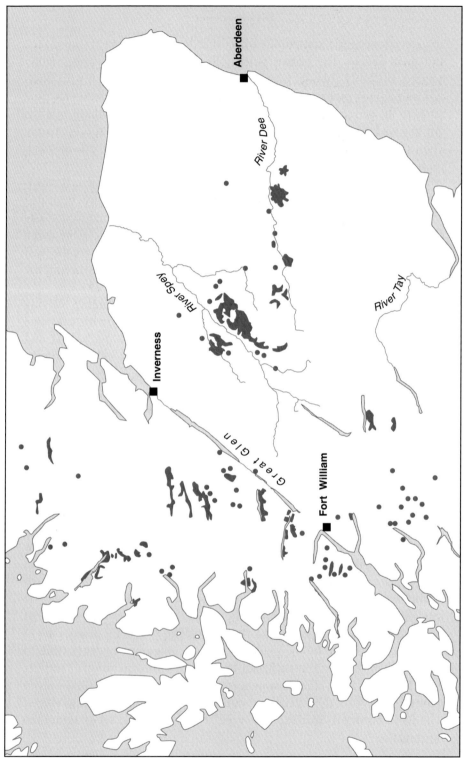

Map 1.1 The ancient semi-natural pinewoods of Scotland. Based on Caledonian Pinewood Inventory, Forestry Commission.

INTRODUCTION

The central question is this – were the Scottish woods used wisely by our ancestors: were they used in a sustainable way? To pose the question is to invite another – what do we mean by sustainability? Michael Jacobs has likened the concept to those older Enlightenment and nineteenth-century ideals of liberty, social justice and democracy: almost everyone is in favour, but deep conflicts remain about how they should be understood. 'In the 1990s it will be hard to find anyone to oppose the ideal; but this will still leave much room for disagreement over what it entails.'[1]

Sustainability is a notion that has already changed much since it originated as the essentially utilitarian creed of 'maximum sustainable yield', with the American Progressives in the presidency of Theodore Roosevelt, 1901–1909. Gifford Pinchot, forester and friend of the equally conservation-minded president, argued in his autobiography, *Breaking New Ground*, that prosperity could never be secure in a society that squandered its wealth. Natural resource management was to become a tool to achieve a large, sustained and repeatable crop:

> The purpose of Forestry, then, is to make the forest produce the largest possible amount of whatever crop or service will be most useful, and keep on producing it for generation after generation of men and trees ... A well-handled farm gets more and more productive as the years pass. So does a well-handled forest.[2]

A century later, that ideal seems inadequate and insensitive, however much an advance at the time on the devastating robber fellings of the timber barons in the American west 'in the traditional frontier style of grab, gut and git out'.[3] It entirely ignores the ecosystem of the forest. In Scotland, for example, this mode of thinking could be (and was) used to justify the clear-felling of native oak or Scots pine and their replacement with successively more productive crops of Sitka spruce or lodgepole pine. Pinchot and his school would have seen this as a natural resource – that is, land capable of growing wood – sustained and enhanced to produce new levels of timber production. We would be more likely to see it as a natural resource destroyed, that is a native wood, in all its biological complexity, replaced by something different and in every respect poorer, apart from the market value of the timber produced.

---

[1] M. Jacobs, *The Green Economy* (London, 1991), p. 60.
[2] Quoted in D. Worster, *Nature's Economy: a History of Ecological Ideas* (Cambridge, 1977), p. 267.
[3] *Ibid.*, p. 266.

Contemporary thinking in respect to sustainability, as enshrined in the Brundtland Report of the World Commission on Environment and Development in 1987, is at first sight hardly less fixated on human needs than Pinchot. Sustainable development it defines as 'development that meets the needs of the present without compromising the ability of future generations to meet their own needs'.[4] Thus, in promoting improvements in the standard of living, 'economic policy has to ensure that the environment is sustained for the sake of future people's welfare'.[5] An international consensus of rich and poor countries quickly appeared around Brundtland precisely because the report so willingly accepted the possibility of future economic growth while warning about the consequences of environmental misuse: 'We can have our cake and eat it, it seems to say ... everyone can be both rich and green'.[6] We would all like that.

Five years later, in 1992, at the Convention on Biological Diversity at Rio de Janeiro, the Brundtland Report was effectively glossed in two significant ways. Firstly, sustainability was seen as necessarily involving the maintenance of biodiversity, and secondly as involving the participation in and acceptance of environmental planning by local populations. The first was justified on the grounds that biodiversity is a resource that contributes to the healthiness of the planet as a whole, and that 'future practical needs and values are unpredictable';[7] the second on the grounds that, without local democratic ownership, conservation is doomed. Nothing can last if it is forced on a population against its will.

These recent definitions, of course, especially the first, would appear to rule out as 'sustainable' the destruction of a native oak or pine wood and its replacement by non-native conifers, as that process presumably alters the local biodiversity of the site. Biodiversity, it should be added, is defined as including 'diversity between and within ecosystems and habitats; diversity of species; and genetic variation within individual species'.[8] It is therefore diminished when, for example, a distinctive habitat is lost or reduced, a bird, a moth or a lichen is locally exterminated, or the genetic range within very variable species like Scots pine or juniper is lessened.

Differences in emphasis about the interpretation of sustainability, however, will remain, partly depending on whether one takes a local view, of the

---

[4] World Commission on Environment and Development, *Our Common Heritage* (Oxford, 1987), p. 43.
[5] Jacobs, *Green Economy*, p. 59.
[6] *Ibidem*.
[7] Department of the Environment, *Biodiversity: the UK Action Plan: Summary Report* (London, 1994), p. 2.
[8] *Ibidem*.

fate of an individual site or resource, or a wider view encompassing the fate of a total environment, in which a site is only one piece in a jigsaw. Thus one environmental economist has argued (albeit pre-Rio):

> A 'sustainably managed' forest... is one that can maintain its output over time. Strictly speaking, however, the relevant capacity for terrestrial resources is only that of the soil... If it is found that a forest is less useful than cultivated cropland, then it should be possible (subject to the requirement that sufficient trees are preserved to retain water and to absorb carbon dioxide, and valuable habitats are not destroyed) to switch from one to the other. The fertility of the soil would have to be maintained in this process, however... [so that] a future generation will have the opportunity to revert to the original cultivation pattern if it so wishes: no irreversible change is brought about... The actual balance of crops, trees, other vegetation and livestock should not be unduly constrained.[9]

That is a view that seems to put less emphasis on biodiversity, since it is virtually impossible not to 'destroy a valuable habitat' when a native wood is destroyed, and undoubtedly 'irreversible change is brought about' when an ancient forest is uprooted. This point seems to have been lost on the Department of Transport and the Regions when it authorised the 'translocation' of ancient woods on the route of the Channel Tunnel Rail Link in Kent, an action hailed by the minister as the way forward in reconciling planning and conservation needs, but derided by the Woodland Trust as 'simplistic at best and arrogant at worst'. As Oliver Rackham has shown, even medieval plantations of broadleaf trees on old forest soils in southern England have yet to develop the diversity of flora of woods descended (in altered form but in unbroken tenure of the ground) from the prehistoric wildwood.[10]

It is also useful to consider what has been described as the 'authenticity' of a wood, a concept close to its 'naturalness'. A forest high in authenticity will be composed largely of the original species of trees and other plants and animals that were there before human interference – it may not have them all, but it will not have many exotics; it will be patterned with different ages and sizes of trees, have a varied understorey, a multi-layered canopy,

---

[9] Jacobs, *Green Economy*, p. 89.
[10] G. Roberts, 'The relocation of ancient woodland', *Quarterly Journal of Forestry*, 24 (2000), pp. 305–12; Woodland Trust, 'Ancient woods and translocation', *Position Statement 19* (2000); O. Rackham, *The History of the Countryside* (London, 1986), pp. 67–9, 154.

open space and much dead timber; it will function naturally, with appropriate nutrient cycling, food webs, relationships between species including parasitism and symbiosis, and natural soil relationships; and processes such as regeneration and ageing, dying out in one place and spreading in another, replacement of one tree species with another, will be allowed to proceed unhindered.[11] Such a forest will of course, be entirely delightful. It will be like the Rothiemurchus pine forest on Speyside or the mixed birch and pine of Glen Affric, not unchanged from the past but high in all these qualities inherited from the past. The maintenance of authenticity, as well as the count of species implied in biodiversity, is a critical part of sustainability in woodland history.

It is, however, also true that any habitat that replaces or modifies the authentic or original wood will have its own ecological character. Tests in a thirty-year-old Norway spruce plantation near Oxford apparently found a density and variety of invertebrate life exceeding that of a native semi-natural oakwood, and work in Scotland has found considerable biodiversity in mature Sitka spruce plantations, notably for fungi, though the richest sites may have been ancient semi-natural woodlands in the recent past.[12] Whether the new ecosystem of exotic conifers is as 'interesting' or contains as many rare species overall as the old one is another matter. It is certainly less 'authentic', in the sense described. The ecological link to what was originally there in the aftermath of the last Ice Age (still present in what ecologists call 'semi-natural' form in the ancient native woods) is either broken or severely damaged. In any case, the replacement of a wood by heather moor or grassland – the usual option in upland Scotland before the twentieth century – would invariably have replaced a more diverse ecosystem by a less diverse one.

In assessing the economic and environmental performance of the past, it is important to appreciate the degree to which its mental horizons were partly, yet by no means entirely, our own. The general ideal enunciated in the Brundtland Report would have been immediately recognisable in earlier centuries. As Martin Holdgate has expressed it, 'the concept of "sustainable development" is less novel than has often been made out. It is, in fact, a synonym for "rational development", because it is a process of making the

---

[11] N. Dudley, 'Authenticity as a means of measuring forest quality', *Biodiversity Letters*, 3 (1996), pp. 6–9; N. Dudley, S. Stolton, J.-P. Jeanrenaud, *Towards a Definition of Forest Quality, a WWF Colloquium* (Godalming, 1993).

[12] C. Hambler and M. R. Speight, 'Biodiversity conservation in Britain: science replacing tradition', *British Wildlife*, 6 (1995), pp. 137–47; J. Humphrey and C. Quine, 'Sitka spruce plantations in Scotland: friend or foe to biodiversity?', *Alien Species: Friends or Foes?*, *Glasgow Naturalist*, 23 (2001), supplement, pp. 66–76.

best practicable use of natural resources for the welfare of people'.[13] It embodies no more than what at peasant level would have been seen as wise ancestral customs, 'adhered to because traditionally they were the only guarantee of survival'.[14] At proprietorial level it would have been seen as good stewardship of the land, part of the long view that resonates, for example, with laws of entail that limited the rights of heirs to sell property acquired by their forefathers. The eighth Earl of Lauderdale, the only Scottish aristocrat to enter the canon of the classical political economists, knew all about sustainability when in 1804 he wrote that 'the common sense of mankind would revolt at a proposal for augmenting wealth by creating a scarcity of any good generally useful and necessary to man'.[15]

To say this is not in any way to assert that in the past peasants were always wise users and landowners good stewards: merely that they would have found understandable the notion that nothing should be done by the present generation that would imperil the chances of the next. On the other hand, the gloss from Rio would have been incomprehensible in an earlier Europe. The notion of preserving biodiversity in particular would have been almost meaningless in a society that believed, with scripture in Genesis, that man had been given dominion over every growing plant and living creature, to use or extirpate them at pleasure, or with Francis Bacon that 'the world is made for man, not man for the world'.[16] Similarly, while the notion that change had to be acceptable to the local population might have worked in some peasant cultures, it scarcely determined the behaviour of the all-important landowning class. Scotland was a country where the rights of property were so strongly entrenched in law and practice as to be able easily to override traditional customs of their social dependants, to a degree exceptional elsewhere in Britain or Western Europe.[17]

On the other hand 'authenticity' would have struck a chord, though contemporaries would not have called it that. To the Romantics, but also to earlier generations, great natural woods were seen as obviously delightful, aesthetically pleasing and sometimes preserved for this alone.[18] Thus the

---

[13] M. Holdgate, 'How can development be sustainable', *Royal Society of Edinburgh Journal*, 143 (1995), p. 25.
[14] M. Redclift, *Sustainable Development: Exploring the Contradictions* (London, 1987), p. 150.
[15] James Maitland, eighth Earl of Lauderdale, *An Inquiry into the Nature and Origin of Public Wealth and into the Means and Causes of its Increase*, 2nd edn (Edinburgh, 1819), p. 44, quoted in H. E. Daly, *Steady State Economics*, 2nd edn (Island Press, Washington DC, 1991), p. 246.
[16] Worster, *Nature's Economy*, p. 20.
[17] T. C. Smout, 'Landowners in Scotland, Ireland and Denmark in the Age of Improvement', *Scandinavian Journal of History*, 12 (1989), pp. 79–97.
[18] T. C. Smout, 'Trees as historic landscapes', *Scottish Forestry*, 48 (1994), pp. 244–52.

pines round Loch-an-Eilein in Rothiemurchus were preserved from felling by the intervention of the laird's family who would not allow their common patrimony to be destroyed (see below p. 304). Yet in a contest between aesthetics and economics, at least in this period, economics was certain to win in nine cases out of ten.

We might therefore say that if early Scottish woodland management practices met the human needs of the time without compromising those of future generations, that was probably an intended consequence, but if they happened to preserve biodiversity or to be socially acceptable to the commonality, that was an accidental by-product. Furthermore, to be realistic, it is perfectly possible and even likely that a management regime which maintained and increased the 'maximum sustainable yield' of timber products even operating within the basic structures of ancient woodland would, to some degree, both damage biodiversity and offend the local population. As we shall see, this was certainly true in late eighteenth-century Scotland.

It is, however, one thing to identify damage to biodiversity or authenticity and another to measure it. The physical destruction of the woodland ecosystem can sometimes be detected, however loosely in most cases, by maps of woodland cover at different dates: but modifications within the wood, for example by adopting or abandoning coppice management in the case of broadleaved trees, or of tidying out dead timber and birch from a coniferous wood, can seldom be spotted in this way, though such management changes may affect the ecosystem profoundly. The richness of invertebrate fauna, for example, can be greatly damaged by clearing away rotting wood. Of the fate of individual species of any group in the Scottish woods we can know very little: approximate dates within our period for the disappearance of fewer than ten, and even for these we have no way of judging very much about the chronology of decline across Scotland. As for gene pool damage, it can hardly be detected except by chance reference to introductions – for example, the use of English acorns to plant or thicken Scottish oakwoods, or of the introduction of Scots pine of German rather than native provenance. But these problems should not blind us to the fact that biodiversity must have been profoundly and continuously affected by woodland management from earliest times.

With these considerations in mind, we might identify a range of ways of managing natural or semi-natural woodland which range from perfectly sustainable to absolutely unsustainable.

1. Light use: living in the 'wildwood'

Today, some hunter-gathering societies living within the forest allegedly leave it immodified in its primeval state, as exemplified by certain

Amazonian Indian tribes, though even in this case there are suggestions that at one time these rainforests were managed and manipulated. We may imagine that when human beings first arrived in Scotland in the Mesolithic era about 9,000 years ago, following the end of the last Ice Age and the gradual establishment of natural forest cover across Scotland, the wildwood, they were at first too few and too primitive in their technology to leave much of a trace on their environment: as hunter-gatherers, they also partook of nature's considerable primal bounty without much altering the world around them. The structure and extent of that 'wildwood' is still unclear and disputed, but the notion of wall-to-wall high forest extending across the horizon, broken only by marshlands, lochs and the tops of mountains, is too simple. In order to accommodate such grass-loving animals as the great aurochs, and such shade-averse shrubs and trees as hazel, it must have had much open space. Franciscus Vera has recently argued that we should think of a natural park-like landscape in western and central Europe, with grass and heath, isolated trees and 'groves' edged with thorny scrub, though such groves could extend to hundreds of hectares, gradually opening out and thinning in the centre. Such a landscape would be kept open by grazing animals, the prey of Mesolithic peoples (see Fig. 1.1).[19]

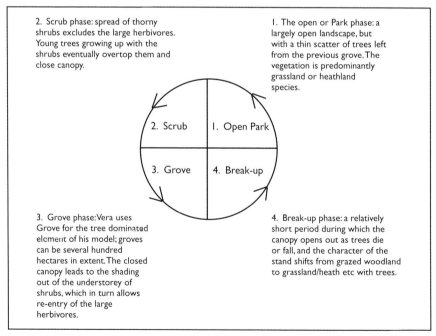

Fig 1.1 Vera's four-stage cycle. Reproduced by kind permission of English Nature.

[19] F. W. M. Vera, *Grazing Ecology and Forest History* (Wallingford, Oxon., 2000).

Others consider that this flies in the face of palaeoecological evidence that cannot find convincing traces of such park-like landscapes.[20] George Peterken, the influential woodland ecologist who favoured a view of high forest predominating and was also sceptical of Vera's conclusions, has recently suggested that the differences between them may be a matter of degree: 'perhaps we can agree that openings lasted longer and formed a higher proportion of the natural forest than most of us have envisaged, without having to accept that wood pasture was everywhere all the time'.[21] (See also pp. 33–4 below.)

We should remember, moreover, that the Mesolithic period lasted for 4,000 years, or twice the length of time since the birth of Christ, and neither the woods nor the habits and numbers of the people remained static. Take, for example, the Caithness Flows, which 7,000 years ago were lightly wooded; charcoal deposits found by archaeologists at the base of the peat indicate numerous woodland fires. About 6,500 years ago an alteration in the climate, making it much wetter and windier, inaugurated a period of peat formation that ultimately inundated the remnants of the woodland and left us with that extraordinary landscape of bog and water that has been so contested by foresters and conservationists in recent times. Mesolithic people here and elsewhere in Scotland may well have come to use fire and cutting systematically, as the Australian aborigines and the American Indians have done in their own environments, no doubt to improve grazing for the larger animals they hunted, and perhaps to encourage its wider colonisation by fruit-bearing shrubs and trees like bramble and hazel.[22] As they moved on, the clearings would revert to nature and the impact would be temporary unless grazing pressures remained high from the aurochs and deer. It is possible that at times and places they were on a scale and of a type that

---

[20] J. C. Svenning, 'A review of natural vegetation openness in north-western Europe', *Biological Conservation*, 104 (2002), pp. 133–48. But a paper by Paul Buckland, 'Holocene woodland history, a palaeoentomological perspective', presented at the conference at Sheffield Hallam University, May 2003, 'Working and walking in the footsteps of ghosts' (to be published), indicated that fossil beetle evidence provided 'some support for Vera's hypothesis on the variegated nature of Holocene forests'.

[21] Quoted in *New Scientist*, 7 Sept. 2002, pp. 225–6.

[22] C. R. Wickham-Jones, *Scotland's First Settlers* (Historic Scotland, London, 1994), pp. 60–1. For an English example, arguing that studies of the development of a peat-covered landscape in Devon 'clearly implicate Mesolithic communities', see C. Caseldine and J. Hatton, 'The development of high moorland on Dartmoor: fire and the influence of Mesolithic activity on vegetation change', in F. M. Chambers (ed.), *Climate Change and Human Impact on the Landscape* (London, 1993), pp. 119–31. For a cautious assessment, see R. Tipping, 'Living in the past: woods and people in prehistory to 1000 BC', in T. C. Smout (ed.), *People and Woods in Scotland: a History* (Edinburgh, 2003), p. 28.

could have interacted with climatic change to trigger the formation of peat: in the absence of trees, the ground would become wetter and more impacted by rain, and what had been traditionally sustainable would now become unsustainable.[23] It is possible – to put it no higher – that in the Caithness Flows and elsewhere, woodland clearance by people had already established the pre-conditions for the irreversible growth of blanket bog when the rainfall began to increase.

2. Deliberately modifying use: the creation and maintenance of 'semi-natural' woodland

With the arrival of Neolithic people, and the beginnings of systematic cultivation and stock-rearing about 5,000 years ago, we can clearly see how the use of fire and domestic animals exploited the existing mosaic of open ground and wood. Over time the extent of agriculture on open ground steadily increased, and that of woodland diminished. As the Neolithic advanced to the Bronze Age, and the Bronze Age to the Iron Age, each with its more sophisticated cutting tools, and the last-named with the first ploughs, the pace of change accelerated, sometimes to a dramatic degree.[24] It seems clear that at least in the Scottish Lowlands deforestation had already occurred on a large scale before the arrival of the Romans, and even in the Highlands the existence of a 'Great Wood of Caledon' appears more myth than reality.[25] The density and impact of the late Iron Age population has probably been underestimated. Indeed, for parts of England it has been argued that the extent of cultivation when the Romans arrived equalled anything in the medieval period, and there is no obvious reason why this should not have been just as true in Scotland.[26] Something of the scale of the human impact can be detected in the Bowmont valley in the

---

[23] I. G. Simmons, 'Vegetation change during the Mesolithic in the British Isles: some amplifications', in Chambers (ed.), *Climate Change and Human Impact*, pp. 109–18.

[24] I. G. Simmons and M. J. Tooley (eds), *The Environment in British Prehistory* (London, 1981).

[25] J. H. Dickson, 'Scottish woodlands: their ancient past and precarious future', *Scottish Forestry*, 47 (1993), pp. 73–8; R. Tipping 'A "History of the Scottish Forests" revisited', *Reforesting Scotland* 8 (1993), pp. 16–21; 9 (1993), pp. 18–21; D. J. Breeze, 'The great myth of Caledon', *Scottish Forestry*, 46 (1992), pp. 331–5, reprinted in T. C. Smout (ed.), *Scottish Woodland History* (Edinburgh, 1997), pp. 47–51. For a different view supporting the older opinion that the Romans themselves had a lot to do with the decay of the woods, see K. E. Barber, L. Dumayne and R. Stoneman, 'Climatic change and human impact during the late Holocene in northern Britain', in Chambers (ed.), *Climatic Change and Human Impact*, pp. 225–36.

[26] B. W. Cunliffe, 'Man and landscape in Britain, 6000 BC–AD 400', in S. R. J. Woodell (ed.), *The English Landscape, Past, Present and Future* (Oxford, 1985), pp. 54–67.

Borders where there was a catastrophic event in the late Iron Age, ca. 400–200 BC, 'of a totally different character to earlier prehistoric events' in the district, when great quantities of soil were washed down the valley into the floor of Yetholm Loch at its foot.[27] This removal of 'virtually all woodland in the pollen catchment, seemingly very rapidly', appears to have resulted in erosion on a scale reminiscent of the effects of the arrival of white farmers and loggers in nineteenth-century Australia and New Zealand. On the other hand, there is local evidence that here the Iron Age farmers learned by their mistakes, and began to adopt contour ploughing as a remedy.

The remaining woodlands, however, became increasingly domesticated. Farm stock grazed in them and modified them; they were used by people for fuel and building materials; traditions of management developed. We know very little about this process, but through it the 'wildwood' (untouched nature) became the 'semi-natural wood', still on the site of the wildwood and its lineal descendant, still with much (but not all) of its ancient flora and fauna, but essentially an environment altered and managed by human intervention and for human needs.

It is impossible to say when most of the wildwood perished in Scotland – that is to say, when the woods (or 'groves' in Vera's parlance) last flourished basically unmodified by human intervention. Even today, among certain of the Atlantic hazelwoods clinging as scrub to the steep sides of Hebridean islands and the adjacent mainland shores, it is possible to feel that little but time and nature has touched them in last 5,000 years: certainly sheep have grazed there, but if people have otherwise used them at all it has been on a casual and unsystematic basis, and a few of the world's rarer lichens remain there as a result.[28] They are, however, a fragment of a fragment. By the eighteenth century it would have been difficult to find any more trace of the wildwood than we can find today, and the original high forest was already gone. Maybe, however, the wildwood high forest had still existed in patches as late as the end of the sixteenth century. There is something about the following description of the woodlands clothing Loch Maree, (also known at this time as the fresh Loch Ewe), probably written by the great Scottish cartographer Timothy Pont in the 1590s, which suggests the magnificent scale and variety of the primeval. Not least is the sense of the

---

[27] R. Mercer and R. Tipping, 'The prehistory of soil erosion in the northern and eastern Cheviot hills, Anglo-Scottish borders', in S. Foster and T. C. Smout (eds), *The History of Soils and Field Systems* (Aberdeen, 1994), pp. 1–25.

[28] A. Coppins, B. Coppins and P. Quelch, 'Atlantic hazelwoods: some observations on the ecology of this neglected habitat from a lichenological perspective', *British Wildlife*, 14 (2002), pp. 17–26.

writer's excitement in seeing something exceptional, though his antennae were naturally tuned to its economic potential rather than its ecological richness:

> Upon this Lochew, do grow plentie of very fair firr, hollyn, oak, elme, ashe, birk and quaking asp, most high, even, thicke and great, all-longst this loch … [it] is compas'd about with many fair and tall woods as any in all the west of Scotland, in sum parts with hollyne, in sum places with fair and beautifull fyrrs of 60, 70, 80 foot of good and serviceable timmer for masts and raes, in other places ar great plentie of excellent great oakes, whair may be sawin out planks of 4 sum tyms 5 foot broad. All thir bounds is compas'd and hem'd in with many hills, but thois beautifull to look on, thair skirts all adorned with wood even to the brink of the loch for the most part.[29]

After that date, however, we are left with the semi-natural woodlands which are the focus of this book, though some of them were undoubtedly more managed and modified than others. Nor is Scotland alone in the loss of the wildwood: it is doubtful if it survived far into the Middle Ages in England. Today it is virtually unknown in Europe. The Białowieza forest of oak, spruce and hornbeam on the borders of Poland and Belarus is close to being one remarkable survival, though it, too, has probably changed its character drastically in the last two centuries as a result of human intervention. To see a truly natural coniferous forest it is probably necessary now to travel as far as the Pechora river on the western side of the Urals in north-central Russia; even there its future is very insecure.[30]

The shift from the wildwood to the semi-natural wood is marked by several changes – eradication or severe diminution of the original girdles of scrub; reductions in the diversity of species among the trees; alterations in the age structure and character of trees of the same species; a general reduction in biodiversity among the vertebrate and invertebrate fauna, and among the flora, of the wood. It becomes somewhat less 'authentic', probably less varied; to the human eye, less disorderly. In the Scottish context the wood is likely to become less mobile, upland birch and pine less able to move from one stance in the enfolding open ground to another, Lowland broadleaf woods confined within enclosures among cultivated

---

[29] A. Mitchell (ed.), *Geographical Collections relating to Scotland made by Walter Macfarlane* (3 vols, Scottish History Society, Edinburgh, 1906), 2, pp. 539–40.
[30] Rackham, *History of the Countryside*, pp. 68–73; Vera, *Grazing Ecology*, pp. 247–55, 271–3; D. Jenkins, *Of Partridges and Peacocks* ([Aboyne] 2003), pp. 264–70.

ground. Characteristics of management regimes in the semi-natural woods in Scotland include:

(a) coppicing, for example of oak or hazel, often on rotation.
(b) rotational clear-felling cutting by 'hags' or coups, without organised coppice, but often with regrowth.
(c) coppice-and-standard, where selected 'maiden' trees are allowed to remain among the coppice-stools.
(d) selective cutting of marketable trees, leaving the remainder standing.
(e) clear-felling of all trees except for selected seed sources, for example 'granny pines'.
(f) forms of wood pasture that allow animals into the wood, usually on a seasonal basis.

All six methods (some are not mutually exclusive) are capable of being reproduced from generation to generation in a perfectly sustainable way, and will come to be associated with relatively stable and diverse ecosystems within the woods. For them to be stable and successful, however, it will be necessary either to enclose the woods for a period against stock, 'haining' as it was called, or to conduct operations with a relatively low density of grazing and browsing animals controlled by strict shepherding and herding. Otherwise these methods become, perhaps gradually and almost imperceptibly, fatally destructive of the wood. The details and the success of semi-natural woodland management may vary even on the same site from one period to another, depending on the quality of the managers themselves, on the varying pressures of human and animal populations, and on the market for timber produce compared to that for other potential produce from the same ground.

The introduction of one regime to an area where it was not known before, for example oak coppicing accompanied by the weeding out of less profitable trees like birch, holly and willow, may damage biodiversity and make that wood less 'natural' than before. It has usually been assumed, however, that coppicing, once introduced, is a perfectly sustainable regime of woodland management that can be continued indefinitely, with its own characteristic and diverse biodiversity related to that of its predecessor, though not identical. This has been questioned, with doubts thrown on the long-term viability and on the richness of the biodiversity of coppiced woods.[31] This is most likely to be the case under conditions like those of the

---

[31] Hambler and Speight, 'Biodiversity conservation in Britain'. For a reply, see R. J. Fuller and M. S. Warren, 'Management for biodiversity in British woodlands – striking a balance', in *British Wildlife*, 7 (1995), pp. 26–37.

west of Scotland where the continuous removal of woody material combined with the effects of heavy rainfall on acid soils might have a more deleterious effect than in the very different conditions of, say, the English Weald. On the other hand, if the coppicing was intermittent or short-term, any damage would be proportionately limited.

3. Radical interference: the plantation

When people take it upon themselves to plant trees within an old wood, as opposed to permitting natural regeneration to take place of its own accord, modern forestry begins to replace traditional woodmanship. 'Authenticity' takes a back seat and is probably regarded as a sign of poor silviculture if it is allowed too much prominence. Again, however, there is a spectrum of management.

(a) Minor planting by hand in a semi-natural wood. This may amount to no more than filling in the gaps that have appeared through overuse in one of the management options under 2 above, and scarcely differs from 'deliberately modifying use'.
(b) Replacement of a semi-natural wood after clear-felling by hand-planted native trees of the same species. In some circumstances this may also differ only in degree from the options under 2, though probably the effect on biodiversity and certainly on naturalness will be greater.
(c) Replacement of semi-natural wood with hand-planted exotics. This might be the replacement of the entire wood or of a part of it. In the eighteenth and nineteenth centuries, beech, larch or Norway spruce would be the non-native species involved: in the twentieth century, typically American species such as Sitka spruce, lodgepole pine or Douglas fir. Depending on how much of the original wood was affected, this could have either a fairly minor impact, or turn the semi-natural woodland into a substantially different ecosystem, and therefore in the latter case be unsustainable from our perspective. It would be perfectly possible, of course, to maintain on the same site a sustainable plantation of the new trees, with its own distinctive biodiversity: but the native semi-natural woodland would be lost. On the other hand, once the exotics are felled, it might be possible to restore the land to semi-natural woodland, especially if specimens of the earlier wood were allowed to stand among the exotics and if the seed bank of the earlier flora persists. Woods of this kind are known as PAWS (Plantations on ancient woodland sites) and represent an

opportunity for restoration following clear-fell, but not one that will last for ever. After fifty years, little of the original seed bank will remain. The Forestry Commission has restored such woods at Glenmore and Glengarry with considerable initial success, as part of a programme of restoring native pinewoods, though the restored woods are likely always to have a proportion of regenerated Sitka. Nevertheless, there is an important window of opportunity which, if used in time, will now permit the recovery of a good deal that was lost in the twentieth century. It will not be open for long.

(d) Replacement of semi-natural wood with exotics after deep ploughing and fertilising. In this case the relics of the old semi-natural wood have been even more thoroughly erased from the landscape, as such traces of the earlier ground flora as might have survived earlier hand-planting would be less likely to survive mechanical working and fertilising. However, such practices only became possible in the decades after the Second World War, and even then they were scarcely practicable on recently felled semi-natural woodland because of pre-existing roots. They were best suited to the open moor, and only used on old woodland sites with rotted stumps which had been felled decades earlier and prevented from regrowth by grazing animals.[32] In this case, the damage to the original ecosystem had already been largely done.

In effect, therefore, it is the first three phases of this spectrum which were significant.

4. Extractive use: woodland becomes moor or arable

If a wood is destroyed by overgrazing, felling or burning, and no attempt is made to permit its regeneration, its management is obviously not sustainable. This can also occur in different ways:

(a) When a wood is allowed to decay standing, and grazing pressure prevents its natural regeneration before it finally collapses. In this case the unsustainability of its management may not be evident for generations, and indeed it could be brought back from the brink of extinction after a century or more of neglect, providing a few seed trees continue to bear and the ground cover permits their germination. Nevertheless, in the long run the tooth of too many animals is

---

[32] D. Foot, *pers. comm.*

the surest way to destroy a wood. Traces of the original flora may remain on the open moor long after the wood has gone, as witnessed by the wood anemones on many Highland braes.
(b) When a wood is cut or burnt and the pressure of grazing animals around is too great to allow for natural regeneration, even though no deliberate conversion to pasture is intended. The same traces of the original ground flora may again persist, though if the grazing is heavy they may be few and far between.
(c) When a wood is deliberately grubbed up to convert to arable or pasture, the land thereafter being generally ploughed, sown and treated with chemicals so that no further traces of its origins remain.

In the above schema, therefore 1–3(b) can be regarded as sustainable ways of managing natural or semi-natural woodland: light hunter-gathering activity in the wildwood and wise use of the managed semi-natural wood, along with plantation as simple patching of semi-natural wood. When the line is crossed into more intensive plantation 3(c) it is unsustainable in our terms but very possibly reversible. When it moves into the extractive use of the wood, whatever the commercial or other economic justification, the management regimes will no longer sustain in any considerable or recognisable way the ecological characteristics of the original wood: in our terms 3(d)–4 are 'unsustainable'. Even so, in many circumstances it may still be possible by planting native trees ultimately to restore the vanished wood to some semblance of its former self though not to its original richness. In 4(c) and in the most overgrazed conditions in 4(b), however, this will be next to impossible except with a time horizon of many centuries, as native woodland flora will not flourish on chemicalised ground and only returns extremely slowly even to ground under unfertilised pasture, unless there are pre-existing seed banks. There is a fine line between most of the management regimes in 2 and those in 4: between sustainability in the semi-natural wood, and its extractive exploitation. Moreover, many practices in the management of semi-natural woods will reduce sustainability (as understood at Rio), at least by damaging biodiversity. It is the circumstances under which this fine line is crossed or avoided, and more generally the impacts of management and the causes of their introduction and abandonment, which are the special foci of this book.

CHAPTER 2

The extent and character of the woods before 1500

The notion of an ancient Great Wood of Caledon that in Roman times covered most of Scotland, or at least the whole of the Highlands, runs ineradicably deep in the Scottish mind.[1] The story is based partly on the positioning of 'Caledonia Silva' in Ptolemy's second-century AD geographical account, available on printed maps in western Europe from 1475, though it occupies variable space: Waldseemüller's edition of 1513 has it in less than half the area of Blaeu's *Atlas* of 1654 (see Fig. 2.1).[2] Ptolemy's account was in turn based on information supplied by the Roman invaders and also written up by Tacitus, Dio and other historians of their inconclusive campaigns, and recent critical commentary has shown how extremely vague the authors of antiquity were about the location of the Caledonian Wood or woods. Perhaps the notion has to be seen against a Roman tradition of explaining away military defeat, as again in Germany, by the fearsome terrain.[3]

Stories of a vast and terrifying forest were revived again in the Renaissance by the Aberdeen historian Hector Boece, who in 1527 put his own spin on them: for him, the Roman wood had stretched north from Stirling and covered Menteith, Strathearn, Atholl and Lochaber, and was full of white bulls with 'crisp and curland mane, like feirs lionis'. There is no evidence in classical sources for this embroidery, but in any case the wood was located firmly in the past by Boece. It was referred to in similar terms by the Elizabethan, William Camden, who added witches and bears for good measure, and by the late seventeenth-century Scottish topographer, Sir Robert Sibbald, who said that nothing remained but inconsiderable

---

[1] T. C. Smout, *Nature Contested: Environmental History in Scotland and Northern England since 1600* (Edinburgh, 2000), pp. 37–46.
[2] Royal Scottish Geographical Society, *The Early Maps of Scotland* (edn Edinburgh, 1973), pp. 6, 20; H. M. Steven and A. Carlisle, *The Native Pinewoods of Scotland* (Edinburgh, 1959), p. 48.
[3] D. J. Breeze, 'The great myth of Caledon', *Scottish Forestry*, 46 (1992), pp. 331–5.

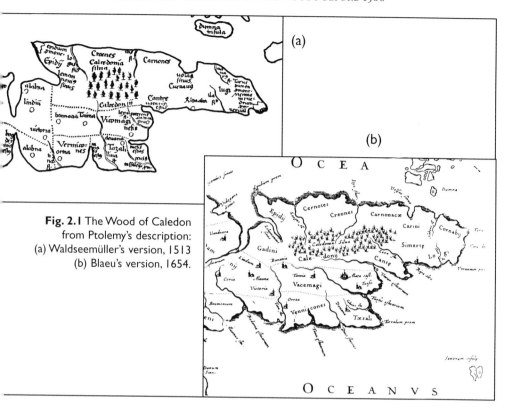

Fig. 2.1 The Wood of Caledon from Ptolemy's description: (a) Waldseemüller's version, 1513 (b) Blaeu's version, 1654.

vestiges.[4] Interestingly, Coronelli's map of *Le Royavme D'Ecosse*, published in Paris in 1708, showed several large woods rather uncertainly placed, and named one of the biggest (very roughly centred on Glenorchy and the Black Mount) the 'Caledonia Forest'. Travellers and antiquarians casually mention the Caledonian Wood from time to time in the later eighteenth and earlier nineteenth centuries, as something of which remnants might occasionally be found, either as tree-trunks in the bogs bearing, as they fancied, traces of Roman fire or felling, or as circumscribed standing woods and groves. Sir John Sinclair's *General Report* of 1814 placed the wood in southern Scotland, stretching to the English border with Callander wood and the Torwood as its remnants, recounting stories of 50,000 men lost by Severus in its depths and of 24,000 axes used against it by John, Duke of

---

[4] P. Hume Brown (ed.), *Scotland Before 1800 from Contemporary Documents* (Edinburgh, 1893), pp. 80–1; W. Camden, *Britannia: or a Chorographical Description of the Flourishing Kingdoms of England, Scotland and Ireland* (edn by R. Gough, London, 4, 1806) p. 107; R. Sibbald, *Scotia Illustrata* (Edinburgh, 1684).

Lancaster in his medieval campaigns against the Scots.[5] On the other hand, it struck no particular chord either with the Romantic poets or with Sir Walter Scott.

The Great Wood took on an altogether new lease of life in 1848, when the main ingredients of the modern myth were supplied by the Sobieski Stuarts in their best-selling *Lays of the Deer Forest*. These two brothers (in reality named Allan) were of Anglo-German descent and claimed to be legitimate grandsons of Prince Charles Edward. Even more preposterously, they claimed to be experts in the Gaelic past, and devised a compilation of poetry, aristocratic and romantic hunting stories, alleged clan histories, supposed folklore and natural history. The tone was Teutonic, and their talk of the woods entirely in the German romantic tradition of the link between the forest and the *Volk*. They spoke of a *Caledonia Silva*, 'the great primeval cloud which covered the hills and plains of Scotland before they were cleared', and its 'skirt', a great forest filled with game and wolves that had occupied the Province of Moray, much of it surviving until recent times, when ruthless and greedy modern man had swept it away.[6]

The tale was taken up by the academic historian, W. F. Skene, when he invited his readers to compare Severus's campaign against the Caledonians to that of General Sir Harry Smith's against the Kaffirs of Cape Colony, and averred (without evidence) that much of Roman Scotland 'presented the appearance of a jungle or bush of oak, birch or hazel'. Skene had surely been encouraged to think in these terms by Patrick Fraser Tytler's earlier *History of Scotland*, which spoke of the face of the country even in medieval times 'covered by immense forests chiefly of oak', even though he made no reference to *Caledonia Silva*.[7]

Tytler and Skene in turn influenced David Nairne, who gave a generally scholarly and much cited address to the Gaelic Society of Inverness in 1891. Nairne began by declaring himself not among those who believed, with the historians, that when the Romans arrived 'Scotland was one dark and dreary forest, as impenetrable as that of Central Africa, and inhabited by a race only a little bigger and scarcely less savage'. He pointed out that the natives

---

[5] J. Sinclair (ed.), *General Report of the Agricultural State and Political Circumstances of Scotland* (5 vols, Edinburgh, 1814), 4, pp. 465–7. More generally, see e.g. G. Chalmers, *Caledonia, or a Historical and Topographical Account of North Britain from the Most Ancient to the Present Times* (edn Paisley, 1887); J. E. Bowman, *The Highlands and Islands, a Nineteenth-century Tour* (Gloucester, 1986), pp. 161, 163 (Bowman travelled in 1827); L. Shaw, *History of the Province of Moray* (edn Glasgow, 1882), 2, p. 11.

[6] C. E. Stuart and J. S. Stuart (alias Allan), *Lays of the Deer Forest* (Edinburgh, 1848), 2, pp. 220–1, 256–7.

[7] W. F. Skene, *Celtic Scotland* (Edinburgh, 1876), 1, pp. 84–6; P. F. Tytler, *History of Scotland* (Edinburgh, 1829), 2, p. 199.

fought in chariots, and chariots demand open country. Nevertheless, he said, 'Scotland was then a tree-grown country, with its greatest forest extending into Badenoch and Strathspey, and ramifying into every Highland strath until it spread over Sutherlandshire, and vanished in the sterility of Caithness.' Despite his disclaimer, he then took off into a flight of fancy worthy of the Sobieski Stuarts, inviting the reader to come with him 'in hurried review' into the natural pine forests, the haunt of bears, wolves and reindeer and the happy hunting grounds of Caledonians: 'Here is Scottish freedom in embryo.' In the Roman invasion the natives made their stand as 'the legions cut their way through the pathless tracts of Strathspey... Victorious! But at what a cost... 50,000 men of the invading hosts have fallen as the trees they felled.'[8] That the dominant image of a Caledonian forest today is one of Caledonian pine is due solely to a coincidence of names and to the manner in which Nairne located the heart of the Great Wood in Speyside.

Nairne attributed the destruction of the ancient woods that once covered the Highlands partly to incessant Dark Age and medieval warfare – 'it was always a sweet revenge to see the sky ruddy with the glare of flames in an enemy's country' – partly to the seventeenth- and eighteenth-century Scottish and English speculators in wood for naval and other purposes, partly to the depredations of ironmasters with their blast furnaces in the same period, and partly to the replacement of black cattle by sheep in the nineteenth century.[9] This account was little challenged for over eighty years and indeed received a considerable boost from the writings of Frank Fraser Darling in the 1940s, who retold the story in essentially the same terms as Nairne, though with a greater emphasis on capitalist ironmasters and speculators, and made it the hinge of his explanation for the ecological deterioration of the Highlands, the collapse of a harmonious, wood-covered, diverse ecosystem into an impoverished 'wet desert'. He said 'the imagination of a naturalist can conjure up a picture of what the great forest was like: the present writer is inclined to look upon it as his idea of heaven'.[10] This endorsement by so charismatic a twentieth-century ecologist made it orthodoxy for the Green movement in the 1980s,[11] and the apotheosis of

---

[8] D. Nairne, 'Notes on Highland woods, ancient and modern', *Transactions of the Gaelic Society of Inverness*, 17 (1892), pp. 170–221. The passages cited are on pp. 171–2.
[9] Nairne, 'Notes on Highland woods', pp. 173–5, 193–5, 220.
[10] F. F. Darling, *Natural History in the Highlands and Islands* (London, 1947), pp. 57–8; see also F. F. Darling, *Pelican in the Wilderness* (London, 1956), pp. 20, 180, 353; F. F. Darling, 'History of the Scottish forests' (1949), reprinted in *Reforesting Scotland*, 7 (1992), pp. 25–7.
[11] Scottish Green Party, *A Rural Manifesto for the Highlands: Creating the Second Great Wood of Caledon* (Inverness, 1989).

the story came, incidentally in the centenary year of Nairne's original talk in Inverness, when Hugh Miles made a prize-winning film and wrote an accompanying book with Brian Jackman, *The Great Wood of Caledon*. The original author would have recognised most of this extract, though no doubt he, too, would have cringed at the tangled errors of the final sentence:

> The Great Wood of Caledon . . . had already been standing for at least 2,000 years when Stonehenge was raised . . . there was hardly a glen that was not roofed with trees, the high hills rising like islands from the blue-green canopy. The wolf and the lynx roamed its trackless deeps. Bears and wild boars snuffled among its roots . . . When the Romans came to Britain it became a refuge for the Pictish tribes who waged guerrilla war on the imperial legions . . . Then came the Vikings, and the war on the Great Wood began . . . they torched the forests, and felled the tall trees to fashion masts for their longships . . . For the Highland Scots the trees were fuel, and huge gaps appeared in the forest canopy as feuding clans burned the woods of their enemies. Yet still the Great Wood stretched for miles, a sanctuary for wolves and renegades alike until the English arrived to smoke them out . . . the crushing of Bonnie Prince Charlie's Highland rebels signalled the end of the glory of Caledon. Down came the mighty trees, felled by the impoverished clan chiefs, who, forced to pay off their hated new Hanoverian landlords, sold their timber to English ironmasters.[12]

By then, however, scholars had long begun to reassess the details of the traditional picture, conveyed in turn so compellingly by Boece, Nairne and Fraser Darling. As early as 1861, Cosmo Innes had already called the Great Wood of Caledon a 'myth', deriving from Pliny who 'liked to place his marvels in inaccessible situations'. Hugh Boyd Watt in 1898, though he believed the Romans had found a Great Wood, also believed it much more circumscribed than 'modern writers', who had extended it from Galloway and Selkirk to Sutherland, and from Lochaber to Buchan – 'an instructive example of the growth and development of an historical fact, until the fact sinks into insignificance as compared with its accretions'.[13] First of the modern critics was James Lindsay, in an important Edinburgh Ph.D. thesis

---

[12] H. Miles and B. Jackman, *The Great Wood of Caledon* (Lanark, 1991), pp. 11–12.
[13] C. Innes, 'Some account of early planting in Scotland', *TRHAS*, new series, 9 (1861), pp. 40–53; H. B. Watt, 'Scottish forests and woodlands in early historic times', *Annals of the Andersonian Institute*, 2 (1900), pp. 89–107.

of 1974, unfortunately never published in full, which questioned in particular Fraser Darling's assumption that the prime culprits for the deforestation of Scotland were external timber speculators and ironmasters.[14] Next, paleoecologists, using a whole range of new techniques including palynology (pollen analysis), dendrochronology (tree-ring analysis), and radio-carbon dating, began to suggest that deforestation, due partly to human intervention and partly to climatic change, had already proceeded far at the point of Roman arrival. As late as 1967 when Mark Louden Anderson wrote his *History of Scottish Forestry*, it was still assumed, as it had been for centuries, that the remains of excellently preserved tree trunks and roots in peat bogs all over Scotland were comparatively recent, and therefore compelling evidence of forests in modern times.[15] We now know that most of them are over 4,000 years old.

Recent articles by James Dickson and Richard Tipping have helpfully summarised the present state of knowledge of the Scottish woodlands in prehistory: both authors emphasise the extent of uncertainty in any detailed picture, particularly in respect of the Highlands.[16] No trees, except possibly juniper, appear to have survived the final episode of the last glaciation in Scotland, the so-called Loch Lomond Stadial, which abruptly ended around 8300 BC. The climate and soil conditions then rapidly became suitable for tree cover, but their invasion of Scotland from southern Britain and the Continent (still joined to the British Isles by land-bridges) occurred only gradually over thousands of years. First to emerge as ground cover was juniper, together with species of willow and rowan, followed quickly by birch, which about 11,000 years ago fragmented the cover of juniper, and then, 800 years after that, hazel spread. Elm and oak arrived in the period 6500–6000 BC. Scots pine first appeared in abundance in Wester Ross in a genetically distinct population, around 7000 BC: it may have come from some isolated population from an area to the west of mainland Scotland

---

[14] J. M. Lindsay, 'The use of woodland in Argyllshire and Perthshire between 1650 and 1850' (unpublished University of Edinburgh Ph.D. thesis, 1974). For a summary of his approach, see J. M. Lindsay, 'The commercial use of woodland and coppice management', in M. L. Parry and T. R. Slater (eds), *The Making of the Scottish Countryside* (London, 1980), pp. 271–90.

[15] M. L. Anderson, *A History of Scottish Forestry* (Edinburgh, 1967), 1, pp. 66–71.

[16] J. H. Dickson, 'Scottish woodlands: their ancient past and precarious future', *Scottish Forestry*, 47 (1993), pp. 73–8; R. Tipping, 'A "History of Scottish Forests" revisited', *Reforesting Scotland*, 8 (1993), pp. 16–21; 9 (1993), pp. 18–22; R. Tipping, 'The form and fate of Scottish woodlands', *Proceedings of the Society of Antiquaries of Scotland*, 124 (1994), pp. 1–54; R. Tipping, 'Living in the past: woods and people in prehistory to 1000 BC', in T. C. Smout (ed.), *People and Woods in Scotland: a History* (Edinburgh, 2003), pp. 14–39. Tipping's work has been particularly drawn on in the pages that follow.

now under the sea, that had survived the final glaciation, or perhaps from Ireland. Pine then greatly expanded its range in the following millennia, apparently mainly from a different genetic source from the east or south. It is uncertain from which source the Cairngorm forests were derived.[17]

The forests reached their fullest extent around 4000–3000 BC, grouping into four main types (Map 2.1) which broadly follow the pioneering vegetation classification of McVean and Ratcliffe developed in 1962.[18] Birch and hazel scrub dominated the Outer Hebrides, Orkney, Shetland, Caithness and northern Sutherland. Pine and birch woodland were predominant in the Highlands, except for most of its coastal margins and the southern edge, reaching much higher up the mountains than it generally does today, even to around 2,500 feet in the Cairngorms. Birch and hazel and oak (with less oak in the north and west) dominated the western Highland littoral south of Loch Maree or Loch Torridon, and the Inner Hebrides; the same forest community (again with less oak in the north and east) also was characteristic of the north-east plain, the shores of the Moray Firth and the southern Highland glens in Angus and Perthshire. Finally, oak, hazel and elm dominated the rest of Scotland, sweeping over the Lowland plains and over even the tops of the Southern Uplands: 'there seems little reason to think that any unforested area need have existed south of the Forth/Clyde line at the maximum extent of woodland'. Yet the cover was not as uniform as the map may suggest, as the composition of the woodlands changed constantly over time (plant communities are not static entities) and small differences of climate, geology and topography even over a short distance 'would have introduced an astonishing beauty, richness and diversity of woods within individual valleys'.[19] Nor should we think of the forest cover as anything like 'a roof of trees' or a 'blue-grey canopy'. Such was the influence of grazing, rainfall and fire that it is more likely to have been an area where grassland, heath and marsh at least came close to equalling the extent of the trees. Precise estimates of the amount of wood under forest at the woodland maximum are not practicable, but James Ritchie long ago, by analogy to Scandinavia, hazarded the guess of 54 per cent, and Fraser

---

[17] Tipping, 'Form and fate', pp. 9–10; K. D. Bennett, 'Late Quaternary vegetation dynamics of the Cairngorms', *Botanical Journal of Scotland*, 48 (1996), pp. 51–63; K. D. Bennett, 'Post-glacial dynamics of pine', in J. R. Aldhous (ed.), *Our Pinewood Heritage* (Farnham, 1994), pp. 23–39.

[18] D. N. McVean and D. A. Ratcliffe, *Plant Communities in the Scottish Highlands* (Edinburgh, 1962). Their work was based on 'the present distribution of woodland and on place-name evidence', but modern palynological studies 'have tended to confirm in broad terms McVean and Ratcliffe's reconstruction': Tipping, 'Form and fate', p. 11.

[19] Tipping, 'Form and fate', p. 14; Tipping, 'Living in the past', p. 24.

Extent and character of the woods before 1500

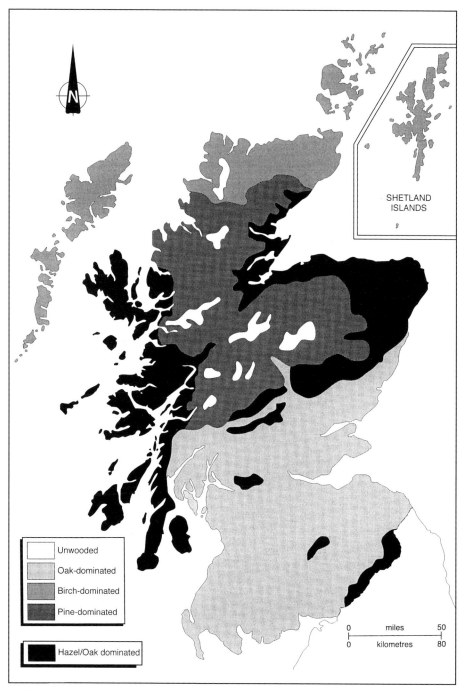

Map 2.1 The major woodland types in Scotland 6,000 years ago. Modified from R. Tipping, *Proceedings of the Society of Antiquaries of Scotland*, 124 (1994), pp. 1–55, with permission.

Darling put it at 'possibly fifty per cent' on the eve of the Neolithic (see also below, pp. 33–4).[20]

People arrived in Scotland 9,500 years ago, perhaps a little before the Scots pine. It has been suggested that people might have followed the hazel, the shells of which often appear in extraordinary quantities in early prehistoric sites. Clear evidence that these Mesolithic hunter-gatherers, whose epoch lasted for some 4,000 years, had a lasting impact on the rich environment from which they sought their subsistence, is hard to discover in Scotland, though in England it is cogently argued that on the North York Moors woodland was cleared by fire to encourage a nutrient-rich sward for wild animals to graze, and that on Dartmoor woodland was transformed into blanket peat, via a phase of acidic grassland, over a period of 600–1,000 years of systematic burning.[21] The degree to which the spread of blanket mires was natural, associated with increasing rainfall, storminess and declining temperatures, or anthropogenic, associated with woodland clearing and the use of fire, is a warmly debated question.[22] Quite possibly it was a combination of the two: as Caroline Wickham-Jones expresses it concerning Caithness, 'The growth of the peat in the Flows must have been encouraged by the increased rainfall but the right conditions may well have been put into place by pre-existing woodland clearance'.[23]

Sorting out anthropogenic from natural causation proves indeed to be a major problem in paleoecology. At approximately the time when Mesolithic hunter-gathering practices began to give way to Neolithic farming ones, there was a major event in woodland history in the south of Scotland, the elm decline, which appears to come at different times in different places, lasting over a period of perhaps a thousand years, commencing at about 3500 BC: it removed the elm as a dominant species from the Scottish Lowlands, as elsewhere in the British Isles. Once thought to be clear evidence for human interference in the forest, it is now thought more likely to have been a consequence of climate change or perhaps of disease

---

[20] J. Ritchie, *The Influence of Man on Animal Life in Scotland* (Cambridge, 1920), pp. 308, 484; F. F. Darling, 'Ecology of land use in the Highlands and Islands', in D. S. Thomson and I. Grimble (eds), *The Future of the Highlands* (London, 1968), p. 39.

[21] Tipping, 'Living in the past', p. 27; I. G. Simmons, 'Vegetation change during the Mesolithic in the British Isles: some amplifications', in F. M. Chambers (ed.), *Climate Change and Human Impact on the Landscape* (London, 1993), pp. 109–18; C. Caseldine and J. Hatton, 'The development of high moorland on Dartmoor: fire and the influence of Mesolithic activity on vegetation change', in *ibid.*, pp. 119–31.

[22] P. D. Moore, 'The origin of blanket mire, revisited', in Chambers (ed.), *Climatic Change and Human Impact*, pp. 217–25; Tipping, 'Form and fate', p. 15; J. Fenton, 'Native woods in the Highlands: thoughts and observations', *Scottish Forestry*, 51 (1997), pp. 160–4.

[23] C. R. Wickham-Jones, *Scotland's First Settlers* (London, 1994), p. 61.

(analogous to the Dutch elm disease of our own time) or a combination of factors. Certainly, disease could have moved faster if farmers were lopping branches from healthy elms to feed to their cattle.[24]

Equally problematic is the decline of the Scots pine over most of its Scottish range. Map 2.1 shows its distribution 6,000 years ago, but later it advanced from its heartland over the now apparently drying peat bogs as far as the north coast of Caithness, and west to Skye, Rum and Lewis (though there never as a dominant tree), and it had even been in Galloway at an earlier period. At about 2400 BC there was a dramatic collapse, not only in the newly occupied areas but throughout the Highlands. In the space of two centuries, pine throughout Scotland shrank back into much the same range as it occupies at the present day: 'this event was one of the more significant and mysterious shifts in vegetation seen in the British Isles in the course of the Holocene'.[25] Where pine declined, it was mostly replaced by blanket peat. In this case climatic change is the favoured explanation, particularly much heavier rainfalls and stronger winds as global atmospheric circulation patterns shifted. On the other hand it was also the period when, in the early Bronze Age, agricultural and technological change may have facilitated woodland clearance, and it is possible that removal of some of the cover increased the physical impact of precipitation on the soil.

Arguably, even the innermost pine forests were not unscathed by human exploitation at an early date. In Strathspey, from about 1700–1900 BC in the case of Loch Garten, and 1000–1200 BC in the case of Loch Pityoulish, there is a decline in woodland and an increase in heather and herbs: the record does not prove clearance, but it has been described as evidence for the forest being opened up by human-related activity. The first archaeological remains in Speyside are from the early Neolithic, there are various monoliths and cup-marked stones presumably from the later Neolithic or early Bronze Age, and half a dozen Iron Age forts between Loch Garten and Laggan Bridge, as well as a number of hut circles, souterrains and crannogs of similar age.[26]

There is no reason to think that this was exceptional for a Highland strath. Crannogs, which used a good deal of substantial timber, are often numerous in Highland lochs. At Balbridie on lower Deeside, the early Neolithic timber hall must have used large numbers of formidably large

---

[24] Tipping, 'Form and fate', pp. 18–22; Tipping, 'Living in the past', p. 35.
[25] Bennett, 'Post-glacial dynamics', pp. 30–3; Tipping, 'Form and fate', pp. 26–7.
[26] Bennett, 'Late Quaternary vegetation', p. 57; P. E. O'Sullivan, 'Vegetation history and the native pinewoods', in R. G. H. Bunce and J. N. R. Jeffers (eds), *Native Pinewoods of Scotland* (Cambridge, 1977), p. 61; N. G. Bayfield and J. W. H. Conroy (eds), *Cairngorms Assets Review* (Cairngorm Partnership, 1995), section 21, 'Historic Aspects'.

trees, to judge from the post-holes remaining. When one considers, in particular, the extensive clearances effected by Iron Age peoples in the Lowlands, it is reasonable to postulate that people also had some substantial impact on the pine and oak forests of the Highlands, well before the birth of Christ, though Richard Tipping thinks that it is 'more likely that climate change and soil deterioration were more significant agents, particularly in northern Scotland'.[27]

Whatever the causes, thinning woodland cover is indicated by palynological study in many places in the birch–hazel–oak zones around the Highland periphery. For example, in Trotternish on Skye, change began about 2200 BC and led to 'a mainly treeless landscape by BC 600'; on Arran, possible anthropogenic influence on the woods, including the spread of heath, begins before 2600 BC, but cultivation appears to come and go as blanket peat spread and climate deteriorated; in Orkney, most of the tree cover went within the first five centuries of Neolithic occupation. Many places in the west suffered from the deepening peat, whatever its cause, as witnessed by the gradual burial of the bottom half of Neolithic monuments at Callanish on Lewis and Kilmartin in Argyll. In the drier eastern Highlands, typified by areas above Loch Tay and above Pitlochry, and within the Caenlochan glens, 'low-intensity grazing pressures, sustained over long periods, seem to have gradually but effectively removed the woodlands' some time before the advent of the Romans.[28]

Yet climate change was all-pervasive in the period. Throughout the prehistoric millennia in northern Europe, long-term cycles involving shifts in the polar wind belts led periodically to episodes of increasing oceanicity, with wet and windy weather at their most damaging on exposed northwestern coasts. Such seem to have occurred in Scotland at around 6200 BC, again at around 2200 BC and at 800–500 BC. In marginal areas of acid soils on ancient rocks such as the Scottish Highlands and the Southern Uplands, this is likely to lead naturally to deepening peat, increasing podsolisation of the ground and conditions successively less favourable to tree growth. In the drier phases of the cycles, as around 3000 BC, trees could sometimes colonise the bogs, only for them to die again in the succeeding episodes. Claims that great expanses of sedge and sphagnum moss in the peat bogs, and not forests of trees, were established as the natural climax vegetation of much of the north and west and of the high ground of the south have much

---

[27] Tipping, 'Living in the past', pp. 36–7.
[28] Tipping, 'Form and fate', pp. 27–30; M. J. Bunting, 'The development of heathland in Orkney', *Holocene*, 6 (1996), pp. 193–212.

to recommend them.²⁹ The activities of early man would only hasten the process.

A very different appraisal to that of the older notions of a Great Wood of Caledon in the Roman Highlands is suggested now by David Breeze, Chief Inspector of Ancient Monuments in Historic Scotland and a leading historian of the period:

> The Highlands of Scotland are certainly an impressive massif. Today, vast and barren, often the only trees are those planted by the Forestry Commission. Their predecessors had been cleared 2,000 years before the Romans arrived, in the Neolithic and early Bronze Ages, leaving only isolated pockets of the original pine and birch woods.³⁰

Possibly, however, the decline of which he writes owed as much to natural causes as to human ones, or at least to a combination of the two.

In the southern half of Scotland – broadly the zone of primal oak–hazel–elm dominance, covering the Lowlands from Angus and Strathclyde to the Southern Uplands – the impact of man seems clearer and easier to date, as the evidence here is more plentiful for judgement to be made. The first indications of farming do not suggest permanence, at most lasting for a few decades or a century in one place before the forest returned. The northern Cheviots, for example, show much evidence for early Neolithic farming of this sort, but by the later Neolithic, at about 2800 BC, semi-permanent areas for pasture and crops were being established. Around 2000–1850 BC there was a quite widespread, almost synchronous, increase in the number of such places both in the lowlands and uplands: they appear to have been quite small in scale, perhaps around discrete farmsteads or groups of hut circles, but also to have persisted for hundreds of years. Then, well into the Iron Age from about 500 BC, in Richard Tipping's words 'a large number of sites quite clearly depict a major clearance episode, the first clearance of any great spatial extent', in different places at different times.³¹ It was accompanied in some places by quite serious soil erosion, analogous, perhaps, to what happened in New Zealand and Australia on the arrival of white farmers in the nineteenth century.

---

[29] R. M. M. Crawford, 'Ecological hazards of oceanic environments', *New Phytologist*, 147 (2000), pp. 257–81; R. M. M. Crawford, 'Oceanicity and the ecological disadvantages of warm winters', *Botanical Journal of Scotland*, 49 (1997), pp. 205–21; H. Tinsley and C. Grigson, 'The Bronze Age', in I. G. Simmons and M. Tooley (eds), *The Environment in British Prehistory* (London, 1981), pp. 211–16; Fenton, 'Native woods'.
[30] D. J. Breeze, *Roman Scotland* (Batsford and Historic Scotland, London, 1996), p. 97.
[31] Tipping, 'Form and fate', p. 31.

This deforestation clearly predates the Roman invasion, though in some places it was still going on when they arrived and in others it had apparently overshot the capacity of the farmers to keep the woods open, and regeneration was occurring. The process was no doubt also rather uneven. Pallisaded settlements of the first millennium BC in the Borders were initially surrounded by timber, later with earth and stone, but in Eastern Dumfriesshire timber appears to have remained the normal material for buildings 'throughout later prehistory and the Roman period', strongly suggesting more wood than elsewhere.[32] Nevertheless, in Tipping's words: 'there seems little doubt that this widespread clearance, almost certainly undertaken to provide vastly increased areas for both crops and grazing, left substantial areas of southern Scotland almost treeless at the time of the Roman advance.' It was 'by far the most substantial anthropogenic alteration of the landscape to have taken place, its scale apparently exceeding later clearance episodes during the historic period'.[33]

Commonsense archaeological observation often confirms the findings of sophisticated palynological investigation. Thus the Antonine Wall used in its construction large quantities of turf, implying open country around, and the Roman signal stations at Findogask above Strathearn, now swallowed up in conifer plantations, could not have worked unless they had been situated within and overlooking open country. Everything points to the invaders having taken over a populous, well-worked countryside in the Lowlands, and confronting a more difficult Highland zone, also thoroughly occupied by a farming people and their animals, modifying the woods, reducing them, and turning them to their own purposes.

Of course the Romans and everyone else made free use of woodland resources and probably did not have to go far to find them. The great legionary fortress at Inchtuthil outside Perth, for example, is thought to have required some 30 linear kilometres of timber-framed wood for its walling, equating to a cleared area well in excess of 100 hectares. But the impact would have been localised, and regrowth will readily occur in a broadleaf forest if the animals are kept out. Similarly, the indigenous population used timber for hut circles, for large individual roundhouses, for the great enclosed hillforts and for the interior of brochs (for the latter,

---

[32] Tipping, 'Form and fate', pp. 31–4. Claims that the Romans themselves cleared large areas of wood for military construction purposes are doubted by Tipping (pp. 33–4). See especially L. Dumayne, 'Iron Age and Roman vegetation clearance in northern Britain: further evidence', *Botanical Journal of Scotland*, 46 (1993), pp. 385–92. For Eastern Dumfries see Royal Commission on the Ancient and Historical Monuments of Scotland, *Eastern Dumfriesshire: an Archaeological Landscape* (Edinburgh, 1997), pp. 17–25.

[33] Tipping, 'Form and fate', p. 32; *Eastern Dumfriesshire*, p. 20.

on the islands, the source was presumably driftwood). Such extensive use of wood combined with a large population of farmers suggests to some archaeologists that the resource must already have been managed.[34]

It is worth considering again at this point the controversy concerning the character and extent of woodland cover at that point in prehistory when human interference was still absent or minimal (see p. 12 above). Until recently, the picture of Ireland and Scotland at the height of the woodland maximum was of densely wooded countries. Jon Pilcher, for example, has described how 'Ireland was covered from shore to shore and to the tops of all but the highest mountains by a dense multi-layer deciduous forest', and the twentieth-century Scottish commentators Ritchie, Darling, Anderson and Tipping have all described or implied a forest covering at its maximum half or more of the land surface – that is, everywhere except the highest mountains and those 'areas of ground at lower elevations where, for geological or topographical reasons such growth was impossible'.[35] The Dutch ecologist Franciscus Vera, however, has recently urged us to revise these pictures of close canopy European forests, and to think rather of an open landscape of scattered trees and groves, maintained by herds of grazing animals including, especially, the great auroch. What proportion of the land would be occupied by the groves is unclear: if it was analogous to the modern New Forest (which Vera considers among the landscapes most resembling the original in Europe), or to the Polish forest of Białowieza as it was mapped around 1830, perhaps between a quarter and a half the natural matrix of forest and grassy heath would be wooded.[36]

This picture, though, does not command universal acceptance. Many palynologists and paleoentomologists are sceptical, which is significant since theirs is the only direct evidence available for the character of vegetation cover in prehistoric times. Woodland ecologists also have reservations, though Peterken perhaps expresses a general view that there may have been more open land than he and others had previously thought. Kirby has also recently suggested a middle way in the controversy, with closed forest dominating on some soils, and groves on others: even in the latter scenario, half the available land surface could still have been covered with wood and another quarter by scrub or 'break-up stands'.[37] The situation across Scotland,

---

[34] I. Armit and I. Ralston, 'The coming of iron', in Smout (ed.), *People and Woods*, pp. 40–59.
[35] J. R. Pilcher, 'A palaeoecologist's view of the Irish landscape', in F. H. A. Aalen (ed.), *Landscape Study and Management* (Dublin, 1996), p. 74; Anderson, *History of Scottish Forestry*, 1, p. 25; Ritchie, *Influence of Man*, Ch. 6; Tipping, 'Form and fate'.
[36] F. W. M. Vera, *Grazing Ecology and Forest History* (Wallingford, 2000), pp. 144, 272.
[37] K. J. Kirby, 'What might a British forest-landscape driven by large herbivores look like?', *English Nature Research Reports*, 530 (2003).

as elsewhere, is likely to have varied considerably, with perhaps less of the land surface being available than in England due to altitude and rainfall.

After the early Bronze Age, in any case, the climate became generally cooler and (even more importantly) wetter and windier. It became more oceanic, influenced constantly by gales and rain moving in from the Atlantic, the ground suffering from waterlogging, leaching of minerals and the build up of peat. Under these circumstances, even if people had not interfered, declining seed-setting and increased windthrow would have further opened up the woods and in places (especially in the west) have eliminated them. The forest would become in many places even more patchy and broken, perhaps difficult in places to discern as wood: woods with clearings would shade into moors with a few trees and scrub, and eventually die out in bogs. In such a landscape the interference of people, even in quite small ways, between the Neolithic and the Iron Age would tend always to increase the openness. In a wet climate, opening the canopy of a woodland grove can create a chain reaction where more rain impacts on the ground and less is absorbed by foliage, and the decline is accelerated. People and nature thus worked together to reduce the proportion of the land under the protection of a leafy canopy.

In these circumstances, we would have greatly to reduce any estimates of woodland cover 5,000 years ago, perhaps by one-half, to arrive at a figure for 2,000 years ago. For a quarter of the land surface to have been wooded then would seem a possible figure. The present woodland cover of Scotland stands at 17 per cent, most of it plantation, and it is frequently urged upon us that the percentage should become much higher. There are many good arguments for planting trees in Scotland – to maintain employment, to give pleasure, to help carbon sequestration and to assist nature conservation. But it seems there may be fewer arguments from history than usually assumed, and none for restoring the fantastical Great Wood of Caledon.

For most of the first fifteen centuries of the Christian era, we know fewer hard facts about the history of the Scottish woods than in the late prehistoric period, partly because few palynological investigations have been carried out. Among those that have, is a study at the Black Loch in north Fife. Here the landscape appears to have become increasingly open at about 2000 BC, showing in the Iron Age all the characteristic signs of intensifying arable as well as pastoral use; these came to an abrupt end at about the time of Roman attack, and were succeeded by five centuries in which woodland of birch, oak and hazel reappeared. The investigators suggest that the native Venicones, faced with at least three Roman camps in the vicinity, were either wiped out or compelled to abandon the area. Only after about AD 520, in the Pictish period, do the indications of arable and pastoral farming reappear. A similar effect has been found in eastern Aberdeenshire, yet further south in the Borders it is not possible to detect conclusively any change to farming or woodland cover arising from the Roman invasion. The Picts have also been discovered practising muirburn elsewhere in Fife, on the sandy edges of Tentsmuir, but that is as likely to have been a standard agricultural practice on heathland as much as an effort to clear out a wood. Another case study in the western Cheviots indicates that from around AD 400 an apparent increase of grassland began in a hitherto partly wooded countryside, and has persisted ever since, kept open by stock.[38]

No doubt throughout the first millennium AD, however, woodland would normally have ebbed and flowed depending on local circumstances, wars, plagues and other demographic events of which we know little. Certainly over the period local populations and incomers alike continued to use it as though it was an abundant resource. At Buiston in Ayrshire, for instance, excavation of a crannog dating from the sixth to seventh century showed it to consist of a roundhouse made of wattle walling with posts of alder and hazel and a floor of alder planks, all on a strong framework of oak and surrounded by an impressive palisade. The dimensions and age of the wood suggested to the archaeologists that the crannog-dwellers had managed a large area of woodland partly by coppice and partly by selective felling, and that the resource had not varied over fifty years, but in the nature of things such conclusions are rather speculative. On Iona, the famous monastery was made of hewn oak and pine and bundles of wattle imported from the mainland, and wooden churches were described by Bede as *modus scottorum* in the eighth century. Both Pictish forts and Anglian halls used large quantities of timber, too.[39] Then there was the intervention of the Vikings. Barbara Crawford has argued that the Norse Earls of Orkney and Caithness had an interest in Easter Ross from the ninth to the twelfth centuries because they could not secure the shipbuilding timber that they needed in the territory that they already controlled (including Sutherland) further north: it also seems that they could not find enough suitable wood in

---

[38] G. Whittington and K. J. Edwards, '*Ubi solitudinem faciunt pacem appellant*: the Romans in Scotland, a palaeoenvironmental contribution', *Britannia*, 24 (1993), pp. 13–25; G. Whittington and K. J. Edwards, 'Palynology as a predictive tool in archaeology', *Proceedings of the Society of Antiquaries of Scotland*, 124 (1994), pp. 55–65; Tipping, 'Form and fate', p. 34; G. Whittington (ed.), *Fragile Environments: the Use and Management of Tentsmuir NNR, Fife* (Edinburgh, 1996), p. 17; C. Campbell, R. Tipping and D. Cowley, 'Continuity and stability in past upland land uses in the western Cheviot Hills, southern Scotland', *Landscape History*, 24 (2002), pp. 111–19.
[39] A. Crone and F. Watson, 'Sufficiency to scarcity: medieval Scotland 500–1600', in Smout (ed.), *People and Woods*, pp. 60–5.

the areas of Pictish settlement along the coasts of Easter Ross and Cromarty, but had to penetrate miles up the river systems such as Strathcarron, Strath Oykell and Strathconon to obtain what they required, leaving behind a string of Norse names such as Amat, Alladale, Eskadale, Dibiedale and Carbisdale.[40]

The Vikings have often been blamed for serious crimes of arson against the wood. Fraser Darling considered the centuries 800–1100 as a particularly disastrous epoch in its history, when 'a destructive and parasitical folk ... set light to the forest to burn out the miserable natives who had taken refuge within it'.[41] There is a widespread Highland tradition of explaining the blackened oaks and pines in the peat bogs as the traces of a burned and ruined forest left when the Vikings ravaged the country. In one version, a witch who was daughter of the King of Norway burned the woods of Lochaber because they were growing so great as to rival the Black Wood of Sweden, 'Coille Dhubh na Suain'.[42] Most of these stories appear to be rationalisations for finding the blackened fossil stumps in the peat. No doubt the Norse would seize any opportunity to burn out their enemies on those rare occasions when their need to do so coincided with dry weather and a suitable species to burn, but fire, far from being an enemy to pine regeneration, encourages it, and excessive modern fire control has been considered as 'maybe the cause of some regeneration problems'.[43] There is no reason whatever to think that the Vikings had any lasting impact on woodland cover.

Dendrochronologists have tentatively drawn attention to a period in the early tenth century when there may have been a sudden reduction in the pressure on the oakwoods, followed by an opportunity for regeneration, but how general it was and what its causes might have been, remain obscure.[44]

In the centuries of the later medieval period (AD 1000–1500), the main sources for the study of Scottish woodland history are documentary and archaeological, though the former in particular, are sparse and ambiguous

---

[40] B. E. Crawford, *Earl and Mormaer: Norse-Pictish Relationships in Northern Scotland* (Rosemarkie, 1995).

[41] Darling, *Natural History*, pp. 58–9.

[42] H. Cheape, 'Woodlands on the Clanranald estates', in T. C. Smout (ed.), *Scotland Since Prehistory: Natural Change and Human Impact* (Aberdeen, 1993), p. 54.

[43] A. Carlisle, 'Impact of man on the native pinewoods', in Bunce and Jeffers, *Native Pinewoods*, p. 70.

[44] A. Crone and C. M. Mills, 'Seeing the wood and the trees: dendrochronological studies in Scotland', *Scottish Woodland History Discussion Group Notes*, 7 (2002), pp. 14–22. The authors (*pers. com.*) have withdrawn their suggestion (p. 15) that Viking raids might have caused a contraction of settlement to safer, inland locations.

indeed compared to the rich record in England or France.[45] From the 1130s, King David I introduced to Scotland the Anglo-Norman concept of *foresta*, supported by a code of forest laws elaborated over the next few centuries, to provide hunting reserves for the king and the nobility.[46] A 'forest' in later Scottish parlance did not have to contain woods, but in the Middle Ages it certainly usually did: thus the forest laws were full of references to 'pannage' (the freedom to feed pigs on acorns in winter) and more generally to rights and restrictions relating to pasturing cattle, horses and sheep in the woods. Clause 5 testified to the unpopularity of trespassing goats:

> Concerning goats found in the forest it is lawful for the forester on each of three occasions to hang one of them by the horns in the trees and on the fourth occasion he ought to kill one of them and leave its entrails there as a sign that they were found there.[47]

Gilbert lists eighty-five known forests, great and small, royal, noble and ecclesiastical, recorded in the thirteenth century, and a further ninety-one recorded after 1296. Most are in the south and east of the country, representing not so much the distribution of wood as the distribution of effective legal administration by the Crown and its officers. Other forests are known later, especially in the north and west.

This information, fascinating though it is, does not solve the problem of how extensive the woods were in medieval Scotland. Over much of the Lowlands, it is, however, clear that large timber, at least, was a dwindling resource, becoming scarce in some areas as early as the thirteenth century, although the problem cannot have been widely felt then, since as late as the mid-fourteenth century the practice of giving abbeys a general right to cut wood in the royal forests continued unabated. In the fifteenth century, however, there is evidence for serious shortage over large areas of Scotland for substantial or good quality building timber, which increasingly came to be imported from the Baltic and Scandinavia. The oak used in Scottish buildings such as castles and churches was now either of foreign provenance or of young native growth. Great beams hundreds of years old, such as those from which the roof of Darnaway Castle had been constructed in 1387, were no longer available (see Fig. 2.2). On the other hand, archaeological

---

[45] O. Rackham, *Ancient Woodland* (Cambridge, 1980); R. Bechman, *Trees and Man: the Forest in the Middle Ages* (trans. K. Dunham), (New York, 1990).
[46] J. M. Gilbert, *Hunting and Hunting Reserves in Medieval Scotland* (Edinburgh, 1979); Anderson, *History of Scottish Forestry*, 1, pp. 90–139.
[47] Translated in Gilbert, *Hunting*, p. 305.

Fig. 2.2 The oaken roof of Darnaway Castle, Morayshire, dated by dendrochronology to 1387. It is the earliest hammerbeam construction in Scotland. Royal Commission on the Ancient and Historical Monuments of Scotland: Crown copyright.

exploration of rural homes and barns, and also of Scottish town houses, shows most of them to have been flimsy constructions of turf, wattle and thatch, utilising pole wood (hazel, ash, willow and so forth), which seems to imply ready enough access to coppice materials.[48]

By the fifteenth century, Parliament began to express concern: an Act of 1424/5 imposed a fine on stealers of green wood and bark peelers, one of 1457/8 ordered landlords to grant leases only if the tenants planted trees, made hedges of live wood and preserved broom, and one of 1503/4, 'considering that the wood of Scotland is utterly destroyed', increased the penalty on stealers of greenwood, and proposed that landowners planted an acre of woodland where there were not already extensive woods.[49] This

---

[48] Crone and Watson, 'Sufficiency to scarcity', pp. 66–81; Crone and Mills, 'Seeing the wood', pp. 14–22.
[49] Gilbert, *Hunting*, p. 237; Nairne, 'Notes on Highland woods', pp. 174–5; Crone and Watson, 'Sufficiency to scarcity', pp. 70–81.

was also the century when the first known enclosures of woodland are made, and all over the Lowlands, lords appear to have clamped down on casual woodcutting. This of course does not mean that the land was bare, only that the woods had become smaller, adequate perhaps for local use but unable fully to supply the increasing needs of the towns and of royal schemes to build grandiose ships and palaces such as characterised the reigns of James IV and James V. When the papal emissary Aeneas Sylvius landed near Dunbar, ca. 1430, he found the country 'scantily supplied with wood' but was told that in reality there were:

> Two distinct countries in Scotland – the one cultivated, the other covered with forests and possessing no tilled land. The Scots who live in the wooded region speak a language of their own, and sometimes use the bark of trees for food.[50]

Many travellers see through a glass very darkly.

Pressure on the woods in southern and eastern Scotland came from various other directions. When it was still comparatively plentiful, wood had been used along the coast of the Firth of Forth to fuel the saltpans below Stirling and in East Lothian, but substitution of coal for wood was easy: even in the seventeenth century the small coals suitable for salt boiling were still known as 'pan-wood'. More irresistible was the growth of human population particularly in the thirteenth century and again in the late fifteenth century. For example, in 1488 the Baron Court of Keith in Aberdeenshire forfeited the lease of a tenant who had grubbed up, ploughed and sown a large part of the woodland of the barony.[51] There is much archaeological and other evidence for 'assarting' (taking land in for arable from the 'forest') often by permission, in Renfrewshire, Roxburghshire, Annandale, Liddesdale and elsewhere, particularly in the Borders.[52] The feuing of the Ettrick Forest by the Crown to a large number of independent small landholders resulted in almost uncontrollable depredation on the woods: the justice ayre at Selkirk in one meeting in 1510 recorded 148 separate woodcutting offences in the forest. Even in the fourteenth century when the pressure of population in the wake of the Black Death should have been less, David II in 1366 made a grant of four oxgangs of arable land in the forest of Plater, Forfarshire, which suggests but does not prove actual

---

[50] P. Hume Brown (ed.), *Early Travellers in Scotland* (Edinburgh, 1891), pp. 26–7.
[51] Gilbert, *Hunting*, p. 234.
[52] P. Dixon, *Puir Labourers and Busy Husbandmen* (Edinburgh, 2003), pp. 42–4.

clearance; by the end of the Middle Ages there was little wood left, as far as we know, either in this forest or in those of Cardenie and Uweth in Fife, or in some of the Lowland Aberdeenshire forests.[53]

The greatest pressure of all in the Borders, however, must have come from commercial sheep farming between the thirteenth and the sixteenth centuries, carried on by Cistercian and other monastic houses, and soon imitated on a large scale by the nobility and the peasants themselves. In the years 1327–1332, for example, 5,700 sacks of wool (360 lbs per sack) were exported annually: the trade peaked at 9,252 sacks in 1372. The animals were small, but with each sheep producing between one and two pounds of wool, such a clip would have been the produce of over 2 million sheep.[54] It is no wonder that there was manifest decline in woodland in the royal and baronial hunting forests in the Southern Uplands, or that, for example, on monastic lands in the south west between the Urr and the Nith, grants of pannage before about 1180 (implying herds of pigs in an oak forest) gave way abruptly to grants of pasture to sheep.[55] There was a more profitable use for the land on which woods were growing, and Cistercians were especially quick to take advantage.

At this time, as earlier, the story is told in Scotland, as it is also (much better attested) in Wales, of military depredations against the forest. For example, in 1316 Jed forest was allegedly hewn down by 10,000 Englishmen to deprive the Earl of Douglas of his hiding place, and again, ca. 1385, John of Gaunt, Duke of Lancaster, brought 80,000 axes (or, in some accounts, 24,000) to fell the woods of the Borders.[56] Apart from the inherent implausibility of the scale of such operations, such felling episodes would destroy standing timber but not eradicate a broadleaf wood, unless the regrowth and regeneration were subsequently grazed away. Of course, if the woods were already subject to intensive grazing pressure anyway, the effect of such an attack might well be to remove the only surviving seed-bearing trees, but the root cause of deforestation would surely be the grazing rather than the felling.

In any case, there are problems with medieval evidence of this sort, because of background expectations. Edward I fancied himself as the very

---

[53] Tytler, *History of Scotland*, 2, pp. 201–2; Gilbert, *Hunting*, p. 238; Anderson, *History of Scottish Forestry*, 1, p. 176.
[54] A. Grant, *Independence and Nationhood: Scotland 1306–1469* (London, 1984), pp. 62, 79.
[55] R. Oram, 'The Lordship of Galloway, ca. 1000 – ca. 1250', (unpublished Ph.D. thesis, University of St Andrews, 1989).
[56] C. O. Badenoch, 'Border woodlands I – Berwickshire', and 'Border Woodlands II – Roxburghshire', *History of the Berwickshire Naturalists' Club*, 46 (1994–5), pp. 115–33, 272–86; W. Linnard, *Welsh Woods and Forests* (Llandysul, 2000), pp. 28–33.

incarnation of a Roman Emperor, and Severus on invading Scotland was said by the classical authors to have employed his legions in felling the trees (slight evidence itself that he did so): it was a flattering touch by chroniclers to attribute, or to exaggerate, tree-felling exploits to Edward's successors. Similarly, heroes conventionally found refuge from their enemies in forests. In a sixth-century Brythonic poem, Merlin, the Arthurian magician, hid in a 'nut rich' wilderness somewhere in the south of Scotland, the extent of which 'he could survey from a mountain top screened by foliage', and in Blind Harry's fifteenth-century poem, William Wallace was, in John Fowler's phrase, 'forever darting in and out of the woods'.[57] The literary conventions that make these such good stories, detract, alas, from their value as historical evidence.

Something of the same problem applies in trying to gauge the situation in the medieval Highlands, where few of the documentary sources are based on first-hand knowledge. Thus John of Fordun in the late fourteenth century speaks, in a much-cited passage, of the foot of the mountains being clothed with 'vast woods full of stags, roe-deer, and other wild animals and beasts of various kinds', where the inhabitants hide their stock, 'for the herds of these parts, they say, are accustomed, from use, whenever they hear the shouts of men or women and if suddenly attacked by dogs to flock hastily into the woods'.[58] But as far as we know, Fordun spent his life as a minor priest in Aberdeen, and was probably only re-telling tales from upper Deeside. Possibly he knew as little about it as Aeneas Sylvius, but both accounts indicate a Lowland view that the Highlands were amply wooded compared to their own area.

Shipbuilding provides some harder, if still limited, evidence of the wooded resource. Clearly there was material on the west coast to support the construction of Highland galleys for the local chiefs, exemplified by the agreement of 1354 between John, Lord of the Isles, and John of Lorn, granting the latter the right to build eight vessels of twelve or sixteen oars. Galley service was indeed the commonest feudal military service all along the western seaboard. Some boats were much larger than John of Lorn's: Robert I demanded service of a boat of forty oars in exchange for a grant of lands in Lochawe and Ardscotnish. According at least to the accounts of the sennachies, they were also numerous: John, Lord of the Isles, was said to have taken sixty 'longships' to the mouth of the Clyde when he went to

---

[57] J. Fowler, *Landscape and Lives: the Scottish Forest Through the Ages* (Edinburgh, 2002), pp. 13–18.
[58] W. F. Skene (ed.), *John of Fordun's Chronicle of the Scottish Nation* (Edinburgh, 1872), 1, p. 36.

his marriage in 1350.⁵⁹ Whatever we make of the last figure, the maintenance of regular naval force by the chiefs presupposes a great deal oak available locally. Again, in 1249 Matthew Paris recounts that a French crusader, Hugh, count of St Paul and Blois, had a 'wonderful ship' built at Inverness for a voyage to the Holy Land, which he is unlikely to have commissioned unless the area already had a reputation in France both for its materials and skills, no doubt inherited, as Barbara Crawford suggests, from the Viking tradition so clearly seen in the galleys.⁶⁰

The Inverness area, and particularly the royal forest of Darnaway in Moray, had some trade within fifteenth- and more in sixteenth-century Scotland supplying large and specialised timber to the Lowlands for palaces and warships,⁶¹ but by then no serious seeker after timber or ships from outside Scotland would have considered the Highlands as a source. At the close of the Middle Ages and in the early Renaissance, the English and the Dutch, and ultimately most of the Scots themselves, looked to Norway and the Baltic countries for their supplies of large wood.

Other indications that the Highlands may have had limited forest, or at least limited wilderness, come from the history of mammals. Norway managed, albeit tenuously in the nineteenth and twentieth centuries, to hang on to a range of impressive large mammals in its forests, including the bear, wolf, lynx, elk and beaver. All these plus the wild boar once thrived in Scotland. The Caledonian bear was known to the Romans, and may have survived until the tenth century but not beyond. The wolf did not need forests in the same way, survived as a pest until the seventeenth century, and possibly lingered as a survival into the eighteenth, in open or scrubby country where domestic stock as well as deer ranged. The lynx is known to have persisted until the third century AD. The elk possibly survived until the ninth century.⁶² The less demanding beaver, which can make do with willows, birch and aspens along lightly wooded streams, survived in the Loch Ness area until the early sixteenth century. Apart from the wolf, they could not cope in a medieval Highland environment, presumably because the woods were already too fragmented and too full of human predators.

---

⁵⁹ Cheape, 'Woodlands', pp. 56–8; J. Munro and R. W. Munro (eds), *Acts of the Lords of the Isles* (Scottish History Society, Edinburgh, 1986). Three charters by John to his brother Celestine all mention oaks (in Skye, Sutherland and Caithness) and in the sixteenth century galleys were built in Loch Kishorn, *ibid.*, pp. xxxix–xl.

⁶⁰ Crawford, *Earl and Mormaer*, p. 16.

⁶¹ Gilbert, *Hunting*, p. 238; N. Macdougall, *James IV* (Edinburgh, 1989), p. 237.

⁶² Ritchie, *Influence of Man*; A. C. Kitchener, 'Extinctions, introductions and colonisations of Scottish mammals and birds since the last Ice Age', in R. A. Lambert (ed.), *Species History in Scotland* (Edinburgh, 1998), pp. 63–92.

The auroch had already perished in the Iron Age although a form of wild white cattle unrelated to the aurochs survived in woodland until the sixteenth century. Some authorities suggest that wild boar may have survived to the sixteenth or seventeenth centuries, but (as in England) it appears hard to find clear evidence beyond the thirteenth century.[63]

This does not of course imply that there was in the Middle Ages as little semi-natural woodland as there is today. There was probably much more. In Strathspey and Badenoch, for example, place-name evidence reveals Kingussic, 'the head of the pines', well beyond the present stretch of the old pinewoods, and several other Gaelic names indicating woods at a higher altitude than they grow now, yet Speyside continued to be well populated with castles, churches and chapels along its length (for place-names see also pp. 389–90 below).[64] The local population was not short of wood in the Highlands, but the area was anything but pristine wilderness.

It has often been claimed that most of the loss of biodiversity in the Highlands was a recent phenomenon. Thus Fraser Darling considers that 'the losses [of species] of the last 200 years are large in proportion to those of the previous 10,000 years'[65] and Carlisle that the pinewoods were 'probably little affected until Lowland timber was exhausted in the sixteenth and seventeenth centuries' and that 'the seventeenth century can be regarded as the time at which the Highland forests began to lose their primeval innocence'.[66] More recently, Mabey has spoken of the late seventeenth century as the time when 'the great forest of native pine and birch' that 'stretched across most of the Highlands, from Perth to Ullapool began to be ransacked, first to provide charcoal for the Lowland iron foundries, then to support the insatiable timber demands of the Napoleonic Wars'.[67] Such views are based on total misconception about the reality and extent of a Great Wood of Caledon in Roman times and on presumptions about its later survival. The distinguished medieval scholar Cosmo Innes was closer to the truth when, in 1861, he expressed what was already becoming an

---

[63] A. C. Kitchener and J. Conroy, 'The history of the beaver in Scotland and the case for its reintroduction', *British Wildlife*, 7 (1996), pp. 156–61; O. Rackham, *The History of the Countryside* (London, 1986), pp. 38–9; Anderson, *History of Scottish Forestry*, pp. 275–6; Ritchie, *Influence of Man*; D. Yalden, *The History of British Mammals* (London, 1999).

[64] For place-names, see H. L. Edlin, 'Place names as a guide to former forest cover in the Grampians', *Scottish Forestry*, 13 (1959), pp. 63–7; for settlement, see J. C. Stone, *The Pont Manuscript Maps of Scotland: Sixteenth Century Origins of a Blaeu Atlas* (Tring, 1989), pp. 51–6.

[65] Darling, *Natural History*, p. 64.

[66] Carlisle, 'Impact of man', p. 72.

[67] R. Mabey, *Flora Britannica* (London, 1996), p. 21.

unfashionable opinion, 'in taking old historical Scotland – Scotland of the fourteenth and to the beginning of the eighteenth century – in respect of wood, to have been very much as at present'.[68] Modern claims by conservation bodies that 'native Caledonian pinewoods now cover only one per cent of their former range'[69] are essentially meaningless, as that 'former range' refers to a period about 5,000 years ago, most of the decline took place before the Christian era, and was probably largely natural. We simply have no idea by how much the pine has declined since, say, the foundation of the Scottish kingdom about a thousand years ago, but it was certainly not by 99 per cent.

---

[68] Innes, 'Some account of early planting', pp. 40–53.
[69] Royal Society for the Protection of Birds, *Time for Pine: a Future for the Caledonian Pinewoods* (n.p., n.d., [1993]), p. 4.

CHAPTER 3

The extent and character of the woods, 1500–1920

It is evident from the last chapter that at the end of the Middle Ages Scotland felt herself to be in some senses short of wood – certainly short of the kind of large timber necessary for great construction projects on land and sea. The wood of Scotland is 'utterly destroyed', declared Parliament in 1503, probably reflecting at least the king's vexation in not being able to get his hands on the supplies he needed for his castles and ships. The tone of legislation in the next two centuries was consistently anxiety-ridden. Acts of 1535, 1607 and 1661 enjoined the planting and protection of timber. Those of 1592, 1594 and 1617 endeavoured to keep royal parks and forests from destruction, the last-named reciting their decay through the presence of shielings, the pasturing of stock and the cutting of wood. The Privy Council in 1564 deplored the widespread misuse of woods in the North, so that the whole resource is 'lyke to pereis', and Parliament in 1609, while stating that large woods had recently been discovered in the Highlands, was anxious that their value should not be frittered away (as they saw it) by the destruction of charcoal burners working for iron smelters.[1] There is no doubt that the greatly increased imports of softwood, especially in the period ca. 1580 to 1640, indicate that the developing towns and industries also could not find sufficient domestic sources of timber for their own concerns, at least not at a competitive price (see Chapter 6 below).

There is a certain amount of evidence (in addition to that of the maps, which we consider later) that some woodlands in the sixteenth, seventeenth and early eighteenth centuries contracted or disappeared. For example, towards the head of Strath Nairn there is an almost bare, rounded hill, with a few isolated Scots pine on it, named Coille Mhór (the Great Wood), local tradition relating that the forest upon it was destroyed to discourage outlaws on the orders of Mary, Queen of Scots. This explanation is unlikely, given her councillors' anxiety about the state of the Highland wood supplies, but

---

[1] *APS*, 2, pp. 242, 251, 343–4; 3, p. 560; 4, pp. 67, 373, 408, 547; 7, p. 263; *RPCS*, 1, p. 279.

the tradition may indicate the period in which it vanished. In 1776 the minister of Weem wrote that 'a great part of the north side' of Schiehallion 'was in the memory of several still alive once covered with a thick birch wood, though now hardly a vestige of it remains'. Its name was also remembered, Coille na Shi, but that has now vanished from the map. Only a few miles to the north another bare slope fronting the Tummel bears the name Coille Kynachan. There are of course many other place-names across the Highlands that appear to denote woods at high altitudes where there is now little but bare heath, but generally there is no indication of what period in the past they disappeared (see also pp. 389–90 below).[2]

More concrete as to causes of woodland loss on lower ground is the witness of Sir Robert Gordon of Straloch, writing around the middle of the seventeenth century and referring to instances of deforestation in order to extend cultivation. He says in his account of Aberdeen and Banff, that at first all the settlements were in villages, but once the woods had been cut down, the farms spread out and became separated: 'I remember seeing instances of this procedure in my early years. The farmers abandoned their villages and removed each to his own possession when any vein of more fertile soil attracted him.' More specifically, he said of the Garioch that 'in former centuries, especially on the banks of the Don, the whole neighbourhood bristled with wood', especially oak, but at the present day nothing remained, so that 'excessive abundance, while no attention is paid to it and there is no thought of the future, degenerated into want'. Furthermore he said of Strathbogie that in the past it had been divided into forty village settlements called davochs, but now the ploughs were all doubled, 'when the whole of the woods have been cut down, and all the land whence there is hope of a crop has been made over to tillage'. He was careful to distinguish this permanent alteration in land use from mere felling episodes, as in Enzie parish where there had been 'a wood clothed with tall oaks when I was a young man; but now having been cut down it flourishes again in a new growth among the hills', or at Birse, where the 'great forest of birch' had been 'entirely cut down through the carelessness of those concerned' but was now 'slowly growing up again without injury to the land, which is very well adapted for this'.[3]

---

[2] We heard of the tradition relating to Strath Nairn at a meeting of the Inverness Field Club. For Schiehallion, see British Library: BM add MS 33977, 65–72 Banks Correspondence, extract of a letter from Rev. John Stuart to Sir Robert Menzies. We are obliged to Mr John Mitchell for this reference. More generally, H. L. Edlin, 'Place names as a guide to former forest cover in the Grampians', *Scottish Forestry*, 13 (1959), pp. 63–7.

[3] A. Mitchell (ed.), *Geographical Collections Relating to Scotland made by Walter Macfarlane* (Scottish History Society, Edinburgh, 1906), 2, pp. 272–6, 282, 284.

The probability is not that the process of agricultural land-claim from the forest was peculiar to the north-east, but that Gordon was a closer observer than most. Lowland woods under pressure from agriculture in the middle of the sixteenth century can be detected at Campsie in Perthshire, even under the relatively firm management of the monks of Coupar Angus (see below pp. 157–8): two centuries later a visiting agricultural expert recommended planting pines to remedy the deficiency of wood in the area.[4] Throughout the sixteenth and earlier part of the seventeenth centuries, population was on the increase and the cultivated area was being extended in most parts of Europe, nowhere more so than in Norway, Scotland's nearest neighbour to the east. It would have been strange if, in Scotland, woods had not also been falling back before the pressures of people and beasts.

Between around 1650 and 1750, population pressure eased in Scotland, as elsewhere, and one would expect to find the rate of woodland extirpation also easing, though not necessarily stopping. Among several hundred woodland sales examined in the Register of Deeds, and in estate papers, 1650–1750, only one relates to the removal of a wood: in 1743, John Erskine sold Gartary Wood near Tulliallan in west Fife to a wright in Leith, on condition that he 'holl out at the root the whole trees and roots of trees within the dykes of the said wood ... and to put the ground thereof in such condition as a pleugh can labour the same as other arable ground'.[5] The rarity of such explicit detail, however, should not obscure the fact that woods could often have continued to disappear or to be nibbled away at the edges and in the centre, without any record surviving.

Within this period, the curtain of uncertainty lifts a little when the first maps showing woods on a local scale begin to appear. They were the work of Timothy Pont, who between 1583 and 1596 travelled the length and breadth of his country making cartographic drawings. Many were in the most distant and difficult parts of the Highlands and Islands, where his travels aroused the same kind of wonder at his intrepidity and resilience that the Victorians were later to reserve for explorers in Africa. Pont's maps were purchased after his death, still unpublished, by Sir James Balfour of Denmilne, following which a fellow Fife laird, Sir John Scot of Scotstarvet, told Joan Blaeu of their potential value to a new world atlas that he was planning in Amsterdam. They were accordingly forwarded to the Netherlands, where thirty-four plates were engraved, and the remainder returned

---

[4] C. Rogers (ed.), *Rental Book of the Cistercian Abbey of Coupar-Angus* (Grampian Club, London, 1879), 1, pp. 220, 222, 237, 242; 2, pp. 69–71; A. Wight, *Present State of Husbandry in Scotland* (Edinburgh, 1778–84), 1, pp. 31–3, 53.
[5] NAS: RD 3/206/3, 9 June 1746.

to Scotland as unusable or illegible. They were then sent to Sir Robert Gordon of Straloch, the leading Scottish chorographer of his age, who despatched new drafts to Amsterdam based mainly on his readings of the Pont manuscripts, but adding new information on three maps of north-east Scotland (he was from that area), and one new map of Fife drawn by his son James Gordon. The historian now has available for consideration everything in volume 5 of Blaeu's *Atlas Novus*, eventually published in 1654, plus about forty manuscript maps by Pont himself, apparently those 'illegible' ones returned from Amsterdam, and a series of draft manuscript maps by Gordon, mainly based on Pont (but possibly with embellishments of his own), many of which were not ultimately published by Blaeu. We also know that other Pont manuscripts have been lost.[6]

This spurt of cartographic activity provides a plethora of valuable information, but it is far from easy or straightforward to use. In the first place, map-makers up to the present day have difficulty in defining a wood. A cartographer has no problem with features like castles, kirks or farmtouns, which may be located by point-spot symbols: a river is linear, and a loch has edges that are clearly defined. A wood, however, particularly an unenclosed wood in montane habitat, may have no clear limits but merge into a zone of low bushes or arctic-alpine shrubs. One senses from time to time an obvious uncertainty of map-makers in these circumstances, even in nineteenth-century and twentieth-century Ordnance Survey work. At the beginnings of cartography the conventions were not even clearly established, and what was entered might be determined only by the personal interest or whim of the individual. In the case of Pont, portraying woods was not at the top of his agenda, and where there are a great many place-names to get in, as on his sketch of Speyside, most of the woods are omitted as symbols, though written comments on the map about woods partly compensate, and Abernethy is shown – 'fyr and other wood with great wilderness'.[7]

Another problem with the maps derived from Pont by Gordon or Blaeu is that his copiers may have left things out or embellished them. For example, in some places in the Highlands, the *Atlas Novus* fails to show woods where we know they existed, as along Loch Tay and Loch Rannoch. In others, it shows very few woods where there were, in fact, very important features, as along Strathspey and the glens of Strathglass. In most of these cases the omissions appear to be due to the fact that Pont himself was

---

[6] J. C. Stone, *The Pont Manuscript Maps of Scotland: Sixteenth Century Origins of a Blaeu Atlas* (Tring, 1989); I. C. Cunningham (ed.), *The Nation Survey'd: Timothy Pont's Maps of Scotland* (East Linton, 2001).
[7] Stone, *Pont Manuscript Maps*, p. 51.

much more interested in settlements than in woods. On the other hand among the unpublished Pont manuscripts is a very clear map of Loch Tay showing blocks of woodlands between the farmtouns, broadly the same impression as in Roy, ca. 1750, or in MacArthur's and Farquharson's estate maps of the 1760s. In this last instance, Blaeu for some reason disregarded the evidence presented to him by Pont. Conversely, an example of mere decoration arises in the Strathnaver area, where the Blaeu plate shows extensive woodland filling most of the space on each bank of Loch Loyal and along the west bank of Strathnaver, meeting in the middle, and stretching well up the hillsides, whereas Pont's original sketch shows two quite constricted and separate woodland areas that do not come anywhere near meeting, and which leave the hills bare (Fig. 3.1). Evidently either Gordon or the Dutch engravers exaggerated the extent of woodland in order to fill bleak empty spaces with a little more interest, rather resembling the latter's treatment of the Ptolemy map (see above p. 21). There is no question of anyone except Pont contributing to information for this map: in Dr Stone's words, 'The manuscript was the source of the printed map and is the more complete historical record.'[8]

Taken overall, however, the Pont maps and their derivatives inspire a good deal of confidence, at least in the Lowlands. Here the overall position as it was in Pont's day and for some time afterwards appears to be unambiguously and relatively accurately represented in Blaeu. The principal woods in Midlothian and East Lothian, for example, appear to be Roslin, Dalhousie, Dalkeith, Ormiston, Keith, Humbie, Pencaitland, Saltoun, Winton, Yester and Pressmennan, all depicted as enclosed, which is very much the same list as appears in woodland contracts of the seventeenth and eighteenth centuries. In Carrick in Ayrshire, woodland is depicted as being restricted to the river valleys which, however, are generously wooded, both with enclosed and unenclosed woods, much as described by Abercrombie's account of the late seventeenth or early eighteenth centuries: 'no countrey is better provyded of wood, for alongst the banks of Dun, Girvan and Stincher there be great woods, but especially on Girvan'.[9]

These early maps are probably a better guide to the location than to the size of the woods involved, though of course, woods may shrink or grow over time. The famous Torwood south of Stirling and west of Airth, the site of 'Wallace's oak' where the hero in legend hid from his enemies, appears in Pont (and subsequently in Blaeu) to cover about half a square mile, but

---

[8] T. C. Smout, 'Woodland in the maps of Pont', in Cunningham (ed.), *Nation Survey'd*, pp. 77–92; Stone, *Pont Manuscript Maps*, p. 20.
[9] Mitchell (ed.), *Geographical Collections*, 2, p. 3.

The Native Woodlands of Scotland, 1500–1920

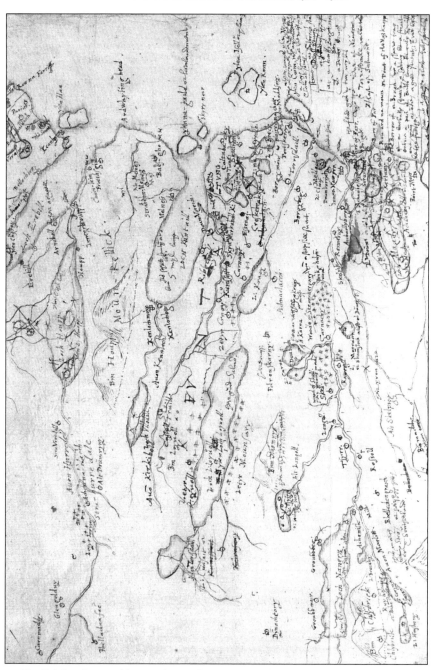

Fig. 3.1 Strathnaver (a) according to Pont, ca. 1590 (b) interpreted by Blaeu, 1654. With no further sources for Blaeu to work on other than Pont's sketch, Blaeu made the map more ornamental by stretching out the depiction of woodland. Reproduced by permission of the Trustees of the National Libraries of Scotland.

Extent and character of the woods, 1500–1920

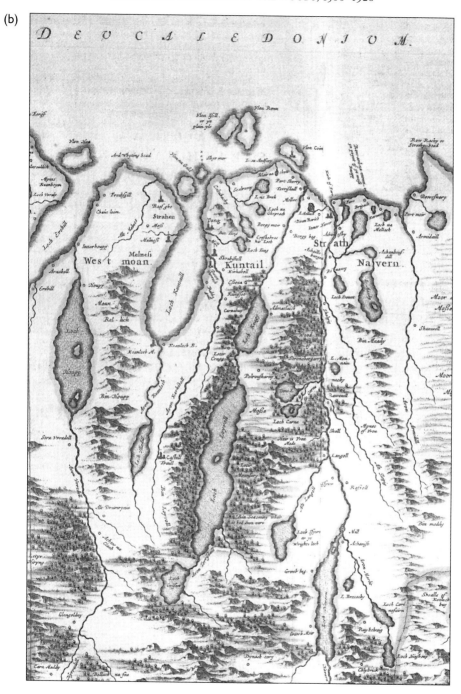

when John Adair made a map of Stirlingshire in 1685 it was only about half that size. As late as 1814 it was reported that 'some acres of natural coppice have been cleared' in the Lower Torwood, and that the land thus cleared was 'as valuable as any in Stirlingshire'.[10]

In the Highlands, the maps of Pont do not at all suggest forests of original splendour surviving throughout the glens and hills until the seventeenth century, as imagined by some authorities (see above p. 43), but, rather, patchy survivals mainly in the same locations as they are in today. There is, however, some ambiguity. Certain of the Gordon maps for areas for which there are no Pont originals use scattered tree symbols over the hillsides in a manner scarcely ever seen in Pont's own surviving sketches, although there is no evidence whatever that Gordon went to the places concerned.[11] Perhaps it is mere embellishment, or he may be trying to convey a sense of some feature about which he had information in these areas.

An interesting example of what may be a series of changes of mind, relates to the high country between the Spey and Ben Nevis, centred on Ben Alder. Gordon has left us two sketches of the area presumably based on lost work by Pont. One has no tree symbols at all, the other has it entirely covered by fairly widely-spaced tree symbols. When the same area is shown on a plate in Blaeu's *Atlas*, it is covered in diminutive, bushy symbols quite different from those used for mature woodlands: possibly Gordon had a third try, and sent another version with the new symbols to Amsterdam which he felt expressed Pont's intention better (Fig. 3.2). Some corroboration for this is offered much later by a comment from John Williams (1784), who speaks of 'a thick stool of oak [that] appears among the heath over great part of that extensive moor, which is situated between Fort William and the River Spey': it failed to grow higher, according to Williams, because of the misuse of farmers, and was cut, burnt and grazed back every year.[12]

Gordon in general shows in his surviving draft maps more woods in upland areas or the north and west than either the Blaeu *Atlas* or most of the Pont manuscript sketches: some of these again appear to resurface in Blaeu as the bushy symbols, especially in the far north-west. Perhaps the reality that we are dimly sensing is a much more extensive layer of montane scrub or thin wood pasture than anything that exists today, but not necessarily a large area covered by high or closed woods.

[10] J. Sinclair (ed.), *General Report of the Agricultural State and Present Circumstance of Scotland* (1814), 2, p. 329.
[11] Stone, *Pont Manuscript Maps*, pp. 6–8.
[12] J. Williams, 'Plans for a Royal forest of oak in the Highlands of Scotland', *Archaeologica Scotica*, 1 (1784), p. 29.

There is, however, available from Pont important and unambiguous evidence of several Highland woods existing in his day that have now disappeared. One good example comes from his manuscript map of the Glencoe area and Mamore (Fig. 3.3), finished to publishable standards but apparently rejected by Blaeu for an unknown reason. It shows (unusually for him) many tree symbols scattered between the hills, and on the south side of Loch Leven are the words 'many fyrre woods heir alongs', stretching from the River Coe to Kinlochleven and beyond. These latter were the woods the remnants of which were being exploited by the Dukes of Argyll in the 1750s, and which are depicted in Jan Dorret's military map of 1750 (which had an eclectic basis) as 'Corrienamor Firwoods'. Nothing is now left of this wood.[13]

Further interesting examples of substantial losses of western woods can be traced on Pont's rougher sketches of Wester Ross. He places a large wood along one half of the southern bank of Little Loch Broom, and another very major wood stretching on both sides of the Gruinard River and its tributaries, clothing both banks of Loch na Sealga and the north bank of Strath na Sealga, as high as Loch an Niel, with an outlier at Loch an Nid – a distance of at least twelve miles, excluding the outlier, now with little more than isolated pockets of birch. At two points along this strath he adds the comment 'fyrwood', east of Loch na Sealga and in Glen Ghiubhsachain, the Gaelic name of the latter indeed confirming that it was a pinewood. Part of an entry written over the area north of Loch na Sealga reads 'mechtie Parck of nature'.[14] This resonant phrase suggests open country with scattered trees, like a gentleman's hunting park on a gigantic scale. All this is unequivocal evidence of heavy losses of natural woodland in the area since the sixteenth century: Roy's map, of about 1750, shows considerably less than Pont, but a good deal more than what is presently there, so the decline has been protracted. Interestingly, Pont did not portray woodland on Gruinard Island, which Dean Munro in the 1540s had described (along with two other inshore islands in Loch Ewe and Loch Broom) as wooded.[15]

Woods also disappeared or changed their location on the east side of the Highlands. Pont's map of Glenesk shows substantial woods at the top of the glen, above Tarfside, around Loch Lee and south of Invermark Castle, all of which have vanished today (except for traces of aspen and birch on the screes) and most of which had already gone in 1750. On the other hand there were apparently few or no woods on the fertile middle stretches of the

[13] See Stone, *Pont Manuscript Maps*, p. 91.
[14] Ibid., p. 38.
[15] M. L. Anderson, *A History of Scottish Forestry* (London, 1967), 1, p. 207.

Fig. 3.2 Ben Alder and its environs according to (a) Robert Gordon and (b) Blaeu, 1654. Gordon was probably the immediate source of both versions, but the published Blaeu appears to be reaching for a depiction of montane scrub. Reproduced by permission of the Trustees of the National Libraries of Scotland.

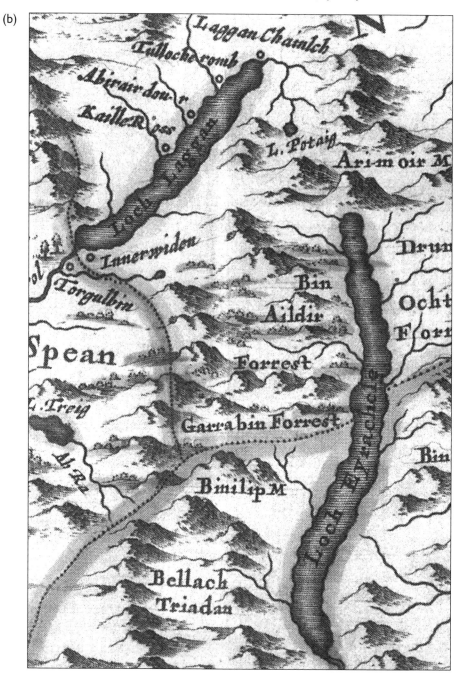

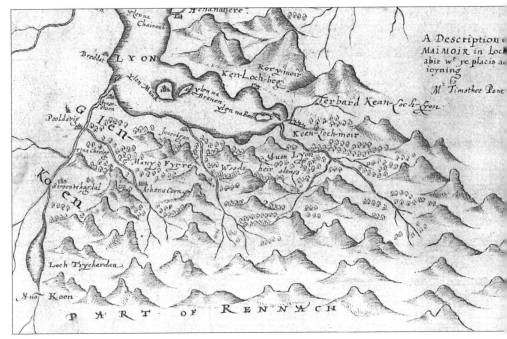

Fig. 3.3 The woods of Glencoe according to Timothy Pont: it is possible that the tree symbols themselves were added by Gordon but the lettering is Pont's.
Reproduced by permission of the Trustees of the National Libraries of Scotland.

glen below Tarfside, where birchwoods perhaps spread as the glen became less populated (Map 3.1).[16] Particularly where change was gradual and undramatic it is singularly difficult to track what was happening. A commentator, probably Pont, around 1600 called Kintail 'a fair and sweet country watered with divers rivers covered with strait glenish woods', and says that 'upon this river of Shiel also is a fair hollyn wood called Letur Choulynn'.[17] Over a century later, in 1719, a Hanoverian army overcame a combined force of Spaniards and Jacobites in Glenshiel, and made a sketch of the field of battle (Fig. 3.4). Small, discontinuous patches and strips of wood are shown in the strath, and a few further small clumps and scattered trees especially on the south-facing slopes, but nothing much either to impede or to assist military action. Today, apart from modern conifers, the slopes of the Five Sisters of Kintail and the course of the River Shiel are even less wooded than they were then.

One should be careful not to assume that all the woodland decline of this

---

[16] Smout, 'Woodland in the maps of Pont', pp. 87–92.
[17] Mitchell (ed.), *Geographical Collections*, 2, p. 544.

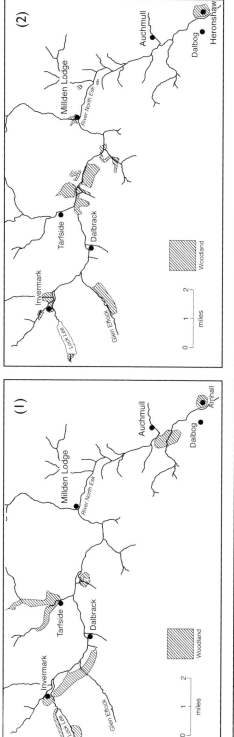
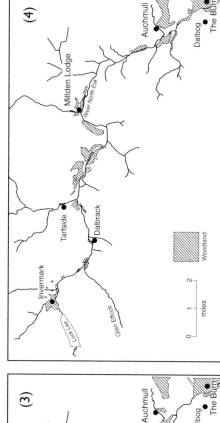

Map 3.1 Woodland in Glenesk, from maps 1590–1946. (1) Timothy Pont, ca. 1590. (2) Roy's Survey, ca. 1750. (3) Ordnance Survey, 1871. (4) Ordnance Survey, 1946. From I.C. Cunningham (ed.), *The Nation Survey'd: Timothy Pont's Maps of Scotland* (2001), with permission.

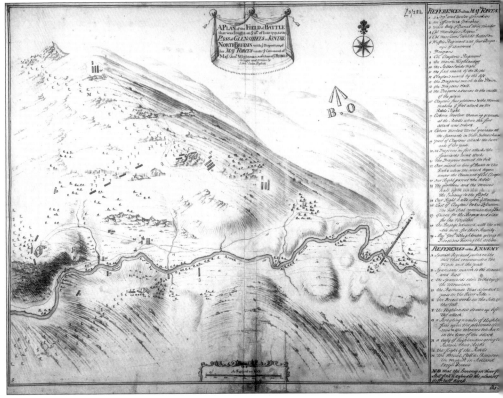

Fig. 3.4 Plan of the battlefield of Glenshiel, 1719. Such wood as is depicted appears then as sparse and stunted, but a century before there had evidently been much more. Reproduced by permission of the Trustees of the National Libraries of Scotland.

period was brought about by people rather than by nature. The sixteenth and seventeenth centuries were the nadir of the Little Ice Age, when temperature fell, precipitation increased, and small glaciers are thought to have appeared again in the high Cairngorms. With accelerated peat formation, especially in the west, woods could simply have failed to regenerate, collapsed and disappeared, as they had done before in earlier periods of climatic deterioration in prehistory.

In 1651, the future Earl of Cromartie, travelling over the neck of land between the head of Little Loch Broom and Loch Broom in Wester Ross, encountered a little inland plain about half-a-mile round, with 'a firm standing Wood; which was so very Old, that not only the Trees had no green Leaves, but the Bark was totally thrown off'. The local people ('the Old Countrymen') told him that 'this was the universal manner in which Firr Woods did terminate', and that within twenty or thirty years the trees would blow over and be carried off by the local people. Fifteen years later he returned, and there was no trace of the trees, only 'a plain green ground, covered with a plain green Moss', into which the wood had disappeared without even being touched by human hand.[18] The death of this wood appears to have been entirely natural, and the manner of its going completely familiar.

Something of the same impression is given by Charles Cordiner describing 'a dale called Derry' near Mar Lodge in 1780, full of huge collapsing pines, 'some thirteen feet in girth', blown over by the wind, and in Glen Lui sinking into the bog.[19] One of the effects of the worst decades of the Little Ice Age in Europe from the Vosges to the Sudeten mountains was to lower the tree line by about 200 metres, and the impact on high ground and the oceanic rim of Scotland might have been considerable. It is easy to imagine that the pinewoods noted by Pont on the south side of Loch Broom and in the twelve-mile stretch up from the mouth of the River Gruinard disappeared in the same way, with the woods on Gruinard Island and its neighbours going first because they were the most exposed. Similarly, fellings such as those at Glenorchy (see below, p. 358) might have removed mature trees, and the woods died out because there was no regeneration following in the growing mat of peat and moss.

A remarkable insight into woodland cover a century and a half after Pont is provided by Roy's Military Survey of Scotland, completed between 1747 and 1755 in the aftermath of the last Jacobite Rising, primarily in order to provide the army with a strategic tool in case of a further rebellion. It has been extensively used by the compilers of the modern Ancient Woodland Inventories in order to help them to locate some of the oldest surviving woods in Scotland, and thus contributes important base-line data for modern conservation initiatives.[20] It exists in two forms, both housed as part of the King's Topographical Collection in the British Library in London: a Protracted Copy, produced in Edinburgh Castle during the winter months, following summers of surveying in the Highlands, and a Fair Copy, into which additional information was fed from the surveyor's notebooks now lost, and sometimes from other sources. For the south of

---

[18] G. MacKenzie, Earl of Cromartie, 'An account of the mosses in Scotland', *Philosophical Transactions of the Royal Society*, 27 (1710–12), pp. 296–301.

[19] C. Cordiner, *Antiquities and Scenery of the North of Scotland* (Banff, 1780), pp. 27–30.

[20] G. J. Walker and K. J. Kirby, *Inventories of Ancient, Long-established and Semi-natural Woodland for Scotland* (Research and Survey in Nature Conservation no. 22, Nature Conservancy Council, Peterborough, 1989); A. J. Roberts, C. Russell, G. J. Walker and K. J. Kirby, 'Regional variation in the origin, extent and composition of Scottish woodland', *Botanical Journal of Scotland*, 46 (1992), pp. 167–89.

Scotland, only maps executed in the same format as the Fair Copy were ever made, and for the islands, no maps were made at all.[21] The Fair Copy, from black and white photographic reproductions, was used to compile the inventories of ancient woodlands.

It has, however, been shown by Whittington and Gibson not only that there are significant differences between the two versions, but that, when compared to contemporary estate maps, both are seriously inaccurate in respect to details such as the layout of farm settlements and unenclosed fields, though in other particulars such as enclosed fields, the Roy maps show a high degree of accuracy. They also suggest that scholars using the Military Survey to study the eighteenth-century landscape would be well advised to use the two versions of an area in tandem.[22]

We therefore undertook a sample comparison of the Protracted and Fair Copies in respect to their portrayal of woodland. In many cases, the depiction of woods in the two versions is so close as to be virtually identical: this was so for Abernethy and Loch Morlich on Speyside, Loch Creran and Benderloch in Argyll, Glen Livet in Banffshire and Glengarry in Perthshire. In other cases there are distinct differences where the more plausible and nuanced version is the Fair Copy. These include the north shore of Loch Rannoch, where the distribution of woodland corresponds better with a contemporary estate map; the woods to the north of Loch Awe, where there is more scatter on the hill; and the woods at Ballachulish, where there is more on the burnside close to the village and those near the ferry are moved several hundred yards back from the water's edge. At the head of Loch Broom, where the Protracted Copy shows a wood on the east bank as a rough circle, the Fair Copy shows a much more irregular shape. In one striking case in our sample, Strathcarron, the Protracted Copy has almost no woods at all, but the Fair Copy has very extended woods that correspond well with contemporary evidence and later maps.[23]

It appears as though the primary purpose of the Protracted Copy was to establish accurate outlines of rivers, lakes, shores and the location of mountains, while the Fair Copy added details from the surveyor's notebooks that often refined first rough impressions of other features. Woods, like enclosures, but unlike unenclosed fields, had a shape and location of potential strategic importance that apparently merited careful depiction,

---

[21] G. Whittington and A. J. S. Gibson, *The Military Survey of Scotland 1747–1755: a Critique*, Historical Geography Research Series no. 18 (Aberdeen, 1986).
[22] *Ibid.*, pp. 58–61.
[23] British Library: Maps C9. b. (the Fair Copy) and K.Top. XLVIII. 25. 11. Tab. 13d. (the Protracted Copy). Also Whittington and Gibson, *Military Survey*, pp. 41, 43, 45, 46, 31, 32, 33.

though this was probably only true where woods were in the neighbourhood of a track that an army might use. Whereas a map-maker like Blaeu might have been tempted to decorate a map with extra woodland for aesthetic reasons, a military surveyor would have found this irresponsible. Similarly, an estate plan might omit depiction of trees of little economic worth, such as trees along a streamside, or scattered in a wood pasture regarded primarily as grassland: Roy's surveyors would be more likely to include any feature that could conceal an enemy unless it was in a remote place.[24] Thus to use the Fair Copy alone as a primary source for identifying woodland cover seems entirely justified.

Nevertheless, the Roy map does not provide anything approaching a full picture of the extent of the native woods of Scotland in the mid-eighteenth century. Firstly, the size of many woods was greatly underestimated, such as the Black Wood of Rannoch, shown on a map made by John Leslie in 1756 for the Forfeited Estates Commissioners as about three times the size depicted on the Fair Copy.[25] Secondly, and more seriously, very many woods were overlooked.[26] The compilers of the Ancient Woodland Inventory counted 6,305 sites (74,165 ha) of ancient woodland still in existence (including PAWS) that were depicted both on the Roy map and on the first edition of the Ordnance Survey a century later. However, they also counted no fewer than 8,263 further sites (73,988 ha) still in existence that appear by their shape and structure to be ancient woodland sites, shown on the first edition Ordnance Survey but omitted from the Roy map. On the other hand, there were a further 2,549 sites (16,941 ha), a few of them originating as plantations, shown on the Roy map but omitted from the Ordnance Survey, but still in existence and mostly showing all the characteristics of ancient woodland.[27]

From this we can perhaps draw two conclusions. First, neither map tells the whole story: the Ordnance Survey omitted woods as well as Roy. Second, if the ancient woods still in existence are an accurate reflection of

---

[24] There is a good example illustrated in Whittington and Gibson, *Military Survey*, pp. 50–2, relating to Glen Livet.

[25] *Ibid.*, pp. 45–8.

[26] J. M. Lindsay, 'Charcoal iron smelting and its fuel supply; the example of Lorn Furnace, Argyllshire, 1753–1876', *Journal of Historical Geography*, 1 (1975), pp. 283–98. See pp. 294–6 in particular.

[27] From the metadata for the Ancient Woodland Inventory in Scotland, kindly supplied by SNH. In addition there were 708 'long-established woods of plantation origin' (LEPO), totalling 7,344 acres identified from the Roy map, and 10,687 such woods totalling 180,328 acres identified from the First Edition Ordnance survey. Records of LEPO from the Borders are incomplete, only 1,500 ha being identified, but some 8,000 ha believed to be present.

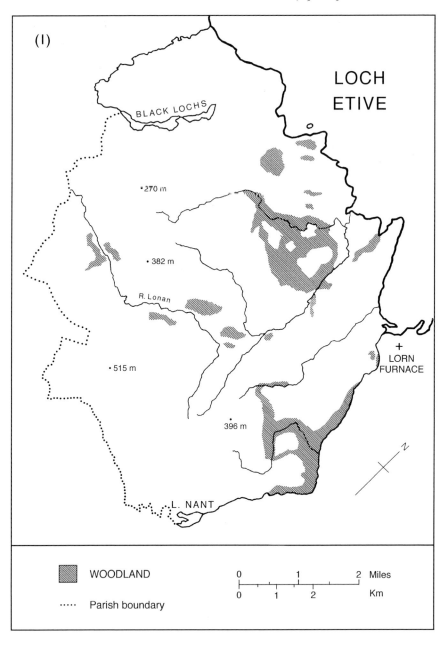

the total of ancient woods 250 years ago, Roy's surveyors omitted to record about one-half of their number. Map 3.2 shows an example of the difference between the military survey and the Ordnance Survey from Argyll. In general, the former tended to miss woods that were off the beaten track or where an army was less likely to march, which also reflected the haste in

EXTENT AND CHARACTER OF THE WOODS, 1500–1920

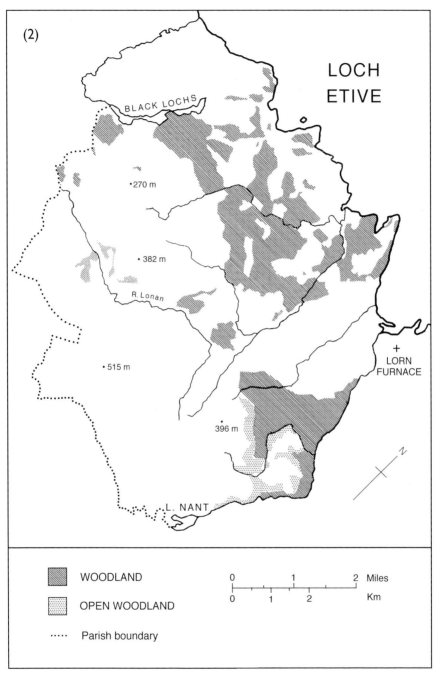

Map 3.2 (above and opposite) Woodland in Muckairn parish, Argyll, according to (1) Roy's survey, ca. 1750 and (2) Ordnance Survey, 1876.
From J. Lindsay (1975), *Journal of Historical Geography*, 1, pp. 295, 297, with permission.

which it was undertaken: Roy himself, significantly, preferred to refer to it as a 'sketch'. The woods omitted also tended to be the smaller ones, though not by very much: of the surviving sites, those depicted on Roy now average 10.2 hectares, those on the Ordnance Survey alone, 8.95 hectares. In other words, there is no way that Roy's map equates to a cartographic census of woodland.

So what percentage of the land surface was covered by native wood in the eighteenth century? James Lindsay sampled the Roy map and found that cover varied from 2 to 3 per cent in Lowland areas and some parts of the Highlands, but rose to 7 per cent or more in other Highland districts, suggesting an overall figure of 4 per cent that has been frequently cited since.[28] If, however, Roy was only showing about half the woods in Scotland and underestimating the size of some, this figure should be doubled.

The first contemporary attempt to estimate the extent of the woods was made for Sir John Sinclair in the *General Report of the Agricultural State and Political Circumstances of Scotland* in 1814, which concluded that the extent of plantations was 412,226 English acres (165,000 ha), and of 'natural' woods, 501,469 English acres (200,600 ha): this equates to about 5 per cent of the land surface, with less than 3 per cent under natural wood. Sinclair, however, was exceptionally cautious in his claims for these statistics: 'in the above statements, many errors will no doubt be found . . . it is only possible to approximate to the truth . . . without actual surveys, accuracy cannot be expected'.[29]

The figures were compiled from the various parish *Statistical Accounts* and country *Agricultural Reports* undertaken in the previous two decades, all estimates made by different people and some with different aims in view. They do, nevertheless, deserve careful scrutiny both for what they conceal and reveal. Table 3.1 shows the original estimates.

Comparison with the extent of ancient woodland surviving in Scotland (whether as semi-natural ancient woodland or as PAWS – see pp. 1, 17 above), and reported in the modern inventories, shows that very often the figures in the *General Report* must have been serious underestimates.[30] Modern Argyll has 22,000 hectares (55,000 acres): Sinclair reported 30,000 acres, which was the lowest of three estimates he had presented to him (the others were 72,000 acres and 98,000 acres). Modern Inverness-shire has 52,800 hectares (132,000 acres): Sinclair reported 45,600 acres. Modern

---

[28] J. M. Lindsay, 'The commercial use of woodland and coppice management', in M. L. Parry and T. R. Slater (eds), *The Making of the Scottish Countryside* (London, 1980), p. 272.
[29] Sinclair (ed.), *General Report*, 2, p. 320.
[30] Roberts *et al.*, 'Regional variation', pp. 188–9.

Sutherland has 5,200 hectares (13,000 acres): Sinclair reported 3,000 acres. Modern Galloway and Dumfries has 7,300 hectares (18,300 acres): Sinclair reported 6,800 acres. Modern East Lothian has 1,000 hectares (2,500 acres): Sinclair reported 400 acres. If these examples are totalled together, and it is assumed that the amount of ancient woodland there today is the same as it was in 1814, they alone would add 135,000 acres to Sinclair's total.

Table 3.1 Sinclair's estimates of natural woodland cover, 1814

County	1,000 acres	County	1,000 acres
Aberdeenshire	74.0	East Lothian	0.4
Angus	5.6	Inverness-shire	45.6
Argyll	30.0	Kincardineshire	0.6
Ayr	6.0	Kinross	0.0
Banffshire	6.0	Lanarkshire	2.15
Berwickshire	0.5	West Lothian	0.2
Caithness	0.6	Nairn	8.0
Clackmannan	0.9	Peeblesshire	0.5
Ross and Cromarty	72.0	Perthshire	118.9
Dunbartonshire	7.0	Renfrew	0.5
Dumfriesshire	3.0	Roxburgh	0.6
Midlothian	3.0	Selkirk	0.0
Morayshire	21.0	Stirling	4.0
Fife	0.0	Sutherland	3.0
Galloway	3.8	*Total for Scotland:*	501.5

Source: J. Sinclair (ed.), *General Report of the Agricultural State and Political Circumstances of Scotland* (Edinburgh, 1814), 2, p. 321.

However, very clearly there has been loss of ancient woodland since 1814 in these counties as in every other. How much might it have been? The estimate for Aberdeenshire was more careful than most, and listed 74,000 acres where now there are about 22,000 acres: Perthshire reported 118,900 acres where now there are 21,000 acres: Ross and Cromarty reported 72,000 acres where now there are also 21,000 acres. If we were to suppose that all these reports in 1814 were accurate and comparable to the modern reports, it would suggest that in these counties ancient wood (including PAWS) covers today only one-quarter of the ground that it did in 1814 – if extrapolated, the natural woodland cover for the whole of Scotland would have been four times what it is at present, or around 7 per cent. Adding another 2 per cent for the plantations, this turns out to be close to the revised estimates for the time of Roy.

A lot of confusion arose, surely, because of uncertainty as to what to include as a wood. The Aberdeenshire reporter said that only about a quarter of his total contained 'trees which are likely to become timber', though some of the rest would have local uses in construction and agriculture: his was an inclusive definition. In Inverness-shire, Sinclair was told that much of the supposed wood was valueless scrub, and the total for that county clearly left it out. In Perthshire, Dunbartonshire, Argyll and Inverness-shire reference is repeatedly made to scrub oak, or stools of oak and other native trees, that put forth leaves seasonally only to be cut back by grazing animals: sometimes such land could be grown on to mature woodland simply by fencing and exclusion of stock.[31] The *General Report* cited examples of this during the Napoleonic Wars when the price of timber and bark repaid the effort and sacrifice of pasture: thus on the Montrose estate oak was spreading over the western shoulder of Ben Lomond and in other localities where only heath and ling had predominated, and west of the bridge at Kilmahog in Callander parish a similar extensive tract of land once moor was becoming covered in trees.[32] This process of alternating regeneration and overgrazing, usually the latter, must have left a lot of land in a condition ambiguous to the eye of the beholder, to be counted as wood by some and as pasture by others.

Such a situation is also reflected at this time in the accounts of travellers, most of whom report the uplands as barren, but a few of whom have a different perspective. Pre-eminent among these was James Robertson, employed by Professor John Hope of Edinburgh University between 1767 and 1771 to carry out a botanical survey of the Highlands. As a professional botanist and a careful observer both of plants and people, his testimony has particular weight.

Robertson described several of the mountains that he traversed as at least partly wooded, even though Roy and other map-makers did not necessarily describe them as such. In 1771 he described the land between Braemar and Ben Avon: 'the tops of the hills produce little but the procumbent azalea; but their sides . . . are covered with several small copses of birch and hazle', and again: 'the hills on each side [of Glen Avon] are covered with hazle, birch, alder, quaking asp, willow, bird's cherry, roses and honeysuckle'. The conjunction of these two passages seems to imply he is talking of lower slopes, and the same may be true when he visited the district from Dingwall up to Loch Monar and down Glen Moriston, saying that 'wood grows naturally upon the hills' and that 'this tract presents a continued succession of lakes, and hills that are covered with birch, hazle, poplar and fir'. He noticed that the sides of Loch Ness were fronted by 'a chain of steep hills, which are covered with birch, hazel, poplar and ash,

---

[31] Sinclair (ed.), *General Report*, 2, pp. 217–22, 306–18.
[32] *Ibid.*, p. 222.

together with some holly, elm and mountain ash'. Between Kinlochleven and Tyndrum he spoke of a tract of land that was either 'forsaken waste' or 'high mountains in many parts bare, in other, covered with birch or fir: common ash and hazel', and a little further on in Glen Dochart, of the hills 'either covered with wood or good grass'. In other stretches of upland he found different and empty landscape: between Spittal of Glenshee and Deeside 'the intermediate ground is hilly, rocky and barren, producing little but heath and bearsberry', between Inverness and Carrbridge 'the ground is very barren', the rising land beyond Blairgowrie was 'unequal, hilly, uncultivated and for the most part covered with heath'.[33]

To summarise so far, it seems likely that by the start of the nineteenth century up to 9 per cent of the land was still in some sense wooded, though probably more than half of this was uncommercial woodland, scrubby pasture or montane scrub, which by some definitions would not count as wood at all. At this point, the only influence that industrialisation had had upon the woods was to encourage landowners to take greater care of them as sources of tanbark, charcoal and building timber, and in places to extend them. Planting of non-native and native species had already begun, and accounted for about a fifth of the total cover. This figure is an upward revision from what has hitherto been accepted for this period by many authors, including ourselves, who have taken the figure of 4 per cent too much at its face value.[34] Nine per cent is still a tentative estimate, and even if correct, would still leave Scotland towards the bottom end of the European league table of countries with woodland cover.

In the course of the nineteenth century, two developments substantially affected the extent and character of the native woods, the coming of modern sheep farming and the arrival of forestry. Their impact is not very clearly revealed by the official statistics, however. No estimates were made between 1812 and 1872, but in the latter year a Board of Trade calculation put the figure of land under wood at 734,490 acres (294,000 ha). At least one contemporary was cynical: 'the accuracy of these figures may well be doubted, as the owners and occupiers of land are often averse to give correct statistics on a matter which might afterwards affect their interests'.[35] These objections should have been removed in later surveys when the

---

[33] D. M. Henderson and J. H. Dickson (eds), *A Naturalist in the Highlands: James Robertson, His Life and Travels in Scotland, 1767–1771* (Edinburgh, 1994), pp. 102, 152, 154, 158, 160, 179, 181, 184–5.

[34] See, for example, T. C. Smout, *Nature Contested: Environmental History in Scotland and Northern England since 1600* (Edinburgh, 2000), p. 46; Walker and Kirby, *Inventories*, p. 9.

[35] A. Smith, 'On Aberdeenshire woods, forests and forestry', *TRHAS*, 4th series, 6 (1874), p. 265.

labours of the Ordnance Survey could be used to cross-check the owners' returns, but the First Edition map was not totally reliable either. Over a quarter of the woods marked on Roy and still identifiable today were not to be found on it, equivalent to 5 per cent of the area but suggesting a larger underestimate. Between 1880 and 1908 a further eight estimates were made, ranging between 811,703 acres (324,681 ha) at the former date to 946,000 acres (129,872 ha) at the latter, or between about 4 per cent and 5 per cent of the land surface (see below, pp. 258–9).

Even the last estimates, however, hardly told the whole story. Neil MacKenzie and Rick Worrell, working back over the twentieth century from the modern statistic of only about 1 per cent of Scotland under ancient semi-natural woodland and observing the documented declines that had taken place over that time, concluded with good reason that around 1900 ancient semi-natural woodland must have covered at least 3 per cent of the land surface.[36] Woods of planted origin were probably as numerous. No-one in the nineteenth century tried after 1812 to discriminate between the two categories, but the compilers of the Ancient Woodland Inventory in the 1980s and 1990s found that from the first edition Ordnance Survey the area of surviving long-established woods of plantation origin exceeded (just) the area of surviving ancient woods of semi-natural origin, and it can be calculated from a survey published in 1885 that 60 per cent of the reported woodland in Ross and Cromarty had been planted, though this may have been above the Highland norm. Perhaps the true figure for Scottish woodland cover around 1900, therefore, was around 6 per cent, some still being missed out of the official figures particularly in the general categories of upland wood pasture or thin scrub.[37] From these figures we can dimly see that half the ancient semi-natural wood surviving around 1800 had probably been lost one hundred years later, the 7 per cent in this category that we identified around 1812 having fallen to perhaps 3 per cent or a little more.

What happened? The woodlands of Victorian Scotland were of course partly ancient semi-natural woods and partly plantation. The former declined as the latter increased and replaced them, but by no means all the decline could be accounted for by conversion. Many ancient woods simply

---

[36] R. Worrell and N. MacKenzie, 'The ecological impact of using the woods', in T. C. Smout (ed.), *People and Woods in Scotland: a History* (Edinburgh, 2003), p. 198, and N. MacKenzie, *pers. comm.*

[37] W. F. Gunn, 'The woods, forests and forestry of Ross-shire', *TRHAS*, 4th series, 17 (1885), pp. 133–202. The early plant ecologist W. G. Smith in 1911 thought that most Scottish woods were 'neither virgin forests, nor plantations, but are the lineal descendants, so to speak, of primitive woods': 'The vegetation of woodlands', *TRSAS*, 24 (1911), p. 11.

vanished, especially those of little commercial value in the uplands, and the main agents of decline were sheep and game management.

Even before the clearances gained momentum there had been instances of woods (for example in Sutherland) disappearing due to overgrazing, sometimes with dire human and ecological consequences (see below, pp. 119–21). From the first quarter of the nineteenth century, however, the sheep-grazing pressures became vastly greater. Robert Dodgshon has calculated that they became between four and eight times as high as they had been in the eighteenth century due to the increase in numbers and in size of animal, and new ways of shepherding to keep the stock longer on the hill.[38] At first, with good shepherding and an estate that valued even its birch woods, the new flocks were compatible with flourishing woods, as experience in Sutherland at Assynt and Strathnaver amply demonstrated: by the 1830s the woods there were 'rising rapidly' and 'now an object worthy of attention': they should be 'put upon a regular system of cutting so as to make use of that which is decaying and at the same time so as not to denude the sheep farms'.[39] More often, however, and increasingly so as profit margins were squeezed and shepherds reduced in number, the woods suffered. Gunn in 1885 gave some examples in Easter Ross. At Balnagown, 'a considerable extent which was once formerly under wood' had been 'reclaimed', and at Tarbat in the same county only seventy acres were left of the old pines 'which had once occupied the whole area': the wood had been grazed by sheep which 'destroyed the natural growth of fir which was springing up as the wood got thinner'. Sometimes there was even conversion to ploughed land in districts where planting turnips for winter feed for stock on the hill was profitable: at Cadboll 'we understand several hundred acres of woodland were reclaimed into arable land of fair quality since 1850', though in general in Ross-shire, 'many tracts of land at present under cultivation . . . should never have been reclaimed'.[40]

Burning the open hill to improve the ground for sheep and grouse was the other great enemy of the wood, sweeping out much ground covered with birch, willow and even oak scrub, while grazing by sheep and deer then prevented any further regeneration. Nairne further condemned the replacement of cattle by sheep on the hill for its effect on the native woods:

---

[38] R. A. Dodgshon, 'The Scottish Highlands before and after the clearances: an ecological perspective', in I. D. Whyte and A. Winchester (eds), *Upland Landscapes* (Society for Landscape Studies, 2004).
[39] M. Bangor-Jones, 'Native woodland management in Sutherland: the documentary evidence', *Scottish Woodland History Discussion Group Notes*, 7 (2002), pp. 1–5.
[40] Gunn, 'Ross-shire', pp. 171, 173–4, 194.

whereas the heavier animals had trodden the seed into the ground and grazed less closely, under the new regime 'the area under wood ceased its natural expansion, the young seedlings being all eaten up, while the herbage got so rough that there was not a suitable bed for it to fall in'.[41] It is easy to see how the woods suffered, particularly those in the ambiguous categories that Sinclair's commentators and later enumerators were uncertain whether to include in their returns or not – the wood pasture, the montane scrub, and those tracts at lower levels which traditionally had come up in summer and been eaten off in winter.

Then there was the question of the new forestry. Increasingly a high proportion of Scottish woods were plantations, established either on new ground, or on the sites of existing semi-natural woods. In 1750, the proportion of planted trees was still very small, despite a long history of legislation to encourage planting, and efforts by noblemen in the seventeenth and eighteenth centuries to plant both for ornament and profit, as at Yester and Tyninghame in East Lothian, Breadalbane in Perthshire, Glamis in Angus and Melville in Fife.[42] Thereafter, the vogue for planting snowballed rapidly under the leadership of magnates like the Dukes of Atholl, the Earls of Moray and the Seafield Grants. By the close of the nineteenth century, the Scots were synonymous with forestry in Britain, and the Scottish landowners were highly influential in putting pressure on London to create a state forest service (see pp. 278–80 below).

What was the impact of this enthusiasm and effort on the character of the semi-natural woods? A vast number of new tree species were introduced by the Victorians, many of the most important by the explorer and plant collector David Douglas, once on the gardening staff of Scone Palace. Sitka spruce, lodgepole pine, grand fir, noble fir, Douglas fir, Wellingtonia and (from Chile) monkey-puzzle were all involved: there are Douglas firs still growing that originated from the seeds sent back by the explorer to his former employer and his friends. But such trees were still confined to policies and arboreta round the great houses, in some cases serving as field trials for species that would be important in the following century. Beyond the policies, non-native species were also extensively used. Sycamore (plane in the documents) had been introduced into Scotland probably during the Middle Ages. Beech had first been tried at Yester and elsewhere during

---

[41] D. Nairne, 'Notes on Highland woods, ancient and modern', *Transactions of the Gaelic Society of Inverness*, 17 (1892), p. 220.
[42] Anderson, *History of Scottish Forestry*, 1; S. House and C. Dingwall ' "A Nation of Planters": introducing the new trees, 1650–1900', in Smout (ed.), *People and Woods*, pp. 128–57; A. C. O'Dell, 'A view of Scotland in the middle of the eighteenth century', *Scottish Geographical Magazine*, 69 (1953), pp. 58–63.

the later seventeenth century. Both were now widely planted for shelter belts and small woods, especially in the Lowlands. Among non-native conifers, larch, Norway spruce and Corsican pine all became popular for commercial planting during the nineteenth century, often in the Highlands. The other main species involved were oak, albeit often the English pedunculate species, *Quercus robur*, planted to fill out and 'improve' existing coppice of native sessile oak, *Quercus petraea*. Above all, Scots pine was planted in very large quantities in the nineteenth century, sometimes on new sites but often on old ones.[43] That said, Victorian planting was on nothing like the scale of that of the twentieth century: it covered only about 3 per cent of the country, unlike the 15 or 16 per cent at the present day.

The activities of the Grant chiefs on their vast Seafield estates are symptomatic of many, though on a much larger scale than others. David Nairne in 1890 spoke of 50,000 acres of Scots pine planted by Sir Francis Grant and his successor, and of close to 200 million trees being planted on Seafield land 'during the last half century'. He called it 'the greatest planting experiment on record'.[44] The famous Forres arboriculturalist and seedsman, John Grigor, considered native seed from Strathspey far superior to German or other strains, and although the latter were sometimes used they produced what he termed 'degenerate plantations'. Most proprietors probably heeded his advice.[45] The Seafield estate divided its Speyside holdings into 'three great divisional forests... Duthil, Grantown and Abernethy', with a twelve-acre nursery at Abernethy (presumably supplying all three), growing, in 1884, 'over two million plants preparing for transference to the hillsides'.[46] Yet the Abernethy Forest is now considered a triumphant example of semi-natural woodland conserved.

Such an opinion is not unreasonable, as the forest would in some respects have been little disturbed. Nineteenth-century forestry was unable to cope with deep peats or high, rocky hillsides where the Sitka spruce reigns supreme today. The choice of the planters for Scots pine was gravelly, moraine ground or well-drained moor at moderate altitude. Planting was not by the crude force of deep ploughing and caterpillar tractor, but (as described by Nairne), by two foresters, assisted by a woman, and using a garden spade.[47] The disturbance of ground flora would have been minimal.

---

[43] Anderson, *History of Scottish Forestry*, 2; L. Mitchell and S. House, *David Douglas, Explorer and Botanist* (London, 1999); House and Dingwall, '"Nation of Planters"'.
[44] Nairne, 'Notes on Highland woods', p. 200.
[45] J. Grigor, 'Report on the native pine forests of Scotland', *TRHAS*, 2nd series, 12 (1839), p. 131; Nairne, 'Notes on Highland woods', p. 200.
[46] Nairne, 'Notes on Highland woods', p. 200.
[47] *Ibid.*, p. 201.

Of course, where exotics were planted, or where Scots pine was planted on open ground that had not been under trees for centuries, there was nothing 'semi-natural' about the ensuing wood. But where native trees were planted within an existing wood, or immediately adjacent to one, or even on a site of an old wood recently clear-felled, the area might well assume some of the characteristics of semi-natural woodland, including parts of its distinctive flora and fauna. This was true even though, in the case of oakwoods, *petraea* and *robur* might hybridise and change their genetic structure, and in the case of Scots pine woods, the plantation was likely to show an unnaturally even age-structure compared to the original. A good example of the latter is Balblair Wood on the north side of Loch Fleet in Sutherland, an even-aged plantation a century old, boasting exceptionally rich northern pinewood flora, including the best population of the one-flowered wintergreen (*Moneses uniflora*) in Britain. It was planted in 1905 on the site of an ancient semi-natural pinewood damaged by windblow but its nature-conservation value is greater than that of many recognised relics of the old semi-natural pinewoods themselves.

Planting on old woodland sites therefore may not always have obliterated their previous ecosystems, but it obviously modified their species diversity in very many cases. Sometimes the woods were given understorey patches of snow-berry, for game shelter and food, or of *Rhododendron ponticum* for ornament, and both often spread far beyond their original coverts, as in the oakwoods of Brodick on Arran or the pinewoods round Achnacarry on Loch Arkaig. Beech and sycamore planted around the perimeter of ancient woods for ornament and use might become invasive, as beech has at Twenty Shilling Wood at Comrie in Perthshire, and sycamore at Flisk in Fife – in each case single instances among many. As oak coppice wore out through excessive or ill-managed cutting, the gaps that appeared might be planted up with specimens of Scots pine, as at Rowardennan on Loch Lomond, or with Norway spruce, as again at Twenty Shilling Wood. As the price of oak bark fell, Gilchrist in 1874 recommended that blank areas among the stools should be filled in with larch: 'they will, even after only twenty years' growth, be found a more valuable crop than planted oak'.[48] McCorquodale had more radically, six years earlier, recommended the general replacement of oak coppice by larch – 'it is a well-known fact, that larch grows better among oak roots than anywhere else'.[49]

---

[48] A. Gilchrist, 'On the treatment and management of oak coppice in Scotland', *TRHAS*, 4th series, 6 (1874), p. 126.

[49] W. M'Corquodale, 'On the conversion of coppice land into a more remunerative crop', *TSAS*, 4 (1868), p. 48.

In an earlier generation, ironically, foresters had strongly advised that all other species be weeded out from among the valuable oaks. Monteath in 1827 taught that 'oak, and nothing but oak, is the only profitable trees for coppice cuttings, and wherever such a plan is intended, nothing else should be reared'. He advised one of his customers to see 'the whole of the birch cut over ... and the birch growths extirpated, and the whole ground filled up with oak', and another to remove birch, alder and planted larch which had been allowed among the oak stools – 'a most injudicious and ruinous method of planting'.[50]

Since there was much profit in oak coppice for tanbark and charcoal in the earlier nineteenth century, we may expect owners to have paid attention to the wisdom of Monteath and his contemporaries. The 'semi-natural' woods of Argyll, Dunbartonshire, Stirling and Perthshire must be far more of a monoculture of oak than they were originally, with more of the oaks having been planted by hand than is discernible today, and many other native species grubbed up. Occasionally we get a glimpse of the more diverse character of these woods before intensive management. In 1751, Dugald Clerk of Bralecken represented to the Duke of Argyll that:

> The whole different species of woods upon my lands such as ash, elm, birch, hazel, alder, rowan, gean, quaking ash [aspen], sallies [willows], haw and sloethorn trees etc. are disponed over to me by your Grace's predecessors and only the oaks, a small part excepted for biggings and other uses of the ground, reserved to your grace.

But, he went on, the oaks in most places made only the smallest share, and the various trees were:

> so intermixed and grow so thick and close together in the form of a hedge or one continued thicket, galling, rubbing upon and smoothing one another for want of due weeding and pruening that it is morally impossible for any person to enter in to these thickets to cutt or weed either any of your Grace's oaks or my other woods without at the same time breaking or cutting less or more of the other in order to get access to those thickets.[51]

It would be hard to find a wood of this character on Loch Fyne today, though woods with a similar mix of species can still be found in woods on

---

[50] R. Monteath, *Miscellaneous Reports on Woods and Plantations* (Dundee, 1827), pp. 102, 121, 137.
[51] NLS: MS.17665, fo. 66.

Islay and Mull. It bears a striking resemblance to the kind of wood that Vera envisages as characteristic of the original cover of Europe after the last Ice Age, very dense, with oak coming up amid thorny scrub.

It is not difficult in other pre-nineteenth-century documents to find accounts of woods with greater species diversity than most modern woods. Thus, the wood of Cruixtoun was sold in 1599 with 'asches, planin, birk, aller, haisill, sauch, chirrie tries' and 'aik treis' – that is, ash, sycamore, birch, alder, hazel, willow, gean and oak. In the same area, on the Arnprior estate near Kippen in Stirlingshire, the Annexed Estate Commissioners in 1782 received a report from consultants who found in the oakwood 'a variety of different trees, such as natural ash, elms, geans, plains [sycamores] and a gray willow and birch, these in such a quantity as to be near or about a third part of the whole wood ... very useful to the countrey for many purposes of husbandry'.[52] On the other side of Scotland, a 'hagg' [division] of wood at Newhall in Yester parish in East Lothian in 1733 was described as 'consisting of oak, ash, birch, plain, thorn, alder and quaikasp [aspen]', and in the same parish in 1755 from the Castle Wood of Yester House were sold 'all the firr trees, all the beech trees [clearly planted] one hundred and ninety seven oak trees, three hundred and thirteen elm trees, one hundred and five ash trees, two saugh trees, nineteen plain trees and four alder trees'.[53]

The species diversity in the Scots pine woods has suffered in a comparable way from the diminution in particular, of birch and aspen. When between 1725 and 1730 Joseph Avery made for the York Buildings Company the first attempt to map the woods of the northern part of the Great Glen, with Strathglass and its tributaries, he described Glen Affric with 'very lage firrs and birches', and 'fine firrs and birches', Glen Cannich with a 'firr and birch wood', Glen Moriston with 'firr and birch' (and an oak wood further down), Loch Garry with 'oak and birch', 'oak, aller and firr' and 'oak and firr' woods. 'Fir', of course, always meant pine to contemporaries. On the south side of Loch Ness he delineated birch woods, and alder, holly and hazel woods.[54] In other words, he does not normally describe single-species woods. Similarly when James Robertson was in Rothiemurchus in 1771 he described the forest as mixed fir and birch, with a good deal of aspen and hazel – not a description that would immediately seem appropriate today. His accounts of the varied woods around Loch Ness we have already noted, and when he came to Abergeldie on Deeside he contrasted the 'plantations of fir' with the 'many natural woods of birch, hazel, poplar

---

[52] NAS: GD 220/6/2004/6; NAS: E 738/59/2.
[53] NAS: RD.13. reg. 26 May 1735; NLS: MS 14754, fos 54–4.
[54] Inverness Museum MS: J. Avery, 'Plan of the Murray Firth, 1730'.

and bird's cherry'.[55] In 1760, the factor on the Lochiel estates explained that 'the firr woods are so interspersed with the other woods that they cannot conveniently be cutt seperatly'.[56]

In most parts of the Highlands today birch remains an integral part of an old Scots pine wood, but not usually as co-equal to the pines. Only occasionally, as in Glenstrathfarrar and parts of Glen Affric, would Avery's phrase 'fine firrs and birches' seem apposite. This is because nineteenth- and twentieth-century foresters for long treated the birch as the 'weed of the forest'. When the last Countess of Seafield took a carriage ride through Abernethy, she always complained to the forester if she saw a birch among the pines.[57] In fact the attempt to extirpate birch probably reduced the fertility of the wood, as its presence enriches the ground rather than otherwise.

Holly and rowan are other species likely to have been relegated to a more minor role in Highland woods, not least because, when young, they are selectively grazed by sheep and deer. One would not now, for example, find in Strathnaver a 'brik [birk] and holyn wood' such as Timothy Pont mapped around 1590, or be able to describe, as he or a close contemporary did, the shores of Loch Maree as growing 'plentie of very fair firr, hollyn, oak, elme, ashe, birk and quaking asp, most high, even, thicke and great, all-longst this loch'. He picked out three dominant components of the Loch Maree woods, 'sum parts with hollyne, in sum places with fair and beautifull fyrrs', and some places with 'excellent great oakes': it is the holly that is not recognisable as dominant anywhere now.[58]

Finally, there has fairly clearly been, over the last two centuries, a severe decline in montane scrub and open wood pasture on the higher land. Little is recorded of when or how it went, but muirburn along with the ceaseless grazing of heavy densities of sheep and deer surely lie at the root of this decline. It is impressive to see within the exclosures of the National Nature Reserve at Inchnadamph in Sutherland how the arctic willows, the holly and the rowan return after two or three decades of close fencing, and there is now similar recovery at Ben Lawers. One has a sense there, for a moment, of seeing what the moorland cover of the Highlands might have been like before the nineteenth century.

If one takes a perspective encompassing the whole sweep of time since the woodland maximum 5,000 years ago, the reduction appears to be of

[55] Henderson and Dickson (eds), *Naturalist in the Highlands*, pp. 155, 167, 181.
[56] NLS: MS.17694, fo. 143.
[57] Roy Dennis, *pers. comm.*
[58] Mitchell (ed.), *Geographical Collections*, 2, p. 539.

long standing and relatively continuous, and, over long periods at least, to be partly natural. Insofar as it was caused by human activity, much of that appears to have taken place at extremely early dates, and to have been carried far by the end of the Middle Ages. If our figures are anything like correct, the present figure of about 1 per cent of the land surface covered by ancient semi-natural woodland today nevertheless represents only about a seventh of the resource even in 1750, maybe about a third of the resource in 1900. There is much to investigate to explain the causes of the decline.

CHAPTER 4

Woodland produce

The economic importance of wood in early modern Scotland was immense. No building from the grandest castle to the smallest home could be constructed or furnished without it. No boat could be built or wheeled vehicle made without wood. No tool or agricultural implement could be fashioned, no fence made to protect crops or stock, no fish or other food packed for keeping, no baskets made of any strength, no mine sunk below the ground, no machinery devised for milling, draining, spinning or weaving, that did not utilise wood. Even nails and locks might be of wood. Wood products were the basis of the very important industry of tanning hides and leather, were used to preserve fishing gear and boats, were important as dyestuffs, were manufactured in some places into ropes or into lights. Wood was burned as domestic firewood, a convenient if rapidly consumed fuel. It was important as kindling in some industrial processes like lime-burning that otherwise used coal; it was vital for bakers who needed a hot fire quickly, and to fish curers and meat curers needing woodsmoke. As charcoal it was almost invariably used in Scotland in the smelting of iron until after 1750, and was the preferred fuel for a high-grade product for long after that.

In the nineteenth century there were substitutions. Coal finally completely replaced charcoal in smelting, and consequently cheap iron and steel largely replaced wood in the manufacture of machinery and ships and in the frames of large buildings. Coal tar replaced wood tar, and chemicals supplemented vegetable tanning. Nevertheless, demand for wood in the economy continued to grow rapidly, partly because there were many uses that were not substituted, particularly in house construction and in such specialised but important industrial uses as bobbin manufacture, and partly because new uses were found. The iron horse ran on iron rails, but countless wooden railway sleepers were needed as their bed. The development of pyroligneous acid manufacture brought small chemical factories into the Scottish woods, though the application after 1880 of wood pulp to paper manufacture, that great consumer of the world's northern forests today, used only foreign supplies. Increased demand for wood does not necessarily equate with increased pressure on the Scottish resource, as, for many purposes, imports had been cheaper and more plentiful ever since the end of the Middle Ages (see below, pp. 124–33). By the Victorian period, the invention of the steamship and the railway, combined with the ideological triumph of free trade, had greatly simplified and cheapened the movement of timber products across the world and within the country, and relatively less use was made within Scotland of native woods than in previous centuries.

Until that time, however, the penetration of the Scottish market by foreign wood was confined to the immediate hinterland of coastal towns and other places where a ship could lie at anchor and unload a foreign cargo. Though this zone covered all the main burghs and many of the smaller ones, much of the countryside was left to supply itself as best it could from local supplies (see below, pp. 134–56).

Given their economic importance, it is not surprising that trees also had a certain place in Scottish cultural life, in the Highlands in particular. Individual trees of great size and antiquity were revered in Lowlands and Highlands alike, such as the Wallace's Oaks in the Torwood, Stirlingshire, and at Elderslie, Renfrewshire (in both of which the hero was said to have hidden from the forces of Edward I), the Birnam oak near Dunkeld (once known as the Hangman's Tree), the Capon Tree, near Jedburgh, another oak, or the great yew at Fortingall, Perthshire.[1] One of the most interesting was the ancient ash at Kilmallie in Lochaber, fifty-eight feet in circumference, 'held in reverence by Lochiel and his kindred and clan, for many generations', and burned by Cumberland's troops in 1746 in a deliberate act of cultural vandalism.[2] Ash in Norse lore was Yggdrasil, the World Tree, Odin's horse, its roots in the underworld, its trunk on earth, its crown holding up the sky. Clasped by the Christian vine, it forms the symbolic motif of the Prentice Pillar in Rosslyn Chapel, the fifteenth-century church of the Sinclair Earls of Orkney in Midlothian. The ash was also the tree most venerated by the Irish Celts: in the Highlands it was the tree of life, powerful to protect against enchantment.[3] Pennant in 1772 described how midwives gave new-born babies 'sap from a green stick of ash held in the fire so that the juice oozed out onto a spoon held at the other end'.[4] The rowan was another and even more universal defence against evil, as its presence around settlements and crofthouses witnesses; apple and birch

---

[1] T. C. Smout, 'Trees as historic landscapes', *Scottish Forestry*, 48 (1994), pp. 244–52.
[2] J. Walker, *Essays on Natural History and Rural Economy* (Edinburgh, 1808), pp. 16–17.
[3] H. Fife, *Warriors and Guardians: Native Highland Trees* (Glendaruel, 1994), p. 164; E. Neeson, *A History of Irish Forestry* (Dublin, 1991), p. 21; T. C. Smout (ed.), *People and Woods in Scotland: a History* (Edinburgh, 2003), pp. 3–7.
[4] T. Darwin, 'Sacred trees in Scottish folklore', *Reforesting Scotland*, 10 (1994), pp. 8–10.

were associated with birth; elder and hawthorn with the spirit of the dead; alder with resurrection; hazel with wisdom; aspen, said to be the wood of the Crucifixion, with evil.[5]

Alexander Carmichael relates some of the elaborations of these Highland beliefs: faggots of 'rowan, yew, oak or other sacred wood' were used on certain saints' days for the fire, but blackthorn, aspen, bird cherry, and other 'crossed wood' were to be avoided in baking cakes. Rowan was used about the house and byres as sprigs and wreaths to keep witches from harming people and animals, and the berries given to stock to keep them safe: if a coffin or the bier on which it was carried was made of rowan, it was treated with special reverence. No aspen might be used in making an agricultural implement, and to use it for a boat, a creel or any fishing gear was to court disaster.[6] The very alphabet 'was itself connected, for mnemonic purposes, with the names of individual trees' – A for *ailm*, B for *beith*, C for *coll*, D for *darach*, and so forth.[7] There were links at all levels between everyday life and trees.

Irish Brehon Law of the eighth century classified trees into four classes with seven species in each: the classes were termed nobles, commoners, lower orders and slaves, reflecting both the Gaelic sense of hierarchy and the economic importance of each group. Oak, hazel, holly, ash, yew, pine and apple were the nobles of the wood; alder, willow, hawthorn, birch, elm and gean the commoners; aspen, juniper and so on, the slaves. Such was the persistence of tradition and the cultural closeness of Irish and Scottish Gaels that, ten centuries later, the symbolic meanings of trees had hardly changed. Around 1721, Sìleas na Ceapaich (Cicely Macdonald), lamented the death of Alasdair of Glengarry. He was:

> The yew above every wood,
> the oak, steadfast and strong,
> you were the holly, the blackthorn,
> the apple rough-barked in bloom;
> you had no twig of the aspen,
> the alder made no claim on you,
> there was none of the elm-tree in you,
> you were the darling of lovely dames.[8]

---

[5] *Ibid.*; Fife, *Warriors and Guardians*.
[6] A. Carmichael, *Carmina Gaedelica* (Edinburgh, 1928), 1, pp. 162, 194, 201; 2, pp. 105, 245, 290.
[7] J. Hunter, *On the Other Side of Sorrow: Nature and People in the Scottish Highlands* (Edinburgh, 1995), p. 61; Neeson, *Irish Forestry*, pp. 28–32.
[8] Neeson, *Irish Forestry*, pp. 27–9; R. Crawford and M. Imlah (eds), *The New Penguin Book of Scottish Verse* (Harmondsworth, 2000), p. 221.

Today, it is the presence of a rowan to keep away witchcraft outside the older Highland homes, standing or ruined, which is the one abiding sign of this corpus of traditional meaning and belief.

Trees also had varied economic values and qualities, according to their species. Oak was considered much the most versatile and valuable of all woods, and many felling contracts of the seventeenth and eighteenth centuries distinguished between oak and other trees, dismissed as 'barren timber'. Oak was the timber of prestige buildings, and when available in beams of sufficient size, used in the dramatic hammerbeam roofs of Darnaway Castle, Morayshire (1387) (Fig. 2.2 above) and its successors at Stirling and Edinburgh Castle, as well as for comparable but surely no less splendid roofs of the medieval cathedrals, abbeys and other large churches that mainly disappeared after the Reformation. Royal builders still drew oak in the fifteenth and sixteenth centuries from a wide area: Lochaber in the Highlands, Darnaway Forest itself, and from the Torwood, Callander, Perthshire and Clackmannan in central Scotland, for use in the ambitious palaces at Stirling, Linlithgow and Holyrood. Native oaks by then, however, were seldom of any great size, and for the biggest beams they now turned to foreign sources.[9]

Fortifications, too, had originally made a great deal of use of oak. The first medieval castles were wooden keeps with wooden palisades, then the keeps and curtain walls became stone, albeit often with wooden tops to the keeps in the form of hoardings, only gradually in the fourteenth century becoming more completely stone on the exterior. Long after this, though, old ways persisted. On both sides of the Borders there were still wooden castles in the sixteenth century. A survey of 1541 spoke of the dwellings of the 'heddesmen' of Tyndale in north Northumberland as 'very stronge houses' with their outer walls made of 'greatt sware oke trees, strongly bounde and joyned together with greatt tenours of the same', so hard as to be almost impossible to destroy, and both walls and roofs covered with turfs and earth so that they could not easily be set alight. A similar building on the Scottish side in Dumfriesshire was attacked by Lord Dacre in 1528: the 'strong pele of Ill Will Armistraunges which was buylded aftur siche a manner that it couth not be brynt ne destroyed, unto it was cut downe with axes'. Of course, much more typical of the sixteenth century were the stone tower houses of the Lowlands, the remains of which can be seen almost

---

[9] M. Baillie and G. Stell, 'The Great Hall and roof of Darnaway Castle, Moray', in W. D. H. Sellar (ed.), *Moray: Province and People* (Scottish Society for Northern Studies, 1993), pp. 163–86; A. Crone and F. Watson, 'Sufficiency to scarcity: medieval Scotland 500–1600', in T. C. Smout (ed.), *People and Woods in Scotland: a History* (Edinburgh, 2003), p. 78.

everywhere. Only two have their original roofs, Alloa Tower and Bardowie Castle in Stirlingshire, both constructed of oak beams of modest size. They would both have had oak boarded floors and possibly (like many castles and churches), oak shingled roofs.[10]

In the period after 1500, big oak trees were very hard to come by for construction purposes, and houses in the main burghs were in any case generally constructed of imported timber. Expense had probably always normally limited oak to buildings of some quality, but where it was plentiful in the countryside it was still sometimes used quite lavishly. In Argyll, the nineteenth-century minister recalled that Cameron of Glendessary had lived in 'a house of very peculiar construction, formed of oak beams placed at regular distances, the intervening spaces being closely interwoven with wickerwork. The outside was wholly covered with heath...'.[11] Small oak timbers were used in tenants' dwellings for beams, planking, doors and panelling, in areas like this where oak was plentiful. Sometimes it was even used to make a plough. Timber of this sort might become a by-product of the tanbark industry and enter tenants' houses as a cheap bargain following the sale of the local wood to a tanner.

For shipbuilding, the qualities of oak for producing strong joints and knees of curved shape always made it desirable for frames, and where strength was a prerequisite, as in warships, it was also valued for planking. We have seen (pp. 41–2 above) that the west-coast medieval galleys were made of oak; and Darnaway and other north-eastern oakwoods were pressed into service to help to manufacture the fleet of James IV, along with a good deal of imported timber. Again, however, post-medieval Scotland was depleted of oaken timber suitable for grand ships, though perhaps we should not take literally Pitscottie's well-known opinion that building the *Great Michael* had 'wasted all the woodis in Fife', except for Falkland wood.

For smaller boats there was no doubt still plenty of modest-sized oak available for local use (Fig. 4.1). It is not clear when the traditional galleys were finally abandoned by the chiefs. Some hint that the west coast was becoming short of suitable oak for their construction comes as early as 1568, when the lord justices in Ireland forbade the export of boards from the ports of Wicklow, Arklow or Dublin in order to prevent the Earl of Argyll from building such ships. Nevertheless, some could still be built from local resources. In 1635, for example Sir Duncan Campbell of

---

[10] Royal Commission on the Ancient and Historical Monuments of Scotland, *Eastern Dumfriesshire: an Archaeological Landscape* (Edinburgh, 1997), p. 215; Crone and Watson, 'Sufficiency to scarcity', p. 72.

[11] *NSA*, 7 (1845), p. 177. For this and several subsequent references to timber building construction in the West Highlands the authors are indebted to Hugh Cheape.

Fig. 4.1 A birlinn, the Norse-influenced galley of the western Highlands, carved on the Macleod tomb at Rodel, Harris. Royal Commission on the Ancient and Historical Monuments of Scotland. Crown copyright.

Glenorchy had a new 'birling' made of oak from 'Glenwiring'. In 1677–80 the Chamberlain of the Duke of Argyll noted yearly in his accounts that the Captain of Dunoon owed an obligation to 'provyde my Lord with a ten oar'd birline or two birlines with six oares each of them but the Cap'n took to considerat'n which of them he would provyde but he did not give ansuer as yet': the annual entry was so formulaic that one senses that Argyll never received the boats, but had not yet given up his claim to them.[12]

On Skye, as late as 1706, Macleod of Macleod built a birlinn of oaken boards caulked with oakum, using thirty yards of white plaiding for its sails: 'it was probably one of the last that he owned'. At the opposite end of the island, however, the MacDonalds of Sleat were still using a boat described as 'the birline' in the 1730s. Boats primarily powered by multiple oarsmen continued to be used on the MacDonald estates of North Uist and Skye into the later 1740s. In 1745, 'the six oar boat with a loadening of timber' was sent to Portree and in 1746 it took another load of timber to Tote in Trotternish. In the same year, the tacksman of the mills of North Uist had the right to take a load of timber to North Uist in Sir Alexander MacDonald's 'ten-oar boat'. When these boats were built is unknown but it is clear enough that MacDonald had oak readily available with which to build them.[13]

It is obscure how far the small, locally built fishing boats constructed up and down the west coast in the eighteenth and nineteenth centuries used oak. No doubt they did sometimes, as they were built with whatever native or foreign materials came to hand: the nineteenth-century boats on Skye had frames of ash and sides of larch plank, and in 1816 Patrick Sellar intended to sell crooked birch from the Wood of Skail as 'knees for boat building' and noted that Letterbeg Wood 'consists entirely of Birch of pretty good size, very much suited for boat building'. Surprisingly, as late as 1895–1900, the timber used for a large type of 'Fifie' built on the east coast for the herring fishing was 'either Scottish oak or Scottish larch, American elm being used for the keel and the gunwale, oak for the stem and stern posts'.[14]

---

[12] R. Loeber, 'Settlers' utilisation of the natural resources', in K. Hannigan and W. Nolan, *Wicklow, History and Society: Interdisciplinary Essays on the History of an Irish County* (Dublin, 1994), p. 274; NAS: Breadalbane Muniments, GD 112/9/6; Inveraray Castle Papers: Chamberlain's Accounts, 1677–80.

[13] I. F. Grant, *Highland Folk Ways* (edn Edinburgh, 1995), pp. 254–5; NAS: Court of Sessions Unextracted Processes, Factors' Account Books, Trotternish and Sleat, CS 96/4261; Clan Donald Lands Trust: Tack of the Mills of North Uist, MS. 4280/7.

[14] Grant, *Highland Folk Ways*, pp. 274–5; R. J. Adam (ed.), *Papers on Sutherland Estate Management, 1802–1816* (Scottish History Society, Edinburgh, 1972), 1, p. 188, see also p. 190; P. F. Anson, *Fishing Boats and Fisher Folk on the East Coast of Scotland* (London, 1930), p. 31.

Oak was also the best tree for making charcoal and for peeling for tanbark, both of which had been important in the woods since the Middle Ages. By the eighteenth century, their production often eclipsed in value that of the oak timber itself, and most oakwoods came to be managed as coppice, the foresters being advised to weed out any other trees that might compete for space. As explained in Chapters 9 and 10 below, the value of the woods rose to a peak in the wars between 1792 and 1815, but competition from coal for iron smelting ultimately knocked the bottom out of the charcoal market, and a combination of free trade and chemical substitution eventually also destroyed the market for oak tanbark. In 1866, the duty on imported timber products was finally removed and the trade in foreign materials 'dealt the home tan-bark industry a crippling blow'. Two years later the price of bark was barely a third of what it had been in working memory, and owners were advised that oak coppice was better replaced by softwoods. In fact, much of it was restored from coppice to high forest, but remained oakwood. Some use of coppice struggled on until the First World War.[15]

Scots pine was used primarily in building construction, where its ability, when properly grown, to produce long, straight beams and planks was highly valued, whether the wood came from Scotland, or, as was more likely, from Scandinavia and the Baltic. A building boom began in Scottish towns in the sixteenth century, and at first the houses made lavish use of external timber cladding, almost all of it pine or spruce from cheap, accessible supplies overseas. The danger of fire led to building regulations that changed the outward appearance of the burghs from wood to stone, but internal joists, floors and panelling were still needed in great quantities: the large, square timbers employed in the New Town of Edinburgh two centuries later were often the famed 'Memel pine', and Adam Smith remarked that the city contained 'not a stick' of Scottish wood.[16]

Scottish-grown pine was not used in most towns both because it was expensive to transport down to the Lowlands and because the quality was often reckoned inferior. In fact, some, though perhaps not much, Scottish pine was equal to anything from Scandinavia, and local houses in the vicinity of the woods made good use of it. The earliest softwood roof structure to have survived anywhere in Britain is at Castle Grant on Speyside, constructed in the sixteenth century from straight-grained timbers of fine

---

[15] W. M'Corquodale, 'On the conversion of coppice land into a more remunerative crop', *TSAS*, 4 (1868), pp. 47–50; M. L. Anderson, *A History of Scottish Forestry* (Edinburgh, 1967), 2, pp. 314, 322.
[16] I. Davies, B. Walker and J. Pendlebury, *Timber Cladding in Scotland* (Edinburgh, 2002), p. 19.

Fig. 4.2 The pine roof of Castle Grant, Speyside, sixteenth century. It is the earliest roof of Scots pine known in Britain. Una Lee.

quality up to 18 feet long and up to 14 inches in diameter, plainly grown from closely spaced trees (Fig. 4.2).[17] Even more impressive must have been the early seventeenth-century tower house of Macdonell of Glengarry built on the shores of Loch Oich, which rose to five storeys and had timber joists and unvaulted ceilings. The main hall was 45 feet by 22 feet on its internal measurements, so probably used single timbers of 24–25 feet in length.[18] Both oak and Scots pine were available to the owner from his own property, but these very long spans must have been of pine. Achnacarry, the home of his neighbour Cameron of Lochiel, was described in 1723 as 'a large house, all built of fir-planks, the handsomest of that kind in Britain': it went up in flames during the Duke of Cumberland's campaign of vengeance in 1747.[19]

---

[17] U. Lee, 'Native timber construction: Strathspey's unique history', *Scottish Woodland History Discussion Group Notes*, 7 (2002), pp. 23–9.

[18] H. Cheape, '"A few summer shielings or cots": material culture and buildings of the '45', *Clan Donald Magazine*, 13 (1995), p. 68.

[19] W. Buchanan, *An Inquiry into the Genealogy and Present State of Ancient Scottish Surnames* (Edinburgh, 1723), p. 129.

The pine would have come from the neighbouring Loch Arkaig woods. Similarly the old Doune of Rothiemurchus contained pine baulks 16 feet long by 12 inches by 3 inches, and the new Doune built towards 1800, baulks 17 feet long and a foot square. It was a sign of the times that when the third Doune was built in 1877 it was panelled in Canadian pine delivered by railway to Aviemore.[20]

A more characteristic and humble building, albeit of unusual size and quality for its type, is the cruck-building at Corrimony, Inverness-shire (also part of the old Glengarry estate), 65 feet long and about 19 feet wide, held up by five large pine crucks, each made from several pieces of wood, with an overall height of about 15 feet (Fig. 4.3).[21] Presumably it was once the combined dwelling house and byre of a substantial tenant or tacksman. The walls of this building were of stone or turf, pine always being used sparingly in proportion to the distance from its source. In Speyside, the vernacular buildings had turf walls and a turf roof, the load-bearing members being pine crucks ('couples') that remained in the ownership of the lairds when a family moved, the earliest written accounts of what was clearly an age-old method of construction dating back to 1585.[22] In some cases the space between the couples would be filled with wattle, as at Inverie on Loch Nevis where Macdonell of Glengarry had another house of stone and lime with one wing 'in the old style', with couples of 'Scotch fir', a clay floor, and infill of hazel basket-work for the walls.[23] For most buildings using native pine, the timber would not be carried above twenty miles unless there was convenient transport downriver, as along the Spey. On the other hand, peat mosses in deforested areas might yield pine timber thousands of years old but still usable for simple crucks and other purposes.

Near the main woods, however, pine was often used lavishly. In Glen Moriston there was around 1700 'ane little parish church of timber... called Millerghead' and in the nineteenth century farmers and landowners used pine for external walls of houses and barns, and for panelling. There were smart timber-clad cottages of ca. 1820 in Speyside, built to the pattern-book designs of J. C. Louden from local pine, and all-timber shooting lodges: Forest Lodge, Abernethy, from 1880, is the biggest to survive in Scotland. Balmoral estate has a large collection of timber ancillary buildings, Aviemore railway station is wooden, and there is another wooden church

---

[20] T. C. Smout and R. A. Lambert (eds), *Rothiemurchus: Nature and People on a Highland Estate, 1500–2000* (Dalkeith, 1999), pp. 65, 75.
[21] G. D. Hay, 'The cruck-building at Corrimony, Inverness-shire', *Scottish Studies*, 17 (1973), pp. 127–33.
[22] Lee, 'Native timber construction', pp. 24–5.
[23] N. M. MacDonald, *The Clan Ranald of Knoydart and Glengarry* (n.p., 1972), p. 134.

WOODLAND PRODUCE

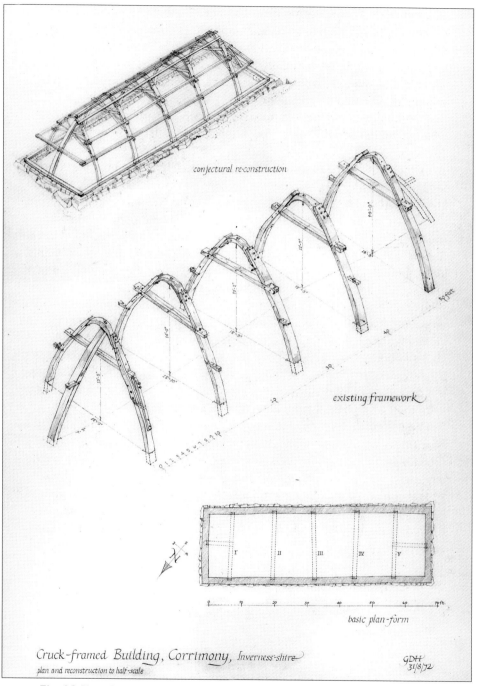

Fig. 4.3 Plan of a cruck-framed building at Corrimony, Inverness-shire, drawn in 1972.
Royal Commission on the Ancient and Historical Monuments of Scotland. Crown copyright.

at Altyre near Forres, dating from the late nineteenth century. Wood was popular, and though many regarded wooden houses as flimsy, the inhabitants of Grantown-on-Spey grumbled when their improving landlord tried to make them rebuild the walls of their homes in stone, complaining that they were cold. Even stone houses, however, had to make much use of pine in their roofs and flooring.[24] From the history of the Doune of Rothiemurchus, however, it cannot be assumed that after the coming of the railways the wood used was always local rather than imported, even in Speyside.

Farmers found many other uses for pine, where it was plentiful. Implements were made of it, like the fine harrows in the Highland Folk Museum at Kingussie, making use of a strong natural cruck. Ropes were made from the roots, even the hawsers and rigging of boats at Gairloch.[25] Fir 'candles' – resinous splinters from living pines, or knots from trees buried in the peat mosses – were widely used. Old people in Strath Dearn where only birches grew, used in the early nineteenth century to walk a dozen miles or more over the Slochd pass to collect pine from Strathspey for making fir candle in the autumn, and the inhabitants of the Black Wood of Rannoch distributed pine stocks for candles far and wide through the fairs of northern Perthshire. Even pine dug out of the moss was widely used for fir candle. James Robertson in 1771 described being at a Highland wedding where the chips were burned by the side of the central hearth on a stone pillar three feet high, 'but when the fireplace is in the end of the house, a stone projects from the wall behind it for the purpose of burning the chips'.[26]

The use of pine in shipbuilding depended on the type and quality of boat under construction. We may imagine that (as with oak) it was long used for local boats where there was access to Scots pine, but usually not on any large scale. For example, around 1748 William Chisholm obtained permission to take as much pine from the Loch Arkaig woods as would build a ferryboat but 'he then got as great a quantity as would have built three or four boats'.[27] Most sea-going trading vessels by the seventeenth century, however, were constructed of softwood abroad. Norwegian-built and Dutch-built ships formed the backbone of the Scottish mercantile fleet on the eve of the Union of 1707.

[24] A. Mitchell (ed.), *Geographical Collections relating to Scotland made by Walter Macfarlane* (1906), 2, p. 171; Ross Noble, *pers. comm.*; Davies, Walker and Pendlebury, *Timber Cladding*, pp. 25–9.
[25] Grant, *Highland Folk Ways*, pp. 205, 258.
[26] *Ibid.*, pp. 184–5; T. Hunter, *Woods, Forests and Estates of Perthshire* (Perth, 1883), pp. 410 ff.; D. M. Henderson and J. H. Dickson (eds), *A Naturalist in the Highlands: James Robertson, His Life and Travels in Scotland, 1767–1771* (Edinburgh, 1994), p. 173.
[27] NAS: E768/44/1.

Because of the straight, strong and flexible nature of the very best pines, these were regarded as indispensable for masts on the warships of the day. The splendid Baltic pines of Riga were preferred, and as a substitute for supplies to the Royal Navy, the white pines of colonial New Hampshire. Nevertheless, Royal Navy purchasers also from time to time looked for timber from Scotland even in the seventeenth century: although they seldom found masts for their liking, Strathcarron and elsewhere were explored to provide what naval timber they could (see below, Chapter 12). In 1785 a naval brig of 110 tons, the *Glenmore*, was built from Scots pine from the Duke of Gordon's Strathspey estate, but it does not appear to have been enough of a success to encourage more military orders. The Royal Navy was never a major consumer of Scottish pine.[28]

Builders of merchant ships were perforce less particular. The Irish, having no pine of their own, were probably among the customers referred to by a mid-seventeenth-century commentator, who spoke of 'manie ships' that came to Ardgour 'to be loadned with firr jests, masts and cutts'. Most shipbuilding concerns were probably small and perhaps used little local wood, but the builders of the *Glenmore*, Dodsworth and Osborne from Kingston upon Hull, started a more heavily capitalised concern at Garmouth at the mouth of the Spey, which under different management became a commercial proposition for the better part of a century, using local and imported wood for planking and masts. Much, but not all, of the pine used at Garmouth was Scottish.[29]

For the rougher and smaller pine there was always a variety of uses in construction and industry, not least as pit props in coal mines, though it is doubtful if it would often bear the cost of transport from the Highlands to the Lowlands or the North of England except in wartime when Scandinavian supplies were cut off. Pine was favoured for constructing canal locks on the Caledonian Canal, and for sleepers for the Highland railways, creating a boom for owners while the construction lasted, but, again, it hardly travelled far.

Pine also provided tar, turpentine and resin, though in Scotland there is little indication that tar was ever made, and if it was, it was only briefly and experimentally (see pp. 197, 335 below). Pine charcoal was perhaps made for iron smelting, briefly if ambitiously, by the York Buildings Company at

---

[28] R. G. Albion, *Forests and Seapower: the Timber Problem of the Royal Navy, 1652–1862* (Cambridge, MA, 1926).

[29] Mitchell, *Geographical Collections*, 2, p. 165; J. Skelton, *Speybuilt: the Story of a Forgotten Industry* (Garmouth, 2nd edn, 1995). Spreull was clear that the Irish were significant importers of west-coast pine in 1705: J. W. Burn (ed.), *Miscellaneous Writings of John Spreull* (Glasgow, 1882), pp. 31, 64.

Abernethy in the early 1730s, but it was considered much inferior to oak charcoal, and Strathspey was in any case poorly situated in relation to transport of iron ore.[30] Walker in 1808 described pine charcoal as the worst that could be bought, worth only 26 shillings for a dozen bags, but oak charcoal was the best and would sell at 36 shillings.[31]

The third of the notable timber trees was ash, widespread yet never very common in Scotland, and always highly regarded. Ashwoods, demanding less acidic ground, were few and far between. Rassal Wood in Wester Ross, a natural ashwood on limestone with some trees over 300 years old, is a rare example indeed. Many ashes were planted, often in hedgerows, but some were found in the texture of mixed natural woodlands. The price of standing ash timber in Pressmennan Wood, East Lothian, in 1797 was 1s. 6d. a foot, appreciably above the 1s. 1d. to 1s. 4d. for oak (and only 6d. for alder), although the overall value of the oaks was much greater because its bark was of good quality as well as the timber.[32]

Ash was valued round the farm and by craftsmen because it was light and highly durable even in slender form: 'from the day its stem is three inches in diameter the wood is equally durable until it shall be three feet or any size whatever'. It was used for construction, for tools, utensils, ploughs, for country furniture, carts, coaches, boat frames, barrels and shinty sticks, and the young trees were particularly useful as poles.[33] These qualities made it vulnerable to the rural pilferer, and as the laird was often particularly proud of his planted ashes, removal could invite retribution by a stiff fine or incarceration in the stocks. In 1755 two tenants in Argyll were pursued for taking an ash from a garden in Stronmagachen to make door posts for a barn. They claimed that they were told at the time that they could take it if they planted two young ash in its place, and that actually they had planted four replacements in the garden: an indication at least that sustainable practice was expected.[34]

The other trees in the wood were commonly known as 'barren timber', but they were almost all extremely useful to the local rural population. Birch was the chief of these, probably the most abundant Scottish tree, especially in the north, and, in Walter Nicol's words it was:

---

[30] D. Murray, *The York Buildings Company, a Chapter in Scotch History* (edn Edinburgh, 1973), p. 64.
[31] J. Walker, *An Economical History of the Hebrides and Highlands of Scotland* (Edinburgh, 1808), 2, p. 286.
[32] NAS: GD 6/1655/a1–3.
[33] W. Nicol, *The Practical Planter, or a Treatise on Forest Planting* (Edinburgh, 1799), pp. 50–1; W. Nicol, *Planter's Kalendar* (Edinburgh, 1812), pp. 76–7.
[34] NLS: MS 17667, fo. 81.

The universal wood of the Scottish Highlanders. They make everything of it; they build their houses of it; make their beds, chairs, tables, dishes and spoons of it; construct their mills of it; make their carts, ploughs, harrows, gates and fences of it; and even manufacture ropes of it.

He went on to say that it was also used in machinery, turnery, wheelwork, for lasts, pattens, wooden shoes, for pit props and for the sleepers for waggonways. It made a good fuel, as it was not smoky, provided the best tanbark after oak (but here it was very much a second best), was favoured above all other woods for curing meat and herrings, and from it birch wine could be made – 'a cooling, agreeable drink'. When J. C. Louden plagiarised this passage in 1842 he added a few details of his own: the value of birch as fuel in whisky distillation, as thatch for houses, as bedding, and as lighting 'when dried and twisted into a rope, instead of candles'.[35] Both authors failed to refer to its extensive use as barrel staves, which was a major factor where birch woods were convenient to the herring fishery, or to its local use in shipbuilding. One of the first commercial ironworks set up in Scotland, described near Aberfoyle in 1724, used birch charcoal.[36]

The uses of birch for turnery, for example in domestic treen ware and furniture, had long been noted, but it was only in the nineteenth century that quantities of birch began to be exported from the Highlands for bobbin manufacture. Bobbin mills, and then pyroligneous acid works and even gunpowder works were set up in the birch woods (see pp. 264–9 below). Until this period, however, it was the cheapness and easy availability of birch which gave it its special place in Highland life. It might not be very substantial or durable compared to oak, pine or ash, but it was readily worked and could quickly be replaced in the home or on the farm, and the landowner probably had less objection to its removal than he had to the taking of most trees.

Quite the most remarkable account of a large birch building was Pitscottie's account of the great hunting lodge built for a royal hunting party by the Earl of Atholl in 1528. It was 'buildit in ane fair medow ane fair pallice of greine tymber wood with birkis that war greine bayth under and above'. It was three storeys high, 'loftit and jeistit', and built in four quarters, and 'in evirie quarter and nuck thairof ane gret round, as it had beine ane bloukhous'. The floors were strewn with flowers and rushes, the walls hung

---

[35] Nicol, *Planter's Kalendar*, pp. 80–1. J. C. Louden, *An Encyclopaedia of Trees and Shrubs* (Edinburgh, 1842), cited in R. Mabey, *Flora Britannica* (London, 1996), pp. 84–5.
[36] Mitchell (ed.), *Geographical Collections*, 1, p. 343.

with tapestry and silk and lit with glass windows, and the whole was surrounded with a wide moat approached by wooden drawbridge and portcullis. On the departure of the court the whole building was set on fire, to the astonishment of the visiting papal ambassador. The king explained that 'it is the use of our hieland men, thocht thaj be nevir so well ludgit, to burne the ludgin when thay depairt'. Such ostentatious consumption (even allowing for Pitscottie's usual overstatements) certainly suggests that large birch was abundant in Perthshire in the early sixteenth century. It is possible that other timber was used in the construction as well, to achieve the necessary height and strength.[37]

Other trees not quite so versatile as birch had important specialist uses. Alder, for example, resisted rot in waterlogged conditions, so it was useful for flooring and roofing sheds and outhouses, for door steps, for lining carts, making gateposts, and for pit props. Along with rowan, it could serve as the footings for building crucks of oak or birch that rested on the ground. It could be used for the main building timbers themselves: the manse of the parish of Kilbuco in Peeblesshire in 1624 was described as 'ane house of sex cuples [crucks] in lenth of birk and aller'. Alder also made good charcoal and tanbark, could be used by turners and pattern makers and was a source of dye in the domestic manufacture of tartan. Its whippy branches were useful for plaiting wattle in house construction. Rowan berries were also used for tartan dye, and the wood used by house wrights for floors, by cartwrights for linings and wheels, and by coopers for making herring casks. Gean (wild cherry) was especially used for furniture, but in the West Highlands was taboo for using in the wattle walls of a house. Elm, according to Nicol, was 'prized next to oak' and used by shipbuilders, block and pump makers, cartwrights, cabinet makers and coach makers: its hollowed logs were used for urban water pipes. Hawthorn was hard enough for mill cogs, aspen (despite its connotations of bad luck) was used for herring casks, and holly was used for fine turnery work and block-making for textile printing.[38]

Two trees of which Nicol says little were nevertheless extremely useful in peasant society. One was the willow or sauch, along with birch a source of 'withies' that could be used to make baskets and ropes, or plaited to form part of building structures. In the regality of Atholl in 1708, the reporter on potential improvements urged that all the tenants should sow hemp seed

---

[37] Pitscottie's account is reproduced under the title, 'Ane fair pallice of greine tymber: how the king passit to the Hieland to the hunting', in *Reforesting Scotland*, 16 (1997), pp. 33–4.
[38] NAS: CH2/295/1/2 Register of the Presbytery of Peebles; Nicol, *Planter's Kalendar*, pp. 73, 77–8, 83, 89, 101, 114–15.

and make rope from hemp rather than chop at the bushes to get withies – 'withs daily booling [breaking] no housband man of a plough gang of land can deny but he bestows a moneth every year upon mudleing at withs'. He thought it cost the regality £10,000 Scots a year.[39]

Willow was a source of dyestuff, and of a particularly active agent in tanning, needed in small quantities. Its charcoal was prized in gunpowder manufacture above all other. It was also convenient for wrights and wood turners: Walker listed its use for platters, dishes, bowls, shoe lasts, heels, clogs, ladders, hay rakes and water wheels, and mentioned that in the Hebrides its roots were plaited into ropes.[40]

The other invaluable tree was the hazel, coppiced since prehistoric times and the subject of systematic exploitation by monastic houses (such as Lindores) in the Middle Ages.[41] Most Highland farms, especially in the west, had their hazel thickets, and the flexible wands were used for constructing temporary fencing and hurdles, for making fishing creels and baskets of all kinds, as the hoops for coopers' barrels, and in places for housing. It was probably hazel that most frequently formed the basis for the simple housing described by a government factor to the Annexed Estates Commissioners in 1753:

> The whole houses of the country are made up of twigs manufactured by way of creels called wattling and covered with turf. They are so low in the roof as scarce to admit a person standing in them, and when these are made up with pains they endure ten or twelve years. They thatch them with rushes.[42]

Contemporaries often called these 'creel houses' (Fig. 4.4). James Robertson on a trip to Ardnamurchan and Moidart in 1768 added the detail that the stakes holding the wattles were nine inches to a foot apart, and the walls four feet or five feet high, thatched with turf, heath or straw. In upper Strathconon he found similar structures, including barns that, because they had no turf pinned to the walls, the wind whistled through and dried their corn. Three years later at the head of Glen Moriston he found further barns of this type, but fourteen feet broad, where the corn

---

[39] Edinburgh University Library: Atholl Papers, DC.1.37.1.
[40] Walker, *Economical History*, 2, pp. 257, 286.
[41] J. M. Gilbert, *Hunting and Hunting Reserves in Medieval Scotland* (Edinburgh, 1979), p. 237.
[42] Quoted in H. Cheape, 'Woodlands on the Clanranald estates', in T. C. Smout (ed.), *Scotland since Prehistory: Natural Change and Human Impact* (Aberdeen, 1993), p. 58.

Fig. 4.4 What may well be a creel house in the central Highlands, ca. 1730. The door is clearly of wicker construction, and the walls probably so. From E. Burt, *Letters from the North of Scotland*. St Andrews University Library.

was deposited every night as it was cut to protect it from 'the deluges of rain to which it is subject'.[43]

Different forms of wattle construction, with or without associated crucks, rafters and posts, were remarkably widespread even in the nineteenth century, as described by Alexander Carmichael in evidence to the Napier Commission:

> In wooded districts throughout the Highlands where materials can be found, doors, gates, partitions, fences, barns and even dwelling houses, are made of wattle-work. In the case of dwelling houses and their partitions, the wattling is plastered over on both sides with boulder clay, and whitewashed with lime, thereby giving an air of cleanliness and comfort to the house.[44]

Hazel was not the only wood used for wattling, but it was admirably suited for it.

In James Lindsay's opinion, house timber was without a doubt the major consumer of large timber in the Highlands.[45] The largest members needed in a timber-framed building were the 'panntries', the horizontal beams resting on the walls and at the ridge of the roof: also substantial were the 'couples', the pairs of crucks forming the apex of the gable and set at intervals at right angles to the panntries: much less substantial were the 'cabers', which rested between the walls and the ridge of the roof, parallel to the couples, on which the thatch was set: most tenants' houses were open to the roof, but might embody smaller cross-members from which hams were hung or where baskets could be placed out of reach of vermin (see Fig. 4.5). More substantial houses, like those of tacksmen, would have upper storeys with wooden ceilings and floors, and wooden box beds and screens would divide the interiors.

Even a small creel house, said an observer in Lochaber in 1767, could use 2,000 of 'the straightest and best of the young wood, as the old does not answer the purpose'. Some idea of the quantities of mature timber cut for larger country buildings comes from the Annexed Estate Papers. In 1749 and 1750 Alexander Cameron of Glen Nevis applied to take wood to rebuild his house and barns, burnt after the rising although he had 'remained at

---

[43] Henderson and Dickson (eds), *Naturalist in the Highlands*, pp. 81, 101, 180.
[44] A. Carmichael, 'Grazing and agrestic customs of the Outer Hebrides', appendix to *The Report of the Royal Commissioners of Inquiry on the Condition of Crofters and Cottars in the Highlands of Scotland* (Parliamentary Papers, 1884), 1, p. 454.
[45] J. M. Lindsay, 'The use of woodland in Argyllshire and Perthshire between 1650 and 1850', unpublished University of Edinburgh Ph.D. thesis (1974), p. 105.

Fig. 4.5 Interior of a byre, Keils, Isle of Jura. Such open construction was also commonplace in homes. Royal Commission on the Ancient and Historical Monuments of Scotland. Crown copyright.

home during the late treasonable commotions'. He cut between eighteen and twenty-four panntries, worth about 6d. each, and 1,000 to 1,200 birch cabers, worth 1/3d. each. Charles Stewart, the ferry keeper near Achnacarry, with a simpler house than a tacksman, cut 600 birch cabers for his home.[46] Birch was plentiful here. In Perthshire, a document dealing with the forfeited estate of the Dukes of Perth declared that a dwelling house needed fifty trees (size and species not specified), a barn or stable 48 feet by 16 feet, thirty trees; a byre 36 feet by 16 feet, fifteen trees, and a shelter only ten trees.[47] In Sutherland, Patrick Sellar in 1816 allowed 126 tenants from Lower

[46] NAS: GD 44/51/743/7, 'A short description of Lochaber' (we owe this reference to Professor R. A. Dodgshon); NAS: E 768/27/11; NAS: E 768/44/14(9).
[47] NAS: E 777/265/5, cited in C. Englund, 'Woodland management on the forfeited estate of Perth, 1746 to 1784', unpublished University of St Andrews M.A. dissertation (1996), p. 14.

Strathnaver a total of 915 birch trees from Letterbeg Wood, as 'required by each for repairing his cabin' – almost seven trees each.[48]

At the other extreme of size and extravagance, Oliver Rackham has suggested that the medieval roof of St Andrews Cathedral might have contained a thousand substantial oaks, and Brian Lavery that an eighteenth-century ship-of-the-line (and perhaps its large sixteenth-century equivalent, the *Great Michael*), 3,000 oak trees, or seventy acres of mature timber. These figures make quite modest the gift to the Duke of Queensberry of 'eleven score of tall stately oaks out of Craignee Wood' for joists for his fine new castle at Drumlanrig, built between 1679 and 1689.[49]

Occasionally it is possible at the sale of a wood to get an impression of the relative values and quantities of different trees. When Saltoun Wood in East Lothian was valued in 1757, the worth of 1,443 oaks was £75, that of 1,468 birches was £34, and there were also 7 ash trees (worth 6d. each) and 52 alders (worth 12 shillings in all). Similarly, when part of Pressmennan Wood was sold in 1797, the contents were reckoned as 5,500 oaks, yielding 8,300 feet of timber (value, £472) and 8,200 stone of bark (value, £410), as well as 3,350 birches (value, £96), 162 ash (value, £40), 16 alders, 15 willows and, surprisingly, 7 hornbeam (total value about £4). The oaks at Pressmennan were probably of better quality and certainly found a better market than those at Saltoun. They were not likely to be of shipbuilding quality but they would certainly have made their way into house building.[50]

The use of wood as fuel in Scotland was generally opportunistic. It was, of course, pleasant to burn, and the well-off preferred it to coal when they were in a position to do so: the Earl of Cassillis in 1642, for example, enjoyed 'twa hundreth faggatis for the fyre' annually from his wood of Dalrymple in Ayrshire. Few industrial processes except for the manufacture of charcoal for iron smelting (see below, Chapter 9) had to have wood rather than coal or peat. Bakers, however, used broom and gorse ('whin') to get a quick hot fire, lime burners needed wood kindling for coal fires, and woodsmoke was excellent for curing meat and fish. In 1761, the 'masters of the easter bake house' in the Lawnmarket of Edinburgh contracted with two wrights in Roslin for a three-year supply of fuel, including broom and whins, to be delivered to their wood-yard and made into a 'sufficient stack ... built and crowned', for which the wrights were to received 9d.

---

[48] Adam (ed.), *Sutherland Estate Management*, 1, p. 188.
[49] Rackham and Lavery in a BBC Radio Scotland broadcast, 'Wallace's Oak', (March 1996); J. Gladstone, 'The natural woodlands of Galloway and Nithsdale', *Forestry*, 34 (1961), 174–80.
[50] NLS: 17150(55); NAS: GD 6/1655/a1–3. Professor Fred Last suggests that the hornbeams were perhaps originally planted as a nurse-tree for young oaks, as recommended by John Evelyn in the seventeenth century.

sterling for each batch of bread baked. Juniper wood was used along with peat for illicit distilling, as a smokeless fuel that gave nothing away to the excisemen.[51]

In few places, however, was there an absolute need to burn wood, and the Scots usually enjoyed plentiful supplies of alternative and more efficient, peat and coal fuels – unlike the eighteenth-century Danes, forced to go to bed in their hats and coats as the price of firewood spiralled.[52] Here and there, however, shortages appeared, particularly when peat supplies were exhausted: in parts of Strathdon, for example, there are extensive plantations of broom on the farms, originally made to supply a need for fuel. Loch Lomondside was an area with little peat, and when in 1758 the tenants of Craigroyston found themselves prevented from taking what fuel they needed from the oakwoods as these became more strictly managed, they petitioned the Duke of Montrose in piteous terms.[53] Where there were plentiful supplies of wood open to the peasantry, the dead branches and fallen trees were allowed for firewood, but not the cutting of green timber. Rules were evaded, of course, but there is normally little indication of overuse. It is possible, however, that holly, a tree evidently once much commoner in the Highlands than at present, was selectively cut out because (according at least to Walker) it made the best firewood.[54]

The most extravagant users of wood for burning were passing armies and garrisons needing fuel in quantity and quickly. Sir Ewen Cameron of Lochiel in 1653 successfully ambushed the Cromwellian garrison at Inverlochy, vulnerable as they attempted to cut a winter's supply of firewood in a wood a mile east of the fort.[55] No doubt as a result of this incident, when Fort William was established in the same locality after 1689, the country people were obliged to bring firewood to the garrison: by 1702 efforts were being made by the commander and by local landowners to get this duty removed, as the local woods were 'so exhausted for severall miles about this garrison, that whose ever lott it falls to provid fyr this year, will have difficulties to gett half year's fyring within reach'.[56] Better and safer arrangements were ultimately made for keeping warm, but soldiers lost none of their old reputation – as it was put by John Matheson in a letter from Inverness in

---

[51] NAS: GD 25/9/78/3/3; NAS: RD 12, reg. 3 May 1762; K. Stewart, *Abriachan: the Story of an Upland Community* (Abriachan, 2000), p. 95.
[52] T. Kjærgaard, *The Danish Revolution, 1500–1800: an Ecohistorical Interpretation* (Cambridge University Press, 1994), p. 97.
[53] NAS: GD 20/6/583/22. We are grateful to John Mitchell for this reference.
[54] Walker, *Economical History*, 2, p. 286.
[55] NLS: MS. 3193, opp. fo. 30.
[56] NLS: 1323/37, Delvine papers.

1751: 'The military cannot endure to see any dry timber but what they burn even the house that accomodats themselve.'[57] Yet, in the totality, armies were few and garrisons small. They cannot have made much impression on the Scottish woods by their need for fuel.

Scots also used the wood for food. Hazel nuts, elderberries, blackberries, raspberries, bilberries and rowan berries were collected for human consumption, though in contrast to parts of continental Europe little use was made of woodland fungi. Nuts were especially significant as they could be harvested for winter. In 1642, for example, the Earl of Cassillis expected the forester of Dalrymple wood to supply him with two firlots (perhaps about 50 lbs) of 'good and sufficient rough nuttis', every year when there was a crop.[58] Pollarded, single-stemmed hazels growing in parts of the Highlands, notably at Glen Finglas, may well have been grown in that form to facilitate nutting.[59] In the Middle Ages acorns ('mast') were considered a valuable resource for feeding pigs in the oakwoods of Galloway and the Borders, but the presence of swine in the Breadalbane oakwoods in the early seventeenth century was a vexation to the laird.[60] At Durness in Sutherland in 1726 it was reported that 'there grows a parcell of large and tall hollys, whereof some have no prickles, hence their twiggs and leaves are cut down in time of snow for food to cows', and it is said that in recent times holly was still used for sheep and cattle in this way in Dumfriesshire.[61] This is a rare explicit record of tree branches being cut for fodder in Scotland, though the form of 'very old multi-stem and possibly pollarded hollies' can be seen in other localities.[62] In Norway, by way of contrast, tree pollarding to provide winter feed for cattle and sheep was commonplace until recently, birch being particularly used in this way.[63] Ash and holly in the Lake District were similarly utilised, as various trees were elsewhere in England. As such practices were at one time widespread in Europe, their apparent rarity in Scotland is puzzling. It is not that Scottish animals dislike the food; some Highland cows learn in winter how to lean on birch trees until

---

[57] NLS: 1359/94, Delvine papers.
[58] NAS: GD 25/9/78/3/3.
[59] C.-A. Hæggström, 'Report on Scottish wood pastures visited 17–20 March 2003', unpublished paper for the Woodland Trust (2003).
[60] NAS: GD 112/17/4. This reference is one among many, but it makes clear that in 1618 the keeping of swine (and goats) in the woods was contrary to the acts of the Baron Court.
[61] Mitchell (ed.), *Geographical Collections*, 1, p. 195; R. Mabey, *Flora Britannica* (London, 1996), p. 248.
[62] P. R. Quelch, 'Upland pasture woodlands in Scotland', part 1, *Scottish Forestry*, 54 (2000), pp. 209–14, part 2 (2001), pp. 85–91. The observation cited is in part 2, p. 88.
[63] I. Austad, 'Tree pollarding in western Norway', in H. H. Birks *et al.* (ed.), *The Cultural Landscape, Past, Present and Future* (Cambridge, 1989), pp. 31–45.

they fall over so that they can eat the tops, and after brush-cutting of birch in leaf the animals eagerly eat the foliage.[64]

The first question to consider is whether historians have simply failed to pick up the evidence from documents, as the archaeology of tree forms in some cases suggests that boughs may indeed have been cut in the winter for fodder. In the ancient wood pasture of Glen Finglas not only hazels but alders and other trees, some possibly 300 years old or more, are cut as low pollards, very much like trees cut for fodder in the New Forest or in Scandinavia.[65] Here and in other old wood pastures in the uplands there is evidence of boughs being cut just beyond the reach of grazing animals (that were smaller then than now), and this is reminiscent of eighteenth-century complaints of tenants spoiling the woods by 'cutting high'. Thus James Robertson says of the wood at Loch Ba on Mull:

> Were not the cattle allowed to feed amongst it, and the inhabitants so unskillful as to cut it high above the ground, there would be great abundance of hazel and birch.[66]

The trouble is that when reference is made to 'cutting high' here and in Baron Court records, or where there are references in the latter to 'cutting greenwood', there is no indication as to purpose. Cutting for tool-making or house construction, or even for fuel use, have also to be considered, and it may be no accident that the houses in Glen Finglas were at some distance from a suitable peat deposit. There is so much literature relating to peasant agricultural practice and also to tree management in eighteenth- and nineteenth-century Scotland, that one would expect a use of trees for fodder to be mentioned more often, had it been widely practised.

The absence of records of tree pollarding for fodder in Scotland may well relate to the law of ownership. A Norwegian peasant possessed both his land and his trees, and could do as he chose with either. A Scottish tenant held his land on precarious tenure, and the trees were the laird's: if he was found cutting away at them without permission he was liable to prosecution, and a culture of regular cropping of birch for fodder may never have developed, because the laird would have regarded it as damaging his trees and punished the culprits accordingly. On the other hand it is a simple fact of cultural history that not all peasant societies seize every

---

[64] Roy Dennis, *pers. comm.*
[65] Quelch, 'Upland pasture woodlands', parts 1 and 2; N. Sanderson, 'Veteran trees in Highland wood pasture', *Scottish Woodland History Discussion Group Notes*, 3 (1998), pp. 4–9.
[66] Henderson and Dickson (eds), *Naturalist in the Highlands*, p. 78.

opportunity that the natural environment offers to them. The use of dead leaves from the forest as arable field fertiliser was widespread in Central Europe in the late eighteenth and nineteenth centuries, on a scale sometimes to degrade woodland soils permanently, but seems never to have been practised in Western Europe or in Britain.[67]

---

[67] W. Schenk, 'Forest development types in Central Germany in pre-industrial times', in S. Cavaciocchi (ed.), *L'uomo e la foresta secc. XIII–XVIII* (Prato, 1996), pp. 201–23.

CHAPTER 5

Woodland as pasture and shelter

As some of the examples of wood use in the last chapter remind us, there was more to woodland than the production of wood. For many tenant farmers and cottars, especially in the Highlands and the Uplands, the main use of wooded areas was to provide pasture and shelter for stock. It is quite impossible to find a Scottish wood, either in the Highlands or the Lowlands, at least before the nineteenth century, from which domestic animals were excluded, except sometimes on a temporary basis. The woods evolved with grazing stock, and the distinction between a wood and a wood pasture is functionally meaningless. The latter term (not used in our period) has come to mean a very open wooded area with 20 per cent or less canopy cover where animals are grazed, but almost all woods were essentially pastures, and certainly all woods provided shelter. Of course, if any wood was too thick it might provide valuable shelter in a Scottish winter but not very satisfactory pasture. As Oliver Rackham reminds us, 'most shade-bearing grasses and herbs have little nutritional value, and some are inedible or poisonous. Domestic livestock love tree leaves, but cannot climb for them.'[1] On the other hand the presence of trees on the ground, especially of birch, undoubtedly enriched the soil, and where trees were relatively thin on the ground and there was plenty of light, there was often a flush of excellent grass which contrasted with heathy land beyond the woodland edge. Woods which were under little pressure from animals in summer were often valued as a source of hay.[2]

The presence of domestic stock in a wood modified its character by their grazing and browsing, and by their trampling and pushing, but in the original wildwood there had always been animals, deer, boar and auroch, who kept grasslands open and perhaps, as Vera suggests, restricted the woods to groves and thickets so that they never became an enveloping cover (see above pp. 11–12, 33–4). Over the centuries, the auroch and boar

---

[1] O. Rackham, 'Forest history of countries without much forest: questions of conservation and savanna', in S. Cavaciocchi (ed.), *L'uomo e la foresta secc. XIII–XVIII* (Prato, 1996), pp. 297–326: quote on p. 299.
[2] I. F. Grant, *Highland Folk Ways* (edn Edinburgh, 1995), p. 97.

were extinguished, though the deer remained: cows, horses, sheep and goats replaced them. The character of the woods continued to be modified by their grazing, as the animals ate up or trampled away a greater or smaller proportion of the young saplings. Most woods would thus have come to have rather an open character, despite occasional descriptions of very dense woods (see above, p. 73). This process could of course be checked by enclosure or regulation if the landowner so decided, but very often the maximum advantage to everyone must have appeared to be a regime where there were just enough trees in a wood to provide shelter, local timber and leaf drop to replenish the ground, but not too many trees to shade out the best grasses on the pasture.

Such open woods would have had spreading trees. The tendency of Scots pine in Scotland to form picturesque, open-crowned shapes ('Granny pines'), compared to the tall upright growth more characteristic of Scots pine in Scandinavia and elsewhere in its range, was a consequence of the trees growing unobstructed by neighbours. It has sometimes been assumed that this was due to foresters leaving, after a clear-fell, series of isolated and perhaps already twisted trees to shed seed for the next generation. It is at least as likely to be a consequence of the pines growing well apart in heavily grazed woods.

Under upland conditions there was likely in any case to be a natural continuum from a relatively thick wood, albeit with rocky or boggy openings, on the valley floor, to a few trees thinning out on the open moor of the upper slopes, a tendency that would have been considerably reinforced by the use of the wood by domestic animals. Changes since 1800 have almost everywhere eliminated the gradation to a natural tree line, as can be seen for instance from an estate map of Eliock Wood in Dumfriesshire in 1767 which showed how the wood spread up the deep cut burns to 600 feet in altitude, with an outlier at 650–700 feet, spreading to 900 feet up one burn, after which the draughtsman showed first small trees then scrub to 1,000 feet. By the time of the First Ordnance Survey Map of 1856, the main wood had been fully enclosed, with the upper margin made straight, and those patches of the old wood left outside explicitly described as pasture.[3] As noted in Chapter 4, it often created a problem for cartographers and surveyors as to the point at which they defined such ground as wood rather than as grazing with a few trees. Similar conundrums elsewhere have led to large inconsistencies in classification – Rackham cites such a case in Greece, where 'official statistics of the extent of *dhásos* [open grazed woodland] in

---

[3] R. C. Blackwood, 'An estate's forest history – Eliock, Dumfriesshire', *Scottish Forestry*, 9 (1955), pp. 13–21.

Crete vary from four and a half per cent of the island in 1981 to thirty-three per cent in 1992'.[4]

To the landowner and the tenant, however, the more material question was whether the ground had greater value as pasture or as wood. The answer initially would depend on the interplay of three variables, the density of the trees and the rates of return on the timber and on the grazing, which in turn, would depend on other variables, the extent of local scarcities of wood and pasture, and the strength of external demands for timber products and animal products.

Sometimes it was perhaps possible for land users to enjoy the best of all worlds, if resources were not scarce. Relatively unrestricted grazing might both leave enough grass for animals and timber for people, if the balance was right. There are times in the seventeenth and early eighteenth centuries when local descriptions give a sense of a world delightfully filled with plenitude. Thus at Conaglen in Ardgour (perhaps around 1630) there was:

> Great number of firr trees in this glen. And it is verie profitable to the Superior and Master of the countrie for it is good to feed guids [cattle] therin being of twall mylls of length or therby, and there is a water in the glen which doeth transport great trees of firr and masts to the seasyde.[5]

Similarly Glen Moriston was 'a very profitable and fertill little glen ... plenteous of corne and abundance of butter, cheeses and milk and great and long woods of firr trees'.[6] At Cabrach in Aberdeenshire 'there is a great wood in Old Doveran and a Forrest where there is frequent resort of Deer, Roes, Heathfowl and other Game, and which with Cattle, Sheep, Goats, Butter, Cheese and Wool are the Commodities of the Place'.[7] In the 1680s, the Earl of Airlie had 'a good interest' in Glenisla 'with two great woods called Crandirth and Craigiefrisch he has a large Glen for grassing with abundance of hay meadows with a free forrestrie which in those places they reckone much worth'.[8] These are examples of co-existence between extensive woods and herds of animals, though how it was managed and whether it was sustainable in the long run remain obscure. Perhaps the

---

[4] Rackham, 'Forest history', p. 299.
[5] A. Mitchell (ed.), *Geographical Collections relating to Scotland made by Walter Macfarlane* (Scottish History Society, Edinburgh, 1906), 2, p. 165. Conaglen appears to have been mistranscribed as 'Cowglen'.
[6] Ibid., 2, p. 171.
[7] Ibid., 1, p. 26.
[8] Ibid., 2, p. 36. See also the description on p. 39.

situation was not always as idyllic as it seemed, as at Ardgour the woods had been very much reduced in extent by the end of the eighteenth century.

The viability of any managed system without enclosure would depend on the relative abundance of trees and animals and the quality of herding. The number of animals could vary quite sharply if the human population rose, so that proportionately more subsistence animals were kept, or if, because of a growing external market for meat, hides or wool, more animals were kept with the same or smaller human population. Sutherland was another district where in the mid-seventeenth century trees and animals seemed to thrive together, the straths being 'replenished with woods, grass, corns, cattell and deer, both pleasant and profitable'.[9] But it was here that, around 1800, and therefore before the main clearances or any intensive commercial farming, but under circumstances of a rapidly increasing population of subsistence farmers, there occurred the clearest evidence of woodland collapse with apparently serious ecological and economic consequences (see below, pp. 119–21).

Very frequently, as at Eliock, there was worked out, in the long run, a compromise between the need for pasture and the need for wood, in which areas where the wood was thin were defined as pasture and eventually lost most of their tree cover, and areas of thicker wood nearer to transport or to habitation came to be enclosed and protected from stock, at least for part of the time. The outcome sometimes was a characteristic mosaic of small woods close to houses or farmtouns and scattered trees on the lower part of the hill and other woods or ribbons of tree cover surviving on screes or rocky slopes at a distance. Such a pattern is well exemplified in Assynt.[10] Grazing lowered the natural tree-line as the upper part of the hill was totally abandoned to animals.

A further complicating factor was the existence of hunting reserves, 'foresta', in which game was preserved for the Crown or the nobility and from which domestic stock were often partially excluded, perhaps sometimes to the benefit of tree cover. In practice most 'foresta' would have been regularly grazed, even in the Middle Ages.[11] By the seventeenth century the royal forests had been made over to private ownership. The nobility

---

[9] *Ibid.*, 3, p. 100. See also p. 101, and 2, p. 454 for Strathnaver 'thick with woods' and full of cattle.
[10] R. Noble, 'Changes in native woodland in Assynt, Sutherland, since 1774', in T. C. Smout (ed.), *Scottish Woodland History* (1997), pp. 126–34; R. Noble, *Woods of Assynt* (project report for the Assynt Crofters Trust, [2000]).
[11] J. M. Gilbert, *Hunting and Hunting Reserves in Medieval Scotland* (Edinburgh, 1979). The Bishop of Aberdeen's Forest of Birse contained seventeen shielings, reflecting the seventeen townships of the parish in the twelfth century.

nevertheless often enjoyed defined hunting rights in the hills, for example the Earls of Mar in Deeside, the Earls of Atholl in northern Perthshire, the Earls of Moray in Glen Finglas, and the Earls of Breadalbane in the Forest of Mamlorn, where their interests conflicted with those of local graziers.

The early form of deer hunt involved driving the animals with beaters and dogs through woods and open moor, into an 'elrig', a natural trap or man-made enclosure where they were shot with bows and arrows: James IV on his jaunt in Atholl in 1528 was said by Pitscottie to have killed thirty-score red deer, and 'uther small beistis, as ra and rebuck [roedeer], wolf and fox'. The total may have been inflated for effect on this occasion, but the slaughter (and cruelty) was always great.[12]

Increasingly more fashionable than the chase and ambush of deer on the open hill, however, especially in the Lowlands, was the park close to the great house. At Cadzow and Dalkeith this has produced famous late medieval and seventeenth-century survivals of Lowland wood pasture that were originally hunting parks (Fig. 5.1).[13] There must have been many more. Falkland Wood was a royal deer park until the English let out the animals and cut down most of it in 1653 to build Perth Citadel. It lingered on to be shown in attenuated form on Adair's map of Fife in 1685, but had gone by start of the following century.[14] Its site (on good brown forest soil) is now a substantial farm. There was perhaps a narrow line between a grand hunting park and a modest ornamental park where the laird surrounded his tower house, or later on his mansion, with scattered specimen trees and grass for fallow deer or cattle. When Abercumbrie wrote of Carrick in Ayrshire that 'every Gentleman has by his house both wood and water orchards and parkes' he was probably referring to the ornamental rather than the utilitarian.[15] The maps of Pont, Gordon and Blaeu all show fenced wooded enclosures round the big houses. All parks, however, were wood pastures in which animals and trees existed together in a controlled environment, and the balance between economic use and pleasure could be varied according to circumstance.

From the early seventeenth century there occur other references to parks

---

[12] P. Dixon, *Puir Labourers and Busy Husbandmen* (Edinburgh, 2003), pp. 43–8; Anon. (ed.), 'Ane fair pallice of greine tymber', *Reforesting Scotland*, 16 (1997), pp. 33–4.
[13] M. Dougall and J. Dickson, 'Old managed oaks in the Glasgow area', in T. C. Smout (ed.), *Scottish Woodland History* (Edinburgh, 1997), pp. 76–85; O. Rackham, *Trees and Woodland in the British Landscape* (revised edn, London, 1995), p. 142.
[14] R. Sibbald, *The History Ancient and Modern of the Sheriffdoms of Fife and Kinross* (edn London, 1803), p. 387; G. Whittington and C. Smout, 'Landscape and history', in G. B. Corbet (ed.), *The Nature of Fife* (Edinburgh, 1998), p. 33.
[15] Mitchell (ed.), *Geographical Collections*, 2, p. 3.

Fig. 5.1 Cadzow's medieval hunting park, Lanarkshire, with trees dating at least to the fifteenth century. They appear to have been planted, or allowed to grow, on still older cultivation rigs. Anne-Marie Smout.

that were not obviously ornamental or merely for amusement, but probably implied a calculated combination of economic grazing and timber production. The records of the Campbells of Glenorchy provide the best evidence of this, with a whole series of parks referred to in the records of Baron Courts held at Killin, Finlanrig and 'Crannich' on Lochtayside as early as 1617 and 1618. We have little idea of how they were managed, apart from a determination to keep them free of the tenants' goats and pigs.[16] It is possible that, like other parks constructed in Galloway late in the seventeenth century, they were intended partly as holding and fattening enclosures for black cattle destined for the drove south but we have no direct evidence of that. The most famous of the Galloway parks was that of Sir David Dunbar of Baldoon, Wigtownshire, from which in the 1680s he sold hundreds of cattle into England. It was said to be an enclosure two and a half miles long and a mile and a half broad.[17] No doubt the trees within

---

[16] NAS: GD 112/17/4.
[17] J. A. Symon, *Scottish Farming Past and Present* (Edinburgh, 1959), p. 323.

this park were intended primarily to shelter the animals rather than to produce wood, but with pollarding the two uses were not mutually exclusive.

By the end of the sixteenth century, and probably in many places centuries earlier, the majority of woods in the Lowlands had become enclosed in order to protect timber supplies, and were the objects of carefully considered woodland management (see Chapter 7). Even these, however, were regularly open to grazing by animals, with the ground subject for two-thirds or three-quarters of the rotational cycle at least to some seasonal grazing. As late as the 1790s the haggs of enclosed coppice oakwoods in Argyll and Perthshire, managed on a twenty to twenty-five year rotation, were normally given from five to eight years' exclusion from cattle. In the woods of the Duke of Montrose, the annual value of the grass was normally less than 10 per cent of the annual value of the wood and bark, but at Cruixton and Mugdock it was worth over 30 per cent.[18] Sometimes the modification of a long-enclosed wood in the interests of the stock could be considerable. Alexander Wight, ca. 1780, told of how Mr Steele of Gadgirth in Ayrshire cleared out the undergrowth of a natural wood that had become so overgrown that it 'left no possibility of pasturing cattle in it', and by clearing, burning and thinning, turned it into very good pasture: 'the remaining trees are in a more thriving state than formerly and the field is very beautiful'.[19] There was in fact no wood in Scotland unaffected by the grazing of domestic stock to a greater or lesser degree.

Grazing pressure was of three descriptions – summer only, winter only, and throughout the year. Summer grazing was mainly on the high ground, associated in the Highlands with the shieling system whereby animals were taken in May to summer pastures where herders tended them for several months, eventually returning to the lower fields from which crops had been harvested.[20] Allocation of grazing was by 'souming', whereby each tenant was allowed a certain number of animals on the hill. Edward Burt in the early eighteenth century said of an area near Inverness:

> A soume is as much grass as will maintain four sheep: eight sheep are equal to a cow and a half, or forty goats ... the reason of this disproportion between the goats and the sheep is that after the sheep have

---

[18] J. Robson, *General View of the Agriculture of the County of Argyll* (London, 1794), p. 58; J. Robertson, *General View of the Agriculture in the Southern Districts of the County of Perth* (London, 1794), p. 97; J. Smith, *General View of the Agriculture of the County of Argyll* (Edinburgh, 1798), p. 130; J. M. Lindsay, 'The use of woodland in Argyllshire and Perthshire between 1650 and 1850', unpublished University of Edinburgh Ph.D. thesis (1976), p. 482.
[19] A. Wight, *Present State of Husbandry in Scotland* (Edinburgh, 1778–84), 3, p. 199.
[20] A. Bil, *The Shieling, 1600–1840* (Edinburgh, 1990); A. Fenton, *Scottish Country Life* (Edinburgh, 1976), pp. 124–36.

eat the pasture bare the herbs, as thyme, etc. that are left are of little or no value except for the brouzing of goats.[21]

There was, though, great variation. In other instances, ten sheep were reckoned to a cow, or a horse was considered equivalent either to two cows or eight sheep.[22] It does not follow that there were necessarily about twenty-five goats, or four to ten sheep, for every cow on the hill: cattle were essential to pay the rent, and were often likely to be preferred alone. In late eighteenth-century Assynt, for example, the sheep-to-cattle ratio varied on ten farms from 4:5 to 1:5, with a mean of 1:3 in favour of cattle.[23] Sheep, goats and horses were certainly important, but the eighteenth-century Highlands were seen essentially as a cattle-raising economy. Yet it was widely agreed that the hills were grazed far below what their carrying capacity would have been if the stock could have been supplied with ample feed in winter.[24] As James Robertson expressed it in Glen Avon:

> A man who pays £3 sterling of annual rent will perhaps have 20 black cattle, 3 or 4 horses, 20 sheep and 10 goats. During summer and autumn the pastures could maintain thrice the number, but they would perish during the winter or spring.[25]

This problem was solved when the turnip arrived in the Highland straths in the nineteenth century, and ever since there have been very much heavier grazing levels than previously on the uplands. The switch from a cattle economy to a sheep economy, that began in the southern edge of the Highlands around 1760 and swept the whole area in the first half of the nineteenth century in the period of the Clearances, exacerbated the problem by making dominant on the hills an animal that grazed very closely and destructively.

Shielings had at one time been common in the Southern Uplands as well as in the Highlands, until they were replaced by permanent shepherds managing sheep flocks all year. By 1500 shielings were rare in the Lowlands,

---

[21] E. Burt, *Letters from a Gentleman in the North of Scotland* (London, 1754), 2, p. 155.
[22] Lindsay, 'Use of woodland', p. 139.
[23] R. J. Adam (ed.), *Home's Survey of Assynt* (Scottish History Society, Edinburgh, 1960).
[24] Lindsay, 'Use of woodland', p. 141.
[25] D. M. Henderson and J. H. Dickson (eds), *A Naturalist in the Highlands: James Robertson, His Life and Travels in Scotland 1767–1771* (Edinburgh, 1994), p. 161. When Rob Roy died in 1734 he left thirteen cows, twenty-three goats, twenty-one sheep and five horses, one blind. NAS: Dunblane Commissary Court CC/6/5/24, pp. 145–6. We are obliged to David Stevenson for this reference.

though they may have continued in some areas south of the Forth into the sixteenth century, as on the moorlands between Clydesdale and West Lothian or in the Ettrick Forest.[26] They remained common throughout the Highlands in the seventeenth and eighteenth centuries, and only died out on the Isle of Lewis in the twentieth century.[27] As population pressure increased, especially after 1700, some Highland shielings began to be cultivated, and even to become permanent, if marginal, settlements: this was happening before 1750 in the upper parts of the Dee, in Glen Shee, Glen Tilt, in Atholl and generally in the Tay catchment area.[28] Wherever this took place, environmental change was accelerated by the reinforcement of grazing pressure by cultivation pressure.

Descriptions of the impact of summer grazing on trees are hard to find, perhaps in some cases because it had been going on for so long that in most localities few trees were left on the shielings themselves. They are not, however, completely lacking. In 1596, Sir Duncan Campbell of Glenorchy was given liberty from Privy Council to destroy unauthorised summer shielings in the Forest of Mamlorn in Glen Lochay. Graziers, apparently from Glen Lyon:

> yeirlie in the sommer seasoun cumis and rapairis to the said forrest, biggis sheillis within and about the same and remains the maist parte of the sommer seasoun at the said forrest, cutting and destroying in the meantyme the growand treis within the said forrest and schuting and slaying in grite nommeris the deir and wyld beistis within the same.[29]

This was far from the end of the matter. In 1738 there were nearly 1,200 beasts grazing in and around the forest. The Earl of Breadalbane was determined to keep them out and wrote to his forester: 'I as the king's heritable keeper, am in duty bound to preserve the forrest from all insults and you are to drive back all beasts out of the forrest as often as they come in there.'[30] It seems unlikely that he had any lasting success: in June 1751 there was a 'riot committed by the Glenlyon people in the forest of Mamlorn by pulling down the forester's house'.[31] There are isolated trees in high Glen

---

[26] I. D. Whyte, *Scotland before the Industrial Revolution: an Economic and Social History, c.1050 – c.1750* (London, 1995), p. 142.
[27] Fenton, *Scottish Country Life*, p. 133.
[28] Bil, *Shieling*, pp. 255–77; Whyte, *Scotland before the Industrial Revolution*, p. 134.
[29] NAS: GD 112/1/342.
[30] NAS: GD 112/15/261/6–8.
[31] NAS: GD 112/15/320/1.

Lochay now that show the same sort of evidence for low pollarding or 'cutting high' that is common in Glen Finglas.

Similar usage occurred in the Forest of Mar, though generally without the same levels of confrontation. In the interesting dispute over the forest between the Earl of Fife and Farquharson of Invercauld in 1758–60, which involved the tenant of the latter cultivating the ground next to the woods of the former, it was argued that the pinewoods gradually 'shift their stances' because they propagated by blowing their seeds onto the ground immediately next to the old woods, or into openings 'where they have Freedom of Air'. It was further said that pines growing up on the moors beyond the woods ran little risk from cattle, because:

> There is hardly any pasture on the moor adjacent to the woods, and the cattle are all grazed in the forrests, where there is the best out-pasture for all kinds of cattle to be found any where in Scotland.[32]

The phrase 'out pasture' seems to imply that at least some of the animals in Mar were out among the pines in winter. There were many summer shielings in the Forest of Mar, some indicated in Farquharson's map of 1703, a number at the edge of the pinewoods but most along the edge of watercourses in open ground. More than 1,000 cattle were in the souming for the Forest in 1729.[33] (See Map 5.1.)

The suggestion that cattle would not eat out the seedlings on the moor is undermined by a later eighteenth-century observation from Argyll. Here John Smith noted that the remnants of old pinewoods stood in the higher glens, and shed seed in winter which was driven to a distance by the storms: from these 'a beautiful plantation rises up in the spring, but when the cattle are driven up to the mountains in summer, this precious crop, the hope for future forests, is for ever destroyed'.[34] Yet he himself also mentions that, in the parish of Little Dunkeld in Perthshire, a 'fir wood of three hundred acres has thus arisen from seed driven by the wind from old trees within these thirty years'.[35] No doubt, as always, the truth varied according to the density and character of the animals on the ground.

Woodland on lower ground was potentially even more vulnerable, depending also on its age and degree of enclosure in addition to the other

---

[32] J. G. Michie (ed.), *Records of Invercauld* (New Spalding Club, Aberdeen, 1901), pp. 143, 145, 153.
[33] Royal Commission on the Ancient and Historical Monuments of Scotland, *Mar Lodge Estate, Grampian: an Archaeological Survey* (Edinburgh, 1995), pp. 6, 9.
[34] Smith, *Agriculture of Argyll*, p. 146.
[35] Ibid., p. 148.

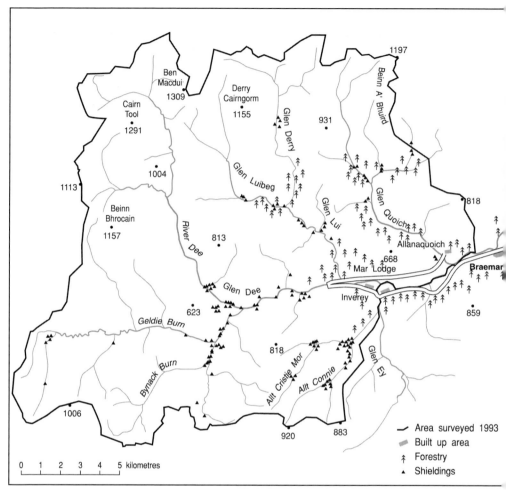

Map 5.1 Distribution of shielings in the Forest of Mar. Royal Commission on the Ancient and Historical Monuments of Scotland, based on OS Map, Crown Copyright, in *Mar Lodge Estate, Grampian: an Archaeological Survey* (RCAHMS, 1995), p. 5, with permission.

factors. Young wood unenclosed was a meal for the taking. 'To plant a wood without enclosing it', wrote William Boutcher in one of the earliest Scottish forestry treatises (dedicated to the Duke of Buccleuch), 'is not only amongst the idlest ways of throwing away money, but is laying up a fund of remorse and discontent that must necessarily occur from the devastation of cattle and sheep.'[36] Bitter experience taught many a landowner that truth. Thus Sir Ludovic Grant of Grant was told by his factor in 1748 that of 100,000

[36] W. Boutcher, *A Treatise on Forest Trees* (Edinburgh, 1775), p. 243.

pines planted on Dulnain Brae, not 4,000 were left undestroyed by cattle and sheep. Sandy Fraser, the miller at Castle Grant had also been letting in his horses at night, but had not been stopped as, if he had been arrested, there would be 'noe body to grind there corn'.[37]

Most of the damage, however, would occur to woods deliberately left beyond any fencing system because it was judged that their value to the tenant (and the rent-roll) as pasture outweighed their value as timber. It was this that stirred the anger of John Williams in 1784, when he spoke of the Gordon lands in Lochaber and the former Jacobite estates in Wester Ross, Inverness and Argyll, as containing 'a great many thousand acres' which:

> form a rich stool of oak in a deep soil, where the most luxuriant shoots are produced in summer, while the goats are in the hills, but they are soon browsed down in autumn, and kept level with the heath, by the goats and other cattle, and if any plant chances to raise its head beyond the reach of the goats, it is soon destroyed by the axe of the Highlandman, who strips off about four feet of the bark quite round, a little above the root, and leaves the young tree standing, to die a lingering death, as a monument of his barbarous greed.

He includes among such lands Coigach, the north side of Loch Broom, Kinlochmoidart, 'several thousand acres' of the north shore of Loch Eil and 'in the glens at the head' of Loch Eil, in Glen Lochy and on the shores of Loch Lochy, around Loch Arkaig, 'beautifully and richly covered with oak, birch and fir, where a good deal of oak is grown out of the reach of cattle'. He adds also the north shore of Loch Leven, with 'a good stoop of oak', Ardshiel ('and a good deal peeps out of the heath on the braes of that estate in summer'), as well as parts of Speyside and 'that extensive moor' between Fort William and the Spey where 'a thick stool of oak appears among the heath'.[38]

This highly important observation, by a man who had obviously travelled over a wide area with critical eyes, identifies the extent of grazing pressure on large expanses of shrubby woodland that was prevented from coming to maturity. Nor does it stand alone. A similar point was made by the Rev. John Walker in 1808 when he observed that the Highlands

---

[37] Quoted in I. D. Grant, 'Landlords and land management in north-eastern Scotland, 1750–1850', unpublished University of Edinburgh Ph.D. thesis (1978), 1, p. 173.

[38] J. Williams, 'Plans for a Royal forest of oak in the Highlands of Scotland', *Archaeologica Scotica*, 1 (1784), p. 29.

contained plenty of vestiges of former woods: 'the forest trees are still to be seen vegetating from large old roots; but what they shoot forth in summer, is eaten down and destroyed by the cattle in winter'. He lists the species: oak, ash, elm, birch, alder, holly, yew, rowan, hazel, whitebeam, goat willow, apple-leaved willow, grey willow, crack willow, bay-leaved willow, hagberry [bird cherry], water elder, blackthorn and whitethorn – all indicative of rich biodiversity. These, he says, have evidently formed in the past extensive woods and could do so again, 'though now nearly obliterated'.[39]

A third commentator, specifically confining his remarks to Argyll, was John Smith, who in 1798 suggested that landowners wishing to cash in on the profitable contemporary market for oak products should turn their attention to 'another kind of ground which ought to be planted, and of which we have large tracts'. This was at present covered with brushwood, such as hazel and birch 'seldom allowed by the cattle to rise above two or three feet high'. Nature, he said, was never mistaken in the suitability of soil for wood, and sometimes such brush also already had sprouts of oak and ash. Besides, 'patches of dwarfish oak, which the cattle never allow to rise one foot from the ground, are also common in many parts of the country'. All that land like this needed was a good dyke and a few extra acorns.[40] It is important to realise that what is being described was not the outcome of extensive sheep farming such as overtook such land in the following century, but of a farming system still based on black cattle. Yet the coming of sheep naturally continued and exacerbated a process already under way. The words of Robert Monteath in 1827 are strikingly similar to those of John Williams forty years before:

> It is as notorious a fact as the sun at noon day, that throughout most of the counties comprising the Highlands of Scotland, particularly the whole of Argyleshire, that there are millions of stools or roots of oak and some other kinds of trees detached throughout almost the whole of the extensive fields now appropriated to sheep pasture; nor is this to be wondered at, when we consider that many thousands of acres of land that was formerly carrying Natural Woods have of late years been left unenclosed and set aside for pasture lands; it is no less wonderful than true, that, the growths of these stools or roots, though devoured and eaten up in winter by cattle and sheep, are never wholly extirpated; as soon as the grass gets up so as to afford a supply of meat for the

[39] J. Walker, *An Economical History of the Hebrides and Highlands of Scotland* (Edinburgh, 1808), 2, p. 277.
[40] Smith, *Agriculture of Argyll*, p. 145.

sheep etc., the growth gets up, and so soon as the grass fails, the growths or saplings are eaten up.[41]

In the end, no doubt, even this scrub would vanish, certainly if muirburn was associated with the management of sheep pasture as it so often was. Yet, providing the pressure was not too intense, it is possible that such a system was sustainable, and a deliberate way of providing some sustenance to stock at a time when the grass had withered. In the absence of rabbits (rare in the Highlands before the nineteenth century) the stools could spring with new shoots year after year, and the soft growth of the season disappear once more in winter, as the writers described. Possibly it fulfilled a similar function to leaf hay in Norway or Sweden, where the winter was too cold for animals to be left out of doors in the Scottish manner. It must, though, always have been vulnerable to misjudgement and change: too many animals, and the stools were sure to die out.

Then there was the impact of grazing on mature woodland that was certainly also valued for timber production, but remained open to cattle in winter and sometimes to a limited degree in summer as well. The Forest of Mar we have already referred to. The arrangements at Comer Wood at the foot of Glen Cannich are explicitly laid down in an agreement of 1686, where the tenants were allowed to pasture the whole woods from 1 November until 5 May, with certain barren ewes allowed between 1 May and 1 September.[42] This was probably fairly typical: Lindsay found that in Argyll and Perthshire most animals traditionally went to the hill grazings between 1 May and 1 November, but milk cows went later and returned sooner.[43] Whether such arrangements prevented regeneration would surely depend primarily on the numbers of animals wintering – those left in summer would not normally be very numerous on the low ground. On the other hand, there were occasions when the summer grazing seemed to be important (Fig. 5.2). In 1758 the tenants of Craigroyston on Loch Lomond petitioned their landlord, stating that for eleven months in the year their only pasture was in the woods that were due to be enclosed, and that without them their tenancy would become impossible. Similarly in 1780 the tenants of Carnghouran in Rannoch claimed that if the Black Wood was enclosed they would lose their best pasture and only wintering ground. In both cases there was reference to summer as well as to winter pasture.[44]

---

[41] R. Monteath, *Miscellaneous Reports on Woods and Plantations* (Dundee, 1827), pp. 53–4.
[42] J. Munro (ed.), *Inventory of Chisholm Writs 1456–1810* (Scottish Record Society, Edinburgh, 1992), no. 560.
[43] Lindsay, 'Use of woodland', p. 137.
[44] *Ibid.*, p. 165.

Fig. 5.2 Summer grazing of cattle among the woods of Lochlomondside, as illustrated in C. Cordiner in 1786. St Andrews University Library.

Agrarian experts often described open woods as pasture. Thus William Marshall explained that the townships on Lochtayside contained on average, within the head dyke, about twenty acres of infield, fifteen acres of outfield, ten acres of meadow, thirty-five acres of green pasture and ten acres of 'woody waste': he further explained that patches that are 'too wet, too woody or too stoney to be plowed, are termed meadow, and are kept perpetually under the sithe and sickle for a scanty supply of hay'.[45] On the estate maps of the area, ca. 1760, small woods are described sometimes as pasture, sometimes as woods, with little consistency or obvious reason.[46] Similarly when George Langlands surveyed Kintyre (1770–7) he detailed all the small patches of woodland; for example, at Barrmolach in Carradale, a township of about 560 acres, he listed five little woods:

---

[45] W. Marshall, *General View of the Agriculture of the Central Highlands of Scotland* (London, 1794), pp. 30, 32.
[46] M. M. McArthur (ed.), *Survey of Lochtayside, 1796* (Scottish History Society, Edinburgh, 1936).

5 acres – mostly birch and some hazel wood, midling grazing.

4 acres, 2 roods – birch, hazel with some alder, a midling good grazing.

11 acres, 1 rood, 10 perch – mostly birch, some hazel and trangent ash and oak – midling grazing.

9 acres – mostly birch with some hazel, tho' a midling grazing.

13 acres, 1 rood, 9 perch – mostly birch with some hazel, not very good grazing.[47]

Land like this in Argyll, if it contained oak, was quite likely to become enclosed and better managed, when, towards the end of the eighteenth century, the returns for tanbark and charcoal rose steeply. Tenants, however, were unlikely to see a benefit to their own pockets. Arrangements varied, but possibly few landlords were as extortionate as James Riddell, the new proprietor of Ardnamurchan, who in 1774 agreed with Donald Campbell, 'tacksman of Ardnamurchan', that if the proprietor enclosed any woods the tenant should receive a rebate equal to the value of lost grazing for five years, but then pay an increased rent at seven and a half per cent of the cost of enclosure when the grazing was restored to him. Nevertheless, Campbell did not own the trees so protected, and was obliged to provide grass for the horses needed to transport the timber produce.[48]

Apparently more consensual was the appointment of two independent assessors on the Breadalbane estate in 1759 to judge compensation to a wadsetter with grazing rights in the wood of Letters near Dalmally when the Earl wished to enclose it. The wadsetter's opinion of the operation nevertheless comes through in his receipt for the money:

> Received by me Donald Mcintyre . . . two years rents of the big inclosur of Leathers which is laid wast for the preservation of the woods.[49]

Thus might a modern farmer receive his set-aside money.

The Duke of Argyll's estate was particularly careful in exploring which woods would repay enclosure and which should be allowed to remain as open pasture, depending on their compactness, species composition and access. Thus a survey of parts of Morvern in 1786 described all the local woods in detail and estimated appropriate rent reductions if enclosure was to be carried out. Some woods, like Ardtornish, were judged to be 'in such rough ground and so difficult to enclose' as not to be worth the expense, or

---

[47] Inveraray Castle Papers: Survey of Kintyre, George Langlands, 1770–1777, no. 200.
[48] NAS: GD 112/10/1/4/68, fo. 6.
[49] NAS: GD 112/15/363/6, fo. 3.

like those at Mechanach, 'mostly good birch, some alder and very few oaks and ashes', which 'should be left open, as barren wood at any distance from the sea will not bear the expence of inclosing and deduction of rent'.[50] Others were more suitable, as at Camusallach, where the woods were mostly ash and oak, with some birch, extensive and thriving, and 'close on the shore': it could form one great enclosure with the neighbouring wood, and rent deductions could be smaller if the tenant were allowed to cut hay in the wood and to labour with spades the patches of arable ground along the shore. Nevertheless, 'it takes in a considerable part of Ardslignish's wintering'.[51] There was some sensitivity to the needs of a working farm: at Gortanbeg it was necessary to leave a walled lane in the enclosure ('a lonie') so that the tacksman's cattle could 'have access to the shore in time of snow and bad weather' – though probably the farmer would rather have had the beasts in the wood than on the shore.[52]

Rebates of rent could be quite substantial. In 1699, when some ironmasters from Cumberland bought certain woods in the parish of Canonbie in Dumfriesshire, their contract specified that the landowner, the Duchess of Buccleuch, would pay damages to the tenants for grazings lost due to the enclosure of newly cut coppice. In the 1720s, such compensations could be as high as £18. 15s. sterling to a single tenant in one year.[53] In the Highlands, the Duke of Argyll's chamberlain's accounts of 1775, list several payments to tenants for loss of pasture due to enclosure for wood, ranging up to £8 a year.[54] Nevertheless, by the exclusions the farms were inevitably made less valuable: even if stock might be allowed back after five to seven years, there was a substantial problem of wintering them in the interim. Scottish practice compares badly with that in Denmark, where by a statute of 1805 all woods were to be enclosed and awarded to one owner, but landowners were also obliged to make available to farmers pasture equivalent to that lost in the forest.[55]

All, however, was not necessarily well for the husbandman even where the tenants got free run of the woods for their stock all winter. There are

---

[50] E. R. Cregeen (ed.), *Argyll Estate Instructions: Mull, Morvern, Tiree, 1771–1805* (Scottish History Society, Edinburgh, 1964), p. 132.
[51] *Ibid.*, p. 130.
[52] *Ibid.*, p. 132.
[53] NAS: RD 4/94, pp. 374–9; NAS: Buccleuch Muniments, Accounts charge and discharge Liddesdale 1722, GD 224/239/24.
[54] Inveraray Castle Papers: Chamberlain's accounts, 1774–5, fo. 242.
[55] T. Kjærgaard, *The Danish Revolution, 1500–1800: an Ecohistorical Interpretation* (Cambridge, 1994), p. 111; B. Fritzbøger, *Kulturskoven: Dansk Skovbrug fra Oldtid til Nytid* (Copenhagen, 1994), p. 341.

some most interesting and significant instances in the northern Highlands, where, in the absence of any attempt to protect the woods, the quality of the land appeared to be deteriorating so notably that it imperilled the viability of the holdings.

The first of these was in Wester Ross, from Inverinate on Loch Duich in 1753, when Torquil Mackay, a wadsetter, wrote to the factor of the Earl of Seaforth in these terms:

> As we in this and the neighbouring countries depend more upon the profite of cattle than that of tillage, we must studdy to have our grasings as good pasturre and as convenient as possible ... I hear from sensible honest men that other places in the country besides my tack have now less wood and more fern and heath than formerly, so that cattle want shelter in time of storm (as we never house any) and their pasture is growing more course and scarce. I know of severall burns that in time of a sudden thaw or heavy rain are so very rapid that they carry down from the mountains heaps of stone and rubbish, which by overflowing their banks, they leave upon the ground next them for a great way and by this means my tack and others are damnaged, and some others more now than formerly.[56]

It was an explicit statement that, in a situation where animals were always outwintered, the woods were diminishing, pasture deteriorating and stream erosion and deposition becoming a bigger problem.

By the end of the eighteenth century there was a wide recognition that, in the north of Scotland, the decline of woodland was an agrarian problem. Sir John Sinclair in 1795, for example, spoke of the 'natural birch woods' as 'much in decline' and likely to give out within a generation or two, which would be a 'melancholy situation' for the people.[57] John Henderson, writing of Sutherland in 1812, gave the fullest account of the phenomenon and its consequences. He spoke of the 'remains of a shrubbery of birch, hazel, aller, willow and some oak bushes, in the straths of the several rivers and burns in the country' that was 'not of so great extent as formerly, and is rapidly decaying in some places'. He continued: 'the natural woods on the several straths in this country to the southern, western and northern coasts are decaying fast'.[58]

[56] NLS: 1359/100, Delvine Papers.
[57] J. Sinclair, *General View of the Agriculture of the Northern Counties and Islands of Scotland* (London, 1795), p. 161.
[58] J. Henderson, *General View of the Agriculture of the County of Sutherland* (London, 1812), pp. 83, 86.

He appeared uncertain as to the cause. 'Naturalists aver', he said, that it was owing to severe winter and spring frosts for many seasons past. At first sight the tree species that he names appear well adapted to far more extreme climates than Sutherland, and a run of freak cold seasons seems an unlikely explanation. On the other hand, if an increase in precipitation had coincided with an increase in grazing pressure, such is the unfavourable oceanic nature of the Scottish environment compared to Scandinavia that regeneration could fail completely. There can be little doubt about the grazing pressure. Henderson himself observed that 'from the constant browsing of black cattle, it is not surprising that the oak is nearly gone'. He also described how until recent years, every farmer had had a flock of twenty to eighty goats.[59]

Whatever the cause, he was clear about the consequences:

> It is a well known fact, that in the straths where these woods have already decayed, the ground does not yield a quarter of the grass it did when the wood covered and sheltered it. Of course the inhabitants cannot rear the usual number of cattle, as they must now house them early in winter, and feed, or rather keep them just alive, on straw; whereas in former times their cattle remained in the woods all winter, in good condition, and were ready for the market early in summer. This accounts for the number of cattle which die from starvation on these straths, whenever the spring continues more severe than usual: and this is one argument in favour of sheep farming in this country.[60]

He cites as support a letter from Alexander Sage, minister of Kildonan (but no advocate of sheep farming), who corroborated the decay of the woods as a 'remarkable alteration on the face of this part of the county in the course of the last 20 years'. Sage adds a few different details: formerly the inhabitants 'out-wintered their cattle till the beginning of January, whereas they must now house them in the beginning of November', and when the wood was wasted away the ground became covered with coarse heather 'in place of the fine strong grass with which the woods abounded'. This led, he said, to 'a degeneracy of black cattle in the parts that were formerly covered with wood'.[61] (See also Fig. 5.3.)

It was, of course, the north of Scotland that was least touched by the external market for wood products in the eighteenth century, partly due to

[59] *Ibid.*, pp. 83, 86, 106.
[60] *Ibid.*, p. 86.
[61] *Ibid.*, p. 176.

Fig. 5.3 A well-wooded glen at Strathmore in Sutherland, with a creel dwelling to the left of the ruined broch of Dun Dornaigil. The decay of the Sutherland birchwoods was a complaint of many at the end of the eighteenth century. From Charles Cordiner's *Views*, 1780. St Andrews University Library.

its geographical remoteness, but also because the woods there were predominantly composed of birch, hazel and other 'barren timber'. Even in Inverness-shire where there was more oak, it was seldom enclosed, but left to the depredations of cattle, 'all shamefully abandoned, after every cutting, in the same neglected condition', in contrast to the situation of enclosed woods further south in Perthshire or Argyll.[62] It was in the north, therefore, that the trees were most vulnerable to an unsustainable regime in using woodland as grazing: interestingly, it appeared as a priority to the new sheep farmers in Sutherland after the clearances that the woods should be rehabilitated, and initially they had some success (see above, p. 69). In the

---

[62] J. Robertson, *General View of the Agriculture of the County of Inverness* (London, 1808), pp. 209–10.

south, forms of woodland protection were, at least for a time, more emphasised, but here the farmers had different problems.

It is not easy to get a long-term perspective on the impact of grazing on the history of Scottish woods. The first point to emphasise is that grazing was a natural component of the woodland ecosystem, ensuring much open land and much space between mature trees even in early prehistory. Auroch, wild white cattle and wild boar ultimately became extinct, but red deer and roe deer persisted and domestic stock replaced the vanished megafauna. In the Middle Ages, however, the numbers of grazing animals, particularly domestic stock, must have been checked in the uplands by the wolf, still reckoned a numerous pest in the fifteenth and sixteenth century.[63] This was a situation in which montane scrub and upland woods, perhaps in the form of very open wood pastures, could have continued to reproduce themselves.

Between 1600 and 1900 a series of changes occurred to this situation. Firstly, climate was less favourable, especially in the seventeenth century, than it had been at the height of the Middle Ages: if it became more difficult for trees to regenerate, a large proportion of seedlings was at risk from being grazed off. Secondly, the wolf, now rare except in the northern Highlands after 1600, was eventually exterminated, possibly before 1700, certainly before 1750. The absence of any predator lifted the ceiling on the number of stock that could be kept, particularly for sheep and goats. The latter became extremely numerous in the seventeenth and eighteenth centuries, until eventually checked by improving landlords concerned at the damage they could do to woods. Thirdly, the population of people and therefore of their subsistence livestock grew steadily in the Highlands in the eighteenth century, and after 1740 the rapid increase in the price of black cattle must have tempted farmers to put more and more stock on the hill.[64] This was the great age of the drove roads, with 60,000 beasts sold at the Falkirk Tryst by 1794.[65] Consequently by the end of the eighteenth century, upland scrub and wood pastures show every sign of being under pressure, eventually becoming vestigial remains almost invisible in the heath. Conversely, woods at lower altitudes, long enclosed in the Lowlands, were now enclosed in Highlands, where access to markets for tanbark, charcoal

---

[63] M. L. Anderson, *A History of Scottish Forestry* (Edinburgh, 1967) 1, pp. 269, 275.
[64] R. A. Dodgshon, *Land and Society in Early Scotland* (Oxford, 1981), pp. 277–320; A. J. S. Gibson and T. C. Smout, *Prices, Food and Wages in Scotland, 1550–1780* (Cambridge, 1995), pp. 196–7.
[65] A. R. B. Haldane, *The Drove Roads of Scotland* (Edinburgh, 1952); H. Hamilton, *An Economic History of Scotland in the Eighteenth Century* (Oxford, 1963), pp. 88–96; J. Mitchell, *The Shielings and Drove Ways of Loch Lomondside* (Stirling, 2000).

and timber justified it: domestic stock were not excluded, but were controlled and limited.

In the nineteenth century, despite the initial experiences in the Sutherland straths, the dominance of sheep and the possibility of keeping many more mouths on the hill once the problem of over-wintering stock out of doors had been solved by turnips, put paid to the chance of any long-term revival of montane scrub and wood pasture, especially when fire was used as the main tool to control and improve the hill grazings. As the price of charcoal and tanbark plummeted, and cheap timber came in from abroad, there was little point in keeping any regulations over stock in enclosed woods either, unless they were to be reserved as game shoots. There was a large drop in the amount of land with scrub and scattered trees on it, and the main cause for that decline must be overuse by cattle and goats in the first instance, and sheep in the second.

By way of postscript, it is intriguing that many of the areas where ancient upland wood pasture survives are in former royal forests, of which Glen Finglas is the prime example. Perhaps this is because in these areas trees were given precedence over stock for longer, since in the minds of their owners the deer that they hunted needed the woodland to survive: 'no woods, no deir' said the factor to the Earl of Moray, impressing upon his employer in 1707 the need to prevent poaching and uphold the need to keep the trees against stock.[66] In districts where there was no such attempt at control, the trees are likely to have gone earlier. By the nineteenth century, the association of red deer with a wooded landscape was much less close, and while deer forest owners might then see some need for low-lying plantations or woods to shelter their animals in winter, stalking itself took place on the open moors, which afforded a clearer line of sight for the rifle if the land was totally unwooded.

---

[66] NAS: GD 50/149. Moray box Lord Doune no. 1/447. Glen Finglas, like many royal and noble forests, had a history of tension between foresters and graziers. In 1580, James VI issued a letter of protection against people who 'compelled the forester to allow them to pasture a great number of their stock within the forest to the destruction of the king's deer and the fouling of the grass' (Anderson, *History of Scottish Forestry*, pp. 194–5) and the factor in 1707 also referred to a previous occasion when the deer had been 'destroyed' and 'your lordship's predecessors were forced to send Callum McGregor to the Muires of Atholl to bring deer out of it to replenish it.' These references may refer to a single event.

CHAPTER 6

Trading and taking wood before 1800

Because of the customs dues charged at Scottish and foreign ports, we know much more about the extent of the import of wood from overseas than we do about the trade in home-grown timber within Scotland. It is worth sketching the main features of this European supply into Scotland, in order to give another context and perspective to our account of how local Scottish wood was used, obtained and marketed.

Evidence from both documentary and archaeological sources shows that as international trade developed and Scottish supplies of good-quality oak became harder to find, timber was imported. In the fourteenth and fifteenth centuries, it was most likely to be oak from the Hanseatic towns of the eastern Baltic such as Danzig or Stralsund, sometimes as beams, sometimes as boards known as 'knappald' or 'knapholt', or 'wainscot'.[1] From the early sixteenth century onwards, however, reference to much larger supplies from Norway, and, to a lesser extent, from western Sweden, become more and more frequent, following the spread of the German water-driven sawmill into Scandinavia. It had reached the Stavanger area around 1550. The eastern Baltic was never entirely replaced: for example, more than one hundred boats coming to Scotland from Königsburg, from 1588 to 1603, carried oak board as part of their cargo. But now it was above all from Norway that Scotland came to draw most of her imported timber over the next two centuries. Most of that was pine or spruce, with a certain amount of hardwood in addition.[2] So critical did these supplies become

---

[1] D. Ditchburn, 'A note on Scandinavian trade with Scotland in the later Middle Ages', in G. G. Simpson (ed.), *Scotland and Scandinavia, 800–1800* (Edinburgh, 1990), pp. 73–85; A. Crone and F. Watson, 'Sufficiency to scarcity: medieval Scotland 500–1600', in T. C. Smout (ed.), *People and Woods in Scotland: a History* (Edinburgh, 2003), pp. 60–81; A. Crone and C. M. Mills, 'Seeing the wood and the trees; dendrochronological studies in Scotland', *Scottish Woodland History Discussion Group Notes*, 7 (2002), pp. 14–22.

[2] S. G. E. Lythe, *The Economy of Scotland in its European Setting, 1550–1625* (Edinburgh, 1960); T. C. Smout, *Scottish Trade on the Eve of Union, 1660–1707* (Edinburgh, 1963); A. Thomson, 'The Scottish timber trade, 1680 to 1800', unpublished St Andrews University

that in 1573 a Scottish statute banning the export of salt specifically excluded Norway, and in 1663 a similar statute banning the export of bullion excluded only coin sent out to purchase grain in time of famine, or sent out to buy Norwegian timber.[3]

The absolute quantities imported are difficult to determine, but they were clearly already substantial by the end of the sixteenth century. In Dundee, for example, one ship in three coming in from abroad was from Norway, almost invariably carrying timber: in 1599 the wood imports into that town were itemised as 12,324 deals, 70,751 spars, 3,700 'knappald', 14,000 'scowis', 26,200 'pipe staves' for barrels, 314 wainscot and a certain number of Swedish boards.[4]

In Norway itself, the impact of Scottish demand could be very significant locally. In Ryfylke, the district in the south-west of which Stavanger is the capital, the export trade in timber was known as the 'Skottarhandel' for 150 years, though in the long run the Scots were replaced by the Dutch as the biggest importers. In the oldest extant list, however, from 1567, twenty-five out of thirty-eight foreign vessels trading there were Scottish. By 1619–20, as many as 115 voyages were made from Scotland in one year to buy timber from this small stretch of the Norwegian coast, and there was also substantial trade to the neighbouring districts of Agder and Sunnhordland, as well as to Telemark and the area round Oslo fjord to the east, and later to the Møre-Trødelag districts to the north.[5] It was a close and personal trade, in which the skippers often came to be well acquainted with the Norwegian farmer-sawmillers, sometimes spending weeks in up-country creeks choosing and loading wood. The most important and valuable items were the sawn boards or deals, of which Ryfylke alone exported 74,000 to Scotland at the recorded peak in 1641–2, and beams or baulks, especially the so-called 'skottebjelker' of nine ells or less in length, of which 28,000 were also exported to Scotland in that one year. The trade in hazel barrel-hoops and in firewood was also significant (peak exports from Ryflke were 35,120 dozen hoops in 1620–1, and 323 cords of firewood in 1641–2). On any individual ship the cargo could be quite miscellaneous. Thus the *Jonas of Aberdeen*, William Walker master, loaded in August 1641 at Boknafjord, 220 boards, 60

Ph.D. thesis (1990); P. H. Winkelman, *Nederlandse Rekeningen in de Tolregisters van Köningsbergen 1588–1602* (The Hague, 1971).
[3] Lythe, *Economy of Scotland*, p. 147; Smout, *Scottish Trade*, p. 118.
[4] Lythe, *Economy of Scotland*, p. 146; I. F. Grant, *The Social and Economic Development of Scotland before 1603* (Edinburgh, 1930), p. 326.
[5] A. Lillehammer, 'The Scottish-Norwegian timber trade in the Stavanger area in the sixteenth and seventeenth centuries', in T. C. Smout (ed.), *Scotland and Europe, 1200–1850* (Edinburgh, 1986), pp. 97–111.

narrow beams of pinewood twelve ells long, 360 'skottebjelker', 180 pieces of birch, and 4 cords of firewood.[6]

In the second half of the seventeenth century, the trade did not slacken, although Ryfylke's particular role declined as the best timber was cut out and the smaller, remaining trees proved less saleable: the boom had not produced sustainable forestry. Nevertheless, between 1680 and 1686 three ships out of every ten arriving in Scotland were still wood-carriers from Norway, and in the customs precincts of Montrose, Dundee and Prestonpans it rose to five or six out of ten. One well-documented example of its use was in the building of Panmure House, Angus, for which in 1685 the Earl of Dalhousie chartered two ships with wood from Norway, and bought further supplies from importers in Dundee. The volume of wood imported to Scotland between 1686 and 1696 averaged per year about 354,000 deals and 'single-trees', probably equivalent to the boards and 'skottebjelker' of 1641–2.[7]

In the eighteenth century, as the Scottish economy developed, the pattern of timber imports underwent a considerable change. After 1750, imports of Norwegian deals, at mid-century accounting for three-quarters of the annual total of some 200,000 deals imported from Scandinavia, began to decline in relative importance. Norway was surpassed at first by Sweden and then, after 1765, by Russia in the inner Baltic. Deal imports nevertheless increased, and reached an eighteenth-century peak of about 600,000 in 1792, before the serious disruptions of the French Wars at the end of the century. However, deals themselves began to decline in relative importance at the same time as 'squared baulks' (the equivalent to the seventeenth-century beams and baulks) dramatically increased their share. These came in particular from the Baltic, and were measured in 'loads' of fifty cubic feet. Between 1755 and 1787, the Scottish imports rose from 1,957 loads to 25,936 loads, much of the demand being met from the Prussian port of Memel. Thereafter, with the wars, the figures somewhat declined.[8]

These statistics, however crude, put the alleged depredations on the Scottish woods into proportion. For example, the famous contract by which

---

[6] A. Lillehammer, 'Boards, beams and barrel-hoops: contacts between Scotland and the Stavanger area in the seventeenth century', in Simpson (ed.), *Scotland and Scandinavia*, pp. 100–6; S. I. Langhelle, 'The timber export from the Tysvær area in the sixteenth and seventeenth century', in [S. I. Langhelle] (ed.), *Timber and Trade* (Lokalhistorisk Stiftelse, Tysvær, 1999), pp. 24–36.

[7] Smout, *Scottish Trade*, pp. 155–7; T. C. Smout, 'The Norwegian timber trade before 1707, from the Scottish perspective', in [Langhelle] (ed.), *Timber and Trade*, pp. 37–57.

[8] Thomson, 'Scottish timber trade', *passim*.

the York Buildings Company in 1727 bought 60,000 trees in Abernethy Forest from Sir James Grant, which was abandoned after ten years without by any means all the trees being felled,[9] seems an insubstantial annual rate of extraction when set against Scottish imports. The imported squared baulk figures for 1787 alone equate to about the equivalent of 100,000 Scottish trees of marketable size, and this in one year, as do the deal imports of 1792.[10] Most Scottish wood sales were for much smaller quantities even than this.

The geographical spread of the imports widened very markedly over time. In the early seventeenth century the west coast played virtually no part, and the north rather little: for example in 1630–1, only two boats out of seventy, and in 1640–1 only five boats out of forty-eight, arriving in Ryfylke from Scotland, came from north of Montrose, all the remainder coming from ports in the Tay area or the Firth of Forth. By the end of the seventeenth century, in the years 1686–96, the north was still importing under 4 per cent of the total deals, but the west coast (primarily Port Glasgow, but stretching as far as the Solway), had come to import nearly one-fifth. Gordon of Straloch, in an account published in 1662, said that Aberdeen and Banffshire were already importing building timber from Norway, 'but there is enough at home for country purposes'. Of Buchan, however, he observed that if Norwegian supplies were to fail, 'they would be in an evil plight'.[11]

The disposition in 1765 is laid out in Table 6.1: by then the west took almost a third of all deals, the north about 10 per cent, and the Firths of Tay and Forth the remainder. Perhaps the most interesting aspect of the imports by 1765 however, is the widespread character of the trade. Every precinct in Scotland from Shetland to Dumfries, including Inverness, had some imports: most came into Leith, but Greenock came a close second, and Aberdeen came fifth after Bo'ness and Alloa. Imported wood was reaching all round the coast, often in substantial quantities.[12]

---

[9] H. M. Steven and A. Carlisle, *The Native Pinewoods of Scotland* (Edinburgh, 1959), pp. 112–13.
[10] One ton = 40 cu feet, and was reckoned to be the equal to three trees of the size commonly cut *per annum* in the Black Wood of Rannoch in the third quarter of the eighteenth century: J. M. Lindsay, 'The use of woodland in Argyllshire and Perthshire between 1650 and 1850', unpublished University of Edinburgh Ph.D. thesis (1974), p. 275. Another estimate from 1757 was that 'seven trees make two tons' in the pine woods of the estates of Lochiel and Kinlochmoidart: NLS: MS 17694 fo. 87.
[11] Lillehammer, 'Scottish-Norwegian timber trade'; Smout, *Scottish Trade*; A. Mitchell (ed.), *Geographical Collections Relating to Scotland made by Walter Macfarlane* (Scottish History Society, Edinburgh, 1906), 2, pp. 268, 298.
[12] Thomson, 'Scottish timber trade'.

Table 6.1 Import of deals into Scotland, 1765

Customs precinct	Number of deals	% of total
Northern area		
Shetland	1,032	0.2
Orkney	3,576	0.8
Thurso	4,020	0.9
Inverness	5,700	1.2
Aberdeen	37,092	7.6
		Sub-total 10.7%
Tay		
Montrose	14,028	2.9
Dundee	26,544	5.6
Perth	17,652	3.7
		Sub-total 12.2%
Forth		
Anstruther	5,460	1.2
Kirkcaldy	17,256	3.7
Alloa	39,209	8.3
Bo'ness	47,220	10.0
Leith	76,092	16.1
Prestonpans	22,308	4.7
		Sub-total 44.0%
Western area		
Stornoway	720	0.2
Campbeltown	4,728	1.0
Greenock	71,376	15.2
Dumbarton	7,032	1.5
Port Glasgow	10,440	2.2
Irvine	24,192	5.2
Ayr	20,160	4.3
Stranraer	3,900	0.8
Kirkcudbright	3,948	0.8
Dumfries	7,944	1.7
		Sub-total 32.9%
Total imports	471,624	

Source: adapted from A. Thomson, 'The Scottish timber trade, 1680 to 1800', unpublished St Andrews University Ph.D. thesis (1990).

The spread of imported wood around Scotland was a consequence of the level of demand on the one hand and the availability of alternative supplies on the other. The firths of Forth and Clyde, with their increasing industrial activity and dense population, were also far from native supplies and likely to need much imported softwood. Orkney and Shetland had no native supplies apart from driftwood; they imported what they needed, but their populations were small and poor. Inverness, and perhaps in the seventeenth century the northern mainland generally, presumably imported little because they had supplies of pine relatively near at hand.

Nevertheless, Norwegian wood appears to have been highly competitive even close to the native pinewoods, and was often regarded as indispensable for the finer work. Thus in 1733, when the Earl of Aberdeen was building his house at Kelly on the Ythan (now Haddo House), he found that the Abernethy wood he had purchased from the York Buildings Company was 'too knotty' and 'blew' for anything except the roughest work of scaffolding and sarking, and for lathing, flooring and lining the garrets; to get good wood for the main construction he was obliged to freight a Dundee ship to fetch it from Fredrickstad in Norway. Similarly, in the 1770s, the architect John Baxter brought imported Norwegian, Swedish, Prussian and Russian deals from Leith into Garmouth at the mouth of the Spey to build Gordon Castle: 'the [local] wood is of the roughest, coarsest kind, I am informed it would not answer your purpose'.[13] On the west coast, when Inveraray new town and castle were being built for the Duke of Argyll, 1,000 pines were cut on the estate for scaffolding, and local oak was used to make the piers of the bridges. All other construction timber was imported. Deals came from Norway, Gothenburg and the Baltic as far east as Riga. The demand for softwood was so great that Norwegian skippers called in at Loch Fyne on their way to the Clyde to see whether their cargoes could not be more profitably sold directly to the Duke. The great new inn alone used 10,000 deals. The castle, in addition to an immeasurable quantity of softwood, used oak wainscot imported via the Netherlands from the Baltic or Germany and 3,000 feet of Jamaica mahogany (Fig. 6.1).[14] During the Napoleonic Wars it was commented in Ross and Cromarty that the 'scarcity of timber is much felt', and that even in Lochbroom and Gairloch 'the timber used for the better sort of houses is partly the growth of the country and partly foreign': the advantage of Scottish timber there was its cheapness (10d. a foot compared to 2s. 6d. a foot), but its quality was too bad for any fine work.[15]

At times, however, even very ordinary buildings in the heart of the Highlands and islands were by preference built of Norwegian deal. John Mackenzie, undertaking a construction job in Stornoway in 1763, told the Commissioners of Forfeited Estates that their Loch Arkaig wood would not compete even in price with Norwegian wood by the time it had been cut, transported to Fort William and freighted across the Minch. Even more

---

[13] *Ibid.*, pp. 143, 147–8.
[14] I. G. Lindsay and M. Cosh, *Inveraray and the Dukes of Argyll* (Edinburgh, 1973), esp. pp. 106–20.
[15] G. S. MacKenzie, *General Survey of the Counties of Ross and Cromarty* (London, 1810), pp. 255–9.

Fig. 6.1 Inveraray Castle, Argyll, ca. 1800, William Daniell. Despite the wealth of native woodland around, virtually all the timber used in the house was imported from overseas. St Andrews University Library.

strikingly, in 1752 George Douglas was building a 'little washing house' for the garrison at Fort William, and hoping to use foreign wood imported via Irvine in preference to Loch Arkaig supplies: it 'will be better timber and better season'd than yours'.[16]

Even locally planted pine in the Lowlands could not compete with imported timber. In Clackmannanshire, the agricultural report told how 188 acres had been planted, mostly with 'firs', in 1730, and came to be felled between 1770 and 1782. The new stone and lime farmhouses had used such 'home-fir, of thirty-five or forty years' growth . . . for the roofs, as it was sold for little more than half the price of foreign wood'. However, it proved a false economy because so much more work was needed in preparing it, and it was so young and unseasoned that 'it scarcely lasted 20 years, whereas foreign wood continued good for a century'. Significantly, the local pine plantation was then replanted with oaks.[17]

---

[16] NAS: E 768/15/4; E 768/44/7(16).
[17] J. F. Erskine, *General View of the Agriculture of the County of Clackmannan* (Edinburgh, 1795), pp. 29, 58.

All the indications are, therefore, that from an early date Scottish timber found it difficult to compete for quality construction with Scandinavian and Baltic pine and spruce, except in wartime. This was cogently expressed by Robert Edward in 1678, writing of his native Angus:

> Here is abundance of timber for labouring utensils and for the houses of the common people: but for the houses in towns, and those of gentlemen in the country, timber is brought from Norway.[18]

Only in wartime when wood prevents trade to Norway, he went on to say, was timber supplied to Angus from woods 'on the west of the Grampians' where sawmills were constantly at work except during frosts. He may have been thinking of timber from the Black Wood of Rannoch and Meggernie in Glen Lyon, which usually had a very limited sale but was sent to wider markets during the Dutch wars.[19]

In the eighteenth century, war occasionally gave an opportunity to the owners of the Strathspey woods to stretch their horizons. Thus in 1746–8, Grant of Grant found it possible to send ships with rough Abernethy timber to Ralph Carr in Newcastle for sale to the mines, but his market there was instantly destroyed when a Danish fleet of timber carriers suddenly arrived from the Baltic.[20] Most famously and most significantly of all, the wars against the French from 1792 to 1815, and the subsequent period of high tariffs on European timber, provided an opportunity for large-scale exploitation in Glenmore, Abernethy, Rothiemurchus and Deeside, as will be more fully explained (see below pp. 214–24, 302–10). But under normal, peace-time conditions, foreign wood easily predominated in the marketplace over native supplies of Scots pine.

From all the foregoing evidence, it might be tempting to conclude that the Baltic and Scandinavia supplied almost all the demand for wood in Scotland. For the urban population this might be true. Even at a comparatively early date, certainly by 1550, the building trades in the east-coast towns south of Aberdeen probably did rely mainly on imported wood, though in the north and west, and perhaps in Perth, this was not yet the case. The burghs of Angus, Fife and Lothian, large and small, must have relied on Norwegian softwood in their building boom of the late sixteenth and earlier seventeenth century, even though their use of timber beams and cladding was

---

[18] R. Edward, *The County of Angus, 1678* (edn and trans., Edinburgh, 1883), p. 21.
[19] T. C. Smout, 'Some problems of timber supply in later seventeenth-century Scotland', *Scottish Forestry*, 15 (1960), pp. 11–12.
[20] Thomson, 'Scottish timber trade', p. 144.

Fig. 6.2 Timber-fronted buildings in old Edinburgh. Lavish use was made of imported softwoods from Scandinavia to construct wooden outshots and foreworks on otherwise stone buildings. This is Hyndford's Close, off the High Street, drawn in 1848. Royal Commission on the Ancient and Historical Monuments of Scotland. Crown copyright.

Fig. 6.3 The interior of Gladstone's Land, Lawnmarket, Edinburgh. Painted pine ceilings made of imported wood were common in prestigious buildings of the early seventeenth century. Royal Commission on the Ancient and Historical Monuments of Scotland. Crown copyright.

often lavish (Fig. 6.2, Fig. 6.3).[21] By the middle of the eighteenth century, as Table 6.1 suggests, probably only Perth (which planted up its boroughmuir with pine and other trees in the eighteenth century)[22] and the burghs of the Moray Firth were still seriously involved in using Scottish building timber, except under wartime conditions, and even Perth and Inverness imported from overseas. Certainly the most famous urban expansion of the later eighteenth century was built using foreign supplies: as Adam Smith described it in 1776, 'In the new town of Edinburgh, built within these few years, there is not, perhaps, a single stick of Scotch timber.'[23]

[21] Crone and Watson, 'Sufficiency to scarcity', pp. 75–8; I. Davies, B. Walker and J. Pendlebury, *Timber Cladding in Scotland* (Edinburgh, 2002), pp. 14–22.
[22] A. R. MacDonald, 'That valuable branch of the common good: the Perth plantation', *Scottish Forestry*, 51 (1997), pp. 34–9.
[23] Adam Smith, *The Wealth of Nations* (Everyman edn, London, 1910), 1, p. 152.

The overwhelming majority of the population before the nineteenth century, however, were country-dwellers, probably over 90 per cent of the population in 1700. For the most part they did not live in grand buildings but in small and humble ones: yet every house needed some wood, albeit of modest dimensions, and timber and timber products for tools, mill machinery and rural craft industries were universally needed. There is little indication that foreign wood normally met these common needs, except sometimes in rural areas in the Lowlands close to importing centres. Indeed, inland transport costs made it impractical for any except the wealthy to use foreign wood far inland, as illustrated by the experience of John Maitland of Eccles in 1773: it cost him 7d. per cubic foot to have his imported baulks carted forty-eight miles to Coldstream from Leith, higher than the 6d. per cubic foot that it had cost to ship them 1,100 miles from Memel to Leith.[24]

The demand for timber for most people in Scotland before the industrial revolution, therefore, must have been met from within the country. How? Probably the most important way that tenants supplied their needs, especially in the north and west where wood was relatively more plentiful, was by helping themselves. 'In its rude beginnings the greater part of every country is covered with wood,' said Adam Smith, 'which is then a mere encumbrance of no value to the landlord, who would gladly give it to anybody for the cutting.'[25] A general assumption that timber was there for the taking is enshrined in the Highland proverb that three things are free: the fish in the river, the deer on the hill and the tree in the wood. Nevertheless, in the eyes of the law and the landowner, custom no more made the taking of wood without permission legal than it did the poaching of fish and deer. However often the removal of wood might be tolerated and sometimes even encouraged by the landowner, the timber in law always belonged to him and not to the tenant. Temporary permission to use wood might on occasion be enshrined in a lease as an 'allowance' to a tacksman or one of the larger tenants, but it lasted only as long as the lease.

Most tenants in the seventeenth and eighteenth centuries held at will and had no written leases, and most leases, even when they existed, said nothing about wood. The landowner could revoke tacit permission at will, or at least fine tenants for taking trees (without seriously attempting to stop them), as a means of realising income from the wood. Even centuries of undisturbed custom in the removal of timber by the tenantry would not

[24] Thomson, 'Scottish timber trade', p. 151.
[25] Smith, *Wealth of Nations*, 1, p. 151.

establish a legal right for them to do so, though in their own eyes it might well establish a moral or customary right.

This situation accounts for innumerable cases of ill-feeling and outrage among the tenants when, particularly in the eighteenth century, improving landowners tried to put a stop to what they regarded as the tenants' misuse of woodland by taking wood in a profligate or careless manner, or without permission. Such controls were not new. They are certainly implied in fourteenth-century forest laws, such as the regulation that if a free man is found stealing green wood, he should forfeit eight cows.[26] Controls are also found, for example, in late medieval monastic administration at Coupar Angus and in sixteenth- and seventeenth-century Breadalbane. A particularly striking case occurred in 1593 when the Duke of Lennox pursued before Privy Council no fewer than 101 individuals, local people, who 'daylie and nichtlie' had cut his timber and peeled his bark in Methven Wood, 'and hes almaist distroyit the same'. They all failed to appear (they probably had no clue as to what was being done) and were declared rebels. Obviously he had failed to prosecute them successfully in his local Baron Court, and the incident shows the anger and impotence of the lord more clearly than it shows legal effectiveness.[27]

A determination to enforce and systematise controls became, however, much more widespread in the eighteenth century, backed in 1766 by draconian legislation. An Act of Parliament of that year made it an offence to damage or steal any 'Timber tree or trees or trees likely to become timber', or to receive stolen timber, or to allow animals to stray into 'inclosed or parked ground ... woods and hedges', under the penalty of a £20 sterling fine or six to twelve months' imprisonment for the first offence, and transportation to the colonies for seven years for repeated offences. The penalties in legislation of this character were so absurdly disproportionate that they were seldom invoked in their full severity in practice, though the Justices of Peace of Aberdeenshire in 1786 reminded the county of the earlier Act by public advertisement, because of a rising instance of trespassers damaging 'young, green and growing trees' and entering parks and plantations without consent.[28]

Nevertheless, landowners lived in the real world. They generally tried to steer a middle course between allowing the tenants freely to take what they pleased and forbidding them to take wood altogether, since they knew as

---

[26] M. L. Anderson, *A History of Scottish Forestry* (London, 1967), 1, p. 153.
[27] *RPCS*, 5 (1592–9), pp. 95–7.
[28] *Advertisement by the Justices of Peace of the County of Aberdeen, Given at Aberdeen, the Twentieth Day of December, 1786.*

well as anyone that wood was essential to every operation of tenant life and further that all woods contained timber of little or no commercial value on the market. Thus the Duke of Argyll in 1758 instructed his factor in Mull and Morvern that no tenant was to cut any wood without application to 'my woodkeepers', and according to their instructions: the implication was not that they would be refused wood, but that it was to be taken under supervision. To encourage them to build houses of stone and lime, 'I allow them for that purpose the liberty of oak couples and pantrees for roofing, but the same is alwise to be cut at sight of one of my woodkeepers.' To use valuable oak for this purpose was a concession, as earlier rules formulated in 1744 had restricted tenants to taking birch, hazel and alder for their houses, 'but these woods to be cut at the sight of my ground officer, who must survey my tennants houses yearly'.[29] On at least seven occasions between 1744 and 1762 the tenants were told that if they failed to take care of the woods or permitted others to steal wood from their holdings, they would be prosecuted by the estate for neglect: the very reiteration suggests that the edicts did not work effectively.

Indeed, even on this ostensibly well-organised estate there were many occasions for misunderstanding and infringement of the rules. Two examples must suffice. In 1751, Dugald Clerk of Braleckan, a tacksman on Lochfyneside who since 1706 had had a mandate from the Duke for 'preserving, weeding and cleanseing of these woods', found himself prosecuted for cutting oaks. He maintained that while the oaks were always acknowledged to be the Duke's property, he had a right 'disponed over to me with the lands by your Grace's predecessors' to cut the other timber on his holding, 'such as ash, elm, birch, hazel, alder, rowan, gean, quaking ash [aspen], sallies [willows], haw and sloethorn trees', which he used for hoops and small timber. Just as the Duke's woodman, when cutting oaks, sometimes damaged the other timber, so his woodmen unavoidably occasionally damaged an oak tree: 'I should think it pretty hard to be subjected to the penaltys prescribed by law for breaking or cutting a few of your Grace's oak shrubs in these thickets for access of cutting or taking out my own from amongst them', especially as the Duke had sustained not a hundredth part of the damage that had befallen himself when the oaks were felled 'these two last times'.[30]

A tacksman, of course, as a holder of a written lease and a gentleman, was a cut above ordinary tenants, and might well be allocated rights over

---

[29] Inveraray Castle Papers: NRA4 (Argyll), Instructions to Airds 1750–62: Instructions, 16 October, 1758: Instructions to Chamberlains, 1744.
[30] NLS: 17665. Memorial for the Duke of Argyll by Dugald Clerk of Braleckan, 11 September 1751.

certain timber in his lease, though he could lose them again at the termination of that lease. Ordinary tenants and cottars held at will, like those in Glen Aray who in 1748 had explicitly been told that they were 'not allowed to cutt any timber whatsoever', except when they make 'reasonable demand for reparation of their houses or other necessary occasions', and then only after their requests had been laid before, granted and recorded by the overseers at Inveraray.[31] That the estate regulation had made little impact even on the doorstep of the great house became clear when the Duke prosecuted 'cutters of wood in Glenaray' in 1755: no fewer than seventy-two tenants were examined, and although many denied any malpractice, others freely admitted to having helped themselves as need arose. One admitted to taking a few cabers for building a peat loft, a few kiln ribs, 'two or three birch sticks ... as fire-wood'; another cut eight young oak trees in the woods of Kilmun for a peat loft and an oak stick for a plough head; a third took 'some small stakes for making divisions between his cows and some small sticks for creels and such other uses for the farm', and so on. The deposition of Donald McIlvoil, illustrates the problem of getting the tenants to keep within either the spirit or the letter of the laird's law:

> Being solemnly sworn depones that he has been in use of cutting timber in the Duke of Argyle's woods every year these three years past, particularly he built a house in summer last of two couples and got the great timber out of a house at Dalchrunnoch by Duncan Campbell late ground officer his allowance who allowed him to cutt pann trees and cabbers for it which the deponent accordingly did. And being interrogate if he had cutt one hundred trees over and above those put in the house depones he cannot tell but was in use to cutt oak and birch [ash, deleted] as he had occasion for them Depones he neither sold or cutt for fireing any timber or shipt for bark any except those he took off the oak trees he cutt as above.[32]

This kind of evidence suggests that the landowner often had little practical control of the wood. We should be careful not to confuse a landowner's prohibition of an activity with that activity actually stopping. Tenants clearly continued to help themselves to wood whenever they could get away with it and perhaps considered that they were, in some sense, within their rights to do so.

---

[31] Inveraray Castle Papers: NRA4 (Argyll), Instructions to John Campbell, Chamberlain of Argyll, 1748.
[32] NLS: 17667. Depositions of witnesses against stealers of wood in Glen Aray, 9 May 1755.

A telling instance of the problems that might arise when tenants had customary 'wood-leave' came again on the Argyll estates, where the inhabitants of the bleak and treeless island of Tiree were wont to sail over to Loch Sunart (forty miles by sea) to fetch wood, sometimes selling it to people who were not even the Duke's tenants. The Duke's chamberlain on Mull and Morvern complained in 1786 that due to 'abuses' committed by the people of Tiree, 'in a few years they will utterly destroy the woods', so the Duke personally instructed his chamberlain on Tiree to take measures 'preventing their getting a single stick without your order and knowing what use it is for'. These orders were complied with, but in 1789 it was found that well over 7,000 trees had been felled in the space of fifteen months on Loch Sunart to supply the perfectly legitimate needs of forty-seven tenants. James Ferrier, the Duke's agent and close adviser, was appalled: he called it 'a most abominable waste and should be stopped', proposing that the Tiree tenants should build their houses of stone and lime and 'buy timber themselves, which will give them a better notion of the value of it than they seem to have at present'. The chamberlain defended the tenants, saying that some had needed new houses because of sand blow, and pointing out that there were no masons to build of stone, and no fuel to make lime. He could also have pointed out that as 6,440 of the trees had been used to make cabers for the roofs, what the walls were made of was immaterial. Consequently no further check seems to have been made, at least for a time, on the access of Tiree people to these woods, though the problem must have continued to escalate as the population of that overcrowded island grew and grew.[33]

Landowners were often quite generous to tenants in permitting free use of wood: it was, after all, frequently in their interest to be so, especially where the wood was of little value and the loyalty and prosperity of the tenant were a matter of concern. William Marshall in 1794 suggested that in the old days when 'a good soldier, or a fool-hardy desperado was of more value than a good husbandman', and when it had been more necessary 'to train the tenantry to war than to rural improvements', the country people had been profligate in using up timber resources. Even 'within the memory of middle-aged man, tenants had been allowed to cut without restraint and with impunity'. On the other hand there were inevitably problems of real or perceived misuse of any general permission to take wood. The Duke of Atholl in 1708 was told that his tenants selected the straightest young coppice shoots as barrel-hoops, which if left would have grown on into a

---

[33] E. R. Cregeen (ed.), *Argyll Estate Instructions: Mull, Morvern, Tiree, 1771–1805* (Scottish History Society, 1964), pp. 3, 7, 11–19.

valuable crop. The factor of Rannoch in 1752 gave permission for three small bridges to be constructed, but then complained when the tenants used marketable pine in place of birch and alder of little commercial value.[34]

Problems like these troubled improving landlords as the eighteenth century proceeded, more than it had done the traditionally minded chiefs of an earlier generation. Thus Cameron of Lochiel reserved certain of his inferior woodlands in the Loch Arkaig group for community use. Witnesses, after the 1745 Rising, spoke of 'the place that was formerly alloted by Locheill for the common use of the country, being apart from the principal wood commonly called the Gusach', and of 'Putichan' in Glen Loy where the woods 'are not good at least not so valuable as that of the firr woods on other parts of the estate', but where 'in Lochiel's time people were allowed to cutt firr for their buildings'.[35] To this day the ancient woodland in Glen Loy bears witness to unsystematic coppicing and pollarding of hardwoods, and irregular lopping of branches of pine, which probably date from this tradition. Such woods reserved by custom or permission for common use were referred to in parts of Scotland as 'bogs', a word that is probably derived from an early form of 'bough' and has nothing to do with marshy land. The 'Bog of Balnagoun', for example grew 'with a southerly exposure on a very sandy soil'.[36]

On another forfeited estate, on Fraser lands in Glen Moriston, witnesses declared that 'the custome of the barrony was to allow wood for repairing the houses that are ruinous', without hindrance or written orders from Lord Lovat.[37] This comment finds a clear echo on another forfeited estate, the Duke of Perth's lands in the south, where the factor remarked that 'it was the old use and wont of the estate, to give ash and other barren timber to the tenants for nothing, for the reparation of their houses . . . [it] was so much established that they went to the woods and took it at their own hand'.[38] In Glen Moriston, the previous factor further maintained to the new owners that the old chief actually 'found fault with the disponent and other gentlemen in Stratherick for not building right houses when they had

---

[34] W. Marshall, *General View of the Agriculture of the Central Highlands of Scotland* (London, 1794), pp. 24–7; J. M. Lindsay, 'Some aspects of timber supply in the Highlands, 1700–1850', *Scottish Studies*, 19 (1975), p. 43.

[35] NAS: E 768/44/7(1), 19(4) (the modern spellings are Giubhsaich and Phuiteachain).

[36] Lindsay, 'Use of woodland', p. 88; W. F. Gunn, 'The woods, forests and forestry of Ross-shire', *TRHAS*, 4th series, 17 (1885), p. 173.

[37] NAS: E 769/35/10(4).

[38] NAS: E 777/86/13(1). Cited by C. Englund, 'Woodland management on the forfeited estate of Perth, 1746 to 1784', unpublished University of St Andrews MA dissertation (1996), p. 17.

liberty of his lordship's woods'.[39] In other words, he apparently considered that his tenants had not used his woods freely enough.

When the Annexed Estates Commissioners took over, these pleasant arrangements ceased at both Loch Arkaig and Glen Moriston; there was, however, at first in both cases wholesale, almost anarchic, plundering of the woods by local tacksmen and others, which the government factors found extremely difficult to control. At Loch Arkaig, the culprits included not merely the supposedly Jacobite tenantry, but such pillars of the Hanoverian establishment as the local customs officer, the local minister, even officers of the garrison at Fort William and the Sheriff Depute of Inverness.[40] It was not easy to draw up new rules for reasonable use. The new factor in Glen Moriston had visited one of the 'gentlemen's' houses and certainly agreed that it was in need of repair: 'I slept a night in it September last and observed it supported with props and pillars . . . I did not think it very safe to sleep under such a roof.' On the other hand, he thought the tacksmen had 'taken a better quality of wood and a greater quantity than I think would have been necessary to repair thair mudd houses, unless they had a mind to build new ones with stone and lime', and in any case they should have made a proper written application for an agreed quantity.[41] Just as in Argyll, the new brooms in estate management wanted to sweep out the informalities of the past, and there was certainly nothing in law to prevent them doing so. In practice, they were seldom able to do so as quickly or as thoroughly as they would have liked.

Proprietors nevertheless persisted in their efforts to bring to an end this tradition of informal assumption of rights to wood, sometimes indeed by severe and exemplary justice. On occasion, the offenders were very respectable people. The Annexed Estates Commissioners in 1779 were incensed to discover that the Rev. James Robertson in the parish of Callander (a man later responsible for several observations in praise of good woodland management) had cut without permission on their own land a very large oak, and damaged several other old trees by cutting off their biggest branches: they not only ordered that he should be prosecuted but that the news of his prosecution should be published in the Edinburgh press.[42] More characteristic and drastic was the harassment of poor tenants for assuming similar rights. An extreme case on Skye was the expulsion in

---

[39] NAS: E 769/35/10(4).
[40] T. C. Smout, 'Cutting into the pine: Loch Arkaig and Rothiemurchus in the eighteenth century', in T. C. Smout (ed.), *Scottish Woodland History* (Edinburgh, 1997), pp. 115–26.
[41] NAS: E 769/35/9.
[42] NAS: E 721/25 fo. 225. Cited by Englund, 'Woodland management', p. 42.

1814 from his holding in Tokavaig in Sleat of Alexander Nicolson by his landowner, Lord MacDonald. Strict regulations had been introduced on the estate around 1800, reserving all wood to the landowner and promising prosecution of any tenant who took timber without permission. The tenant in this case pleaded that he had taken wood only to repair a house 'consumed by a sudden fire' in order to shelter and preserve his family, that the wood he had used for a 'miserable hut . . . for a temporary shade' was mainly 'picked from among the ruins of his former house', and that a former owner of the land had encouraged him or his predecessors to use the wood freely. Despite the tenant's pleas that 'the sentence of dispossession passed against him is so very unsupportable to his distressed family', Lord MacDonald upheld the eviction, being 'determined to make an example of every person who shall presume to cut down wood on any pretence whatsoever without my permission upon my estate'.[43] The Improvers certainly had the law of Scotland and the United Kingdom on their side, but were only too ready to assume that reason or mercy equated with weakness.

The only circumstances in which secure long-term, legally recognised, rights over wood existed, were when one proprietor, often a feudal superior, granted to another heritable proprietor and his tenants a 'servitude', or right of use, over timber that was the first proprietor's property. A servitude of this kind differed from what was generally called 'wood-leave' or an 'allowance' from a proprietor to his own tenants, in that it was not time-limited. Because proprietors inherited property rights, the servitude was normally perpetual unless bought out, or allowed to lapse through forty years' disuse.[44] Such a right was valuable to the tenants of the second proprietor as they were enabled by it to cut wood for their own needs, but if a dispute arose it was fought out in the courts between the landlords concerned. Servitudes of this sort were commonly granted before the eighteenth century, giving the impression that when wood was plentiful and had little market value, it mattered little to the ultimate owner who took it, provided that the privilege was not abused and that the takers were his vassals or others within his sphere of power. The superior might be particularly helpful if there were types of wood that were perhaps not plentiful on their vassals' or allies' lands. A characteristic example was the servitude established in 1686 over woods at Kingairloch in Morvern, allowing

---

[43] Clan Donald Papers, Skye: MS 4241/1.
[44] D. M. Walker (ed.), *The Institutions of the Law of Scotland by James, Viscount Stair, 1695* (Edinburgh, 1981), pp. 452–62; G. Watson (ed.), *Bell's Dictionary and Digest of the Law of Scotland*, 7th edn (Edinburgh, 1890), pp. 1002–4.

tenants of a proprietor in Lismore to collect annually six large boat-loads of any timber except rowan and straight hazel: it was not extinguished until 1844.[45]

Normally a servitude was conveyed to those who would use the timber for local, essentially subsistence, purposes, not for sale on the market, and it might be hedged with conditions to ensure this. For example, in 1707 the Earl of Breadalbane granted a wadsetter of the lands of Auchmore, a servitude of woods that were not enclosed, and then only for repairing buildings. Similarly, in 1675, Cameron of Lochiel granted a wadsetter the right to use any of the laird's woods in Lochaber or Lochiel, but only 'for the proper and necessary use of the said wadset lands'.[46]

Servitude rights could as readily lead to litigation between wadsetters and landlords as between the greater landlords themselves. In the last-mentioned case, the wadset lands passed to Cameron of Fassifern, who in 1736 obtained a new wadset on the same terms as the old, but excluding any woods that themselves were in tack to another. Lochiel lost his estates to the Crown after the 1745 Rising, but Fassifern's wadset remained unaffected by the new ownership. Around 1769, the government factor, with the approval of the Barons of the Exchequer, sold all the woods on the estate (except the pine) for making charcoal. Cameron of Fassifern maintained that his rights were being affected; his lawyer advised applying to the sheriff for an interdict, or taking the case to the Court of Session.[47]

Not surprisingly, in the eighteenth century landowners became much less happy with servitudes. For one thing, new attitudes towards property inclined them to exercise their rights absolutely and to the full, and made them impatient of restrictions and concessions of any kind. For another, freedom to use the wood seemed, as population rose, to occasion damage to the wood more frequently, and as the value of timber increased this resulted in potential loss of income to the ultimate proprietor. The forestry management of the Duke of Gordon, for example, was criticised by an adviser on woods and estates, William Lorimer, in 1763:

> The family of Gordon fewed out Badenoch in their distress during Cromwell's usurpation – they have not the 10th part of the value now

---

[45] Lindsay, 'Use of woodland', pp. 83–4.
[46] NAS: GD 112/2/19/20/; GD 1/736/4. A wadset was a loan in return for rights of possession as long as the loan remained unredeemed.
[47] NAS: GD 1/736/4.
[48] G. A. Dixon, 'William Lorimer on forestry in the Central Highlands in the early 1760s', *Scottish Forestry*, 29 (1975), p. 208.

paid them, and the woods decay daily, partly because every fewar cuts as much as he pleases, and having that servitude, the Duke dares not sell or cut for his own use.[48]

Five years later, the Duke, fearing that 'the whole wood of the lordship will be run out and destroyed', ordered his baron baillie in Badenoch to co-operate with a newly appointed forester to 'stop every body from cutting upon any pretence whatsoever, unless they produce an order from you ... no use is to be allowed but for the necessary uses of building and repairing upon the ground of the Lordship'.[49] This ushered in a campaign right across the Gordon estates against illegal cutters of wood and peelers of bark, the transgressors being brought before the justices of the peace and fined or imprisoned. One list of offenders, from September 1775, contains the names of nearly 150 offenders in Strath Avon and Glen Livet alone. Similar attempts to prevent destruction of wood over the next few years suggest, however, that these efforts were to little avail.[50] They would have been easier to enforce at will on unprotected tenants (as many of those in Strathaven and Glen Livet perhaps were) than on feuars, who were effectively small independent landowners, since servitudes held by the latter could not readily be extinguished by bluster.

There was certainly a good deal of litigation in the eighteenth century about servitudes, none of it more famous than the quarrels in Deeside involving the *nouveau riche* Lord Braco, later Earl of Fife, and the older families on the land, notably the Farquharsons of Invercauld.[51] Braco had acquired the superiority of the old Mar lands in 1735, following the 1715 Rising, and Farquharson had thereby become his vassal instead of being the vassal of the Jacobite earl. The status of vassal, by the eighteenth century, meant little more than having to pay to the superior a very modest annual rent, and as long as the vassal continued to do so he remained in absolute possession of his lands. Farquharson was therefore a 'heritor', a secure and important landowner in his own right, not a tacksman or a tenant. However, in this case, the feudal superior, being in the past extremely keen on hunting in the forest, had from olden times retained the right to hunt game and the ownership of woods, but had granted local gentlemen a servitude to use the pinewoods and grazings for their needs. Even before the forfeiture, there had been extensive 'abuses' in the Mar woods, and

---

[49] NLS: MS 1281, Delvine Papers, Instructions to John Clerk, 11 July 1768.
[50] NAS: GD 44/39/26/1; GD 44/39/27/4, 12, 14 and 15.
[51] J. G. Michie (ed.), *Records of Invercauld* (New Spalding Club, Aberdeen, 1901), pp. 124–53.

anxiety by the former landowner's agents about curbing illegal cutting.[52] To the new landowner, this would seem justification enough for action (Fig. 6.4).

The issues at law between Braco and Farquharson in 1758–60 related to the limitation and extent of such a servitude on the woods. James Farquharson maintained that he had rights 'for the uses of labouring and building'; Braco, although he acknowledged an obligation to leave the defender with enough wood for legitimate needs, argued that he had an absolute right to do as he wished with the rest, and to ensure that enough ground within and without the wood should be left unploughed to allow the wood to spread. The questions resolved into whether Farquharson's tenants were entitled to cultivate land adjacent to, or intermixed with, the pinewoods, and whether they were entitled to take pine from Braco's woods when they had birch enough on their own grounds.

Farquharson argued that Lord Braco wanted the country to be a 'wild uninhabited desert, all overgrown with wood', but that hard-working men like his tenant Calder, about whom the complaint was particularly raised, needed land:

> In the space of two or three years, [he has] taken in so much of this ground with a spade that he is able to subsist himself, and a family of nine or ten... it is most certain that the cultivation is greatly increased in this country within the memory of man, and many tacks and possessions have in that time been newly set down, or greatly enlarged.[53]

As to the second point, birch was ideal for farm implements, but made only crooked, leaky and short-lived roofs: so pine was considered indispensable for tenant building, whenever and wherever it could be obtained.

Lord Braco's reply emphasised the value of the Highland pinewoods, especially those of Mar, and the degree to which they were continually damaged by 'the waste and abuses' of servitudes. The best trees, it was alleged 'four or five feet diameter', were often felled for trifling purposes, only the tops and branches being used for building, the rest being left to rot on the ground. Another universal and unacceptable practice was 'cutting out the hearts of the finest trees to serve for candle-fir, by which the tree

---

[52] These were intense in the period 1712–13, and renewed after the forfeiture (but before the sale to Lord Braco) in 1717–33. See NAS: GD 124/15, especially 1051, 1086, 1089, 1098, 1109, 1176, 1206 and 1406.

[53] Michie, *Invercauld*, pp. 127–8, 130.

Fig. 6.4 Mar, the scene of furious eighteenth-century litigation, at 'Craggan rock (which marks the commencement of the Earl of Fife's so ample forest)' with 'the saw-mill at the opening of the adjacent romantic valley' of Glen Quoich. Charles Cordiner, ca. 1795. St Andrews University Library.

perished and decayed upon the foot'. There was also alleged to be illegal sale of wood cut under cover of servitudes.

On the question of the legitimacy of cultivation near the woods, Braco's lawyers explained how the pinewoods regenerated:

> It is known to your Lordships, that, as fir-woods do not spring from the root, but are propagated by the blowing of the seed in the grounds immediately adjacent to the old woods, or in the openings, where they have freedom of air, these highland fir-woods are not fixed to a particular spot, but gradually shift their stances.[54]

The activities of the tenant Calder and his like, pulling up thousands of young trees to make his holding, would stifle the woods, which 'must be

[54] *Ibid.*, pp. 142–3.

allowed to shift their stances . . . by confining them to those spots where the stool of the wood now stands, they would soon be extirpated'.[55] It was a classic conflict between farming and forestry, and arose at the point where population pressure was growing and the value of the woods also increasing. It also showed an admirable grasp of the ecological requirements of Scots pine, a tree that indeed does not regenerate readily in the shade of other trees even of its own species.

The Earls of Fife continued their legal campaign to restrict servitudes for many years, and also by the obstructiveness of their factors made the rights less useful. As late as 1719 the minister of Crathie had assumed servitude rights to use pine from the Mar woods for the roof, stairs, doors and windows of his manse, but his successors and many local people who had made similar assumptions were resolutely opposed in this by the new owner.[56] The Annexed Estates Commissioners, when they took over the small Deeside estate of Monaltry near Braemar, were informed in 1756 by 'lawiers of the greatest repute' that their servitude over the Braemar pinewoods obliged them to use timber from their own land before going into Lord Braco's woods ('there is a great deal of birch and allar timber upon the lands of Monaltry very fit and more than sufficient to serve all the inhabitants'), and that, anyway, their servitude is 'only for the use of labouring tools for houses and office houses necessary for the farms, but that it does not extend to household furniture, kirks, manses, bridges and miln works'.[57]

Two decades later, in 1773, the Monaltry tenants were in no doubt that the rules of the game had been changed to their disadvantage: under the Earls of Mar they had used his woods 'for all their necessaries without any limitation or restriction', but now they were 'almost quite deprived of the benefit of our servitude'. The Earl of Fife's factor made them go to the remotest part of the pinewood, ten or twelve miles away (the nearest part was within a mile) and 'when we come there, they give us but the crops and grains [branches] of the trees and none of those but within 8 inches broad and sometimes refuse to give any at all'. They had been explicitly refused wood for the repair of the waulk mill, which was now consequently in ruins, or for the ferry boat, or for a house built by resettled soldiers. In the past they had been able to use the pinewood a mile away for 'all their necessaries for more than a hundred years before the Earl of Fife purchased the estate'.[58]

---

[55] Ibid., p. 145.
[56] Ibid., p. 153.
[57] NAS: E 773/55/1–2. Lorimer's statement that on the Mar estate the vassals 'have no Servitude on the Firrs' was misinformed: Dixon, 'Lorimer on forestry', p. 198.
[58] NAS: E 773/32/50(3).

For the local farmers and villagers it was a poor deal, but Lord Braco was admired by professional foresters for the stand that he had made against the vassals and their tenants. It was a clash of cultures, between the improvers and a more traditional society. For William Lorimer, the Earl of Fife's administration on Deeside provided a model for the future treatment of all such servitudes in the Highlands, which others should take care to imitate. He described approvingly how any tenant of a vassal claiming servitude on the Fife woods had to get a ticket signed by the vassal specifying precisely what was wanted, then to present it to the Earl's forester, who allowed only the wood specified and subsequently inspected the tenant's house to make sure that he had made exactly the use of it claimed. Furthermore:

> My Lord allows only the month of June for giving off his Servitude Timber – Advertisement is given at the Church, of the days appointed for certain tenants, and if they don't come on the days appointed, they receive no Timber that year. My Lord frequently allows no Deals or Boards to the Vassals tenants for their Doors or Checks – he services them with Backs and Slabs from the mills.[59]

That is to say, he allowed them only inferior or refuse wood.

By the 1770s, the notion that anyone could expect to be allowed to take wood without paying for it was not extinguished, but it was certainly in retreat, and if it was not possible to obtain timber for nothing, it would have to be bought. There were basically three ways of doing this – purchasing at the local burgh markets, buying at a sale held in the woods, or paying the landowner's factor or forester in money or kind for what you needed. We are not here concerned with the longer-distance movement of Scottish wood within the country, such as the medieval shipments of oak beams from Darnaway to the king's properties in the Firth of Forth, and various seventeenth- and eighteenth-century attempts to profit by shipping pine from the Highlands into the Clyde and Forth. Some of these are discussed in later chapters.

Most of the burghs round the fringes of the Highlands had markets or fairs where timber and bark were sold. Their significance was recognised as early as 1564, when Mary, Queen of Scots expressed concern at the condition of the woods in the north through the unregulated cutting and selling of timber and the peeling of bark from living trees, so that 'the haill polecie in that part is lyke to pereis'. She ordered through Privy Council that all the burghs in the shires of Inverness, Nairn, Elgin, Forres, Banff and Aberdeen

---

[59] Dixon, 'Lorimer on forestry', p. 199.

should announce by public proclamation in Her Majesty's name that none may buy or sell any sort of timber, great or small 'bot in oppin and plane marcattis at the fre burrowis... or within the fredomis thairof' and also that no-one was to sell bark except on the tree trunk and in the market.[60]

It seems most unlikely that such a regulation could have been enforced in either particular, but the vigour of some of the local burghal wood markets is not in doubt. Thus in the 1680s, Kirriemuir was described as having three great fairs and a weekly market 'of all kind of commodities the countrey affoord but especially of timber brought from the highlands in great abundance', and Brechin as having a weekly market that was 'a great resort of highland men with timber, peats and heather and abundance of muirfowl and extraordinaire good wool in its season'.[61] Inverness was long famous for its country produce market: Pennant in 1769 described it as selling deer and roeskins, coarse country cloth brought down by pedlars, meal in goatskin bags, cheese, butter wrapped in seaweed, 'and great quantities of birch wood and hazel cut into lengths for carts, etc., which had been floated down the river from Loch Ness'.[62] He could have added pine planks and spars from Glen Moriston, floated down Loch Ness, or from the Spey forests, either floated down the river to its mouth and transhipped, or carted on pony-back over the Slochd. In 1778 it was an irritation to the Laird of Grant that his wood from Abernethy, though of better quality than that of the Grants of Rothiemurchus, fetched less at Inverness market.[63]

Just who supplied these markets and how, is, however, a complex and sometimes obscure question. Sometimes people helped themselves and sold the proceeds. We do not know the exact circumstances surrounding the events at Cawdor in 1599, when Gilliecallum Mackintosh allegedly raided John Campbell of Cawdor's lands with thirty 'brokin heiland men' armed with bows and arrows, swords and guns, and 'violentlie cuttit doun' 720 young growing trees of the wood at Easter Balcroy 'callit Torgarve', carrying off the timber partly with his own men and horses and partly with those of John Campbell's tenants, whom the raiding party forced to help.[64] The scale of the raid suggests that the wood was perhaps intended for sale at Nairn market, and behind the affair may have lain a disputed servitude.

Rather clearer were the proceedings in 1753 in the Baron Court at Rannoch against a large group of poor cottars and tenants for straightforward stealing

[60] *RPCS*, 1 (1545–1625), p. 279.
[61] Mitchell (ed.), *Geographical Collections*, 2, pp. 30, 40.
[62] T. Pennant, *A Tour in Scotland, 1769* (edn Warrington, 1774), p. 189.
[63] Mitchell (ed.), *Geographical Collections*, 2, p. 171; W. Fraser, *The Chiefs of Grant* (Edinburgh, 1883), 2, p. 459.
[64] C. Innes (ed.), *The Book of the Thanes of Cawdor* (Spalding Club, Aberdeen, 1859), p. 218.

and selling timber from the Black Wood, only in this case they were so destitute that the fines imposed upon them were meaningless. The Duke of Atholl's tenants at Bunrannoch claimed, through him, a servitude over deciduous timber and candle fir in the Black Wood, and certainly large quantities of candle fir from Rannoch entered the local fairs and markets one way or another: Thomas Hunter said 'the whole of Northern Perthshire depended on the stocks of the Rannoch fir alone for the illumination of their dwellings'.[65] The Grants faced similar problems in Abernethy. In 1690 Thomas Mackenzie in Coulnakyle was fined for 'medling with the Laird of Grant's woods and selling thereof without warrand', and in 1763 complaints of abuses in the wood included reference to 'a most pernicious custom... of cuting pieces from the body of large and thriving trees for candles and other uses, which wood for candles has been even sold and sent out of the estate of Strathspey'. The forester was authorised to call upon three substantial and creditable tenants to help him search houses and holdings where stolen wood might be hidden, to prosecute the offenders, pocket half of any ensuing fines, and 'the other half shall be applied to proper uses as Mr Grant thinks fit'.[66]

Landowners frequently organised the exploitation of their own pinewoods and sale to a local market in a direct way, often in association with running a sawmill. Robertson of Struan, for example, converted the rents of the local farmtouns around the Black Wood of Rannoch into services for cutting and delivering timber to his mill, and sold the deals downriver. Dr Lindsay has shown how short-range was the market: in the period of administration of the Black Wood by the Annexed Estates Commissioners, half the timber was sold within a ten-mile radius, and only 6 per cent went more than twenty miles: half was bought by men of tenant status (mostly from the Struan estate) and only a fifth by wrights or timber workers. Some went by water and some by horseback: attempts to compel the tenants to perform carriage services to carry wood to Crieff and Perth were abandoned in 1757. It was essentially a petty trade, yet nearly half the rental value of the estate in 1767 was accounted for by the woods and the sawmill (see Map 6.1).[67]

Along Deeside, the pinewood economy was also well developed at an early date. At Glentanar in 1725, the people lived 'more by traffiquing in

---

[65] Lindsay, 'Use of woodland', pp. 280–1; T. Hunter, *Woods, Forests and Estates of Perthshire* (Perth, 1883), p. 414.
[66] E. Grant, *Abernethy Forest: its People and its Past* (Nethy Bridge, 1994), p. 8; G. A. Dixon, 'Forestry in Strathspey in the 1760s', *Scottish Forestry*, 30 (1976), pp. 40–1.
[67] Lindsay, 'Use of woodland', pp. 267–70, 294–300.

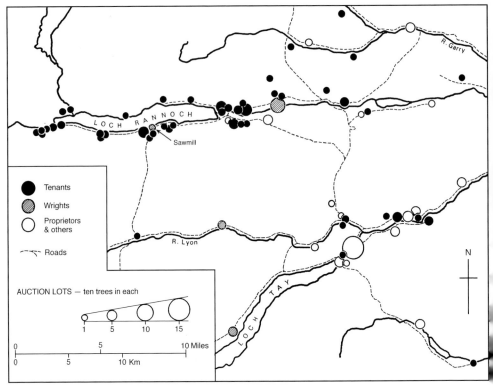

Map 6.1 Destination of pine timber sold in Rannoch, 1779–81. From J. Lindsay (1975), *Scottish Studies*, 19, pp. 39–53, with permission.

timber than husbandry... the whole country round about being served in fir-timber out of it, to the considerable advantage of the Earl of Aboyn who is heretour of it', and in 1797 the inhabitants still 'contrived to live comfortably' by selling wood 'in more distant parts'. By then, the value of the Aboyne estate woods was equivalent to 40–50 per cent of the agricultural rent of the lands, the Earl preferring to sell his trees to local wood merchants without 'the risque, trouble or expense of manufacturing'. He left it to local people to add value by making it into saleable articles.[68] No doubt part of the market here was Aberdeen, which, according to Gordon of Straloch in 1662, had also been served from as far up the Dee as 'the Pannanich wood' (probably Glenchairnich) below Glen Muick.[69] Before the end of the seventeenth century there was a sawmill as far up as Quoich,

---

[68] I. Ross, 'A historical appraisal of the silviculture, management and economics of the Deeside forests', in J. R. Aldhous (ed.), *Our Pinewood Heritage* (Farnham, 1995), p. 141.
[69] Mitchell (ed.), *Geographical Collections*, 2, p. 300.

above Braemar, and in 1726 the successors of the Earls of Mar cleared Glen Lui of the tenants of one of their vassals, Kenneth Mackenzie of Dalmore, in order to facilitate wood extraction and to build a sawmill there: again, the initial sale was local.[70] An account of 1762 describes two water-powered mills at Braemar, cutting deals which 'the country-people buy and sell to the tenants in Perth and Angus shires'. Lord Fife's tenants were said to spend much of the summer buying boards and 'carrying them to Angus' to sell at a better price than they would fetch in Aberdeenshire.[71]

Local timber sales in this part of the Highlands were not confined to pine. On part of the Gordon estates between about 1750 and the later 1770s, a good deal of birch and alder was sold, at a considerable profit, to the tenantry and local craftsmen. From 1750 to 1752 alone, £543 sterling was made from sales ranging from about £17 to over £100. Sales of smaller amounts were more common, however. In August 1760, eighty small lots, each containing about twenty trees, were rouped for less than £50 sterling in total.[72]

On the Spey the situation was much the same. There was apparently at least one sawmill at Abernethy before 1630. An account of the 1680s refers to the great firwoods on both sides of the Spey in the parish of Duthil, 'much timber from thence transported to Inverness ... the people of this much neglect labouring, being addicted to the wood, which leaves them poor': by then there were two sawmills at Abernethy and one at Rothiemurchus.[73] In the eighteenth century the local people seem sometimes to have paid a flat rate to the laird of Rothiemurchus to work in the wood, and sometimes to have bought wood in parcels for ongoing sale. In 1766, for example, there were 106 buyers of wood 'at the back of Lochinellan', all local men apart from one from Cromarty. At Abernethy in the same year the estate sold sawn timber for local demand, delivered at Broomhill of Coulnakyle, 'or upon the buyer's paying the floating at Garmouth': an advertisement listed twelve varieties ranging from spars twenty-four feet long at 26d., through planks and deals of ten to twelve feet selling at 7½d. to 30d. according to quality and thickness, down to 'lath boards and backs' at 3d. each. A lawsuit of 1779 describes how there were by then ten sawmills (as well as two boring-mills making water pipes for the English market)

---

[70] Royal Commission on the Ancient and Historical Monuments of Scotland, *Mar Lodge Estate, Grampian: An Archaeological Survey* (Edinburgh, 1995), p. 8.
[71] Ross, 'Deeside forests', pp. 140–1; Dixon, 'Lorimer on forestry'.
[72] NAS: GD 44/39/42/2; GD 44/39/24/1/14.
[73] J. Munro, 'The golden groves of Abernethy: the cutting and extraction of timber before the Union', in G. Cruikshank (ed.), *A Sense of Place: Studies in Scottish Local History* (Edinburgh, 1988), p. 154; Mitchell (ed.), *Geographical Collections*, 3, pp. 240–2.

operating in Abernethy and Rothiemurchus, and states that the wood trade also extended to the oakwoods at Loch Insh and Ballindalloch: the country people either floated the timber down the Spey to Garmouth, or, in order to 'catch the ready penny' when there was insufficient water in the river, carried it by road to the markets at Forres, Nairn and Inverness.[74]

The northernmost of the catchments with pine resources was Strathcarron, which had attracted exploitation probably at least since Viking times.[75] Gordon of Straloch in the mid-seventeenth century said simply that it 'supplies neighbouring and distant places with timber', and it probably had the same kind of wood-based economy as Glentanar and Strathspey.[76] We know, for example, that the Earl of Caithness bought Glencalvie joists, baulks and deals for his castle in the north in 1661, that there were local sawmills at Bonar Bridge and Spinningdale ('Spainydall') and that there were disturbances in Strathcarron in the 1650s when the English Navy tried to interfere in local business by cutting wood for its own purposes.[77] As late as 1677, the Yorkshire traveller Thomas Kirk needed a guard when he went to view the woods, as the local people threatened to stop his party by force, 'supposing that the King had sent us with orders to cut down the wood'.[78] The wood in Strathcarron was floated down and taken to Dornoch and Tain, but some at least in the side-glens was taken overland. Steven and Carlisle speak of a local tradition 'that until about a hundred years ago there was a large pinewood in Diebidale which supplied timber both for local building and other uses and for export, the timber being carried by ponies along a rough hill track via Strath Rusdale to the Cromarty Firth'.[79] (See also Chapter 12 below.)

North of Strathcarron the pine runs out, but there are signs that local trade in wood still flowed from the more-afforested to the less-afforested areas. In Strath Oykell and towards the mouth of the River Shin, for example, there were in 1726 limited woods of oak and birch growing conveniently on the banks 'so that they may hurle them to the river and they are very soon carried to the sea', and in Strathnaver around the start of the seventeenth

---

[74] T. C. Smout, 'The history of the Rothiemurchus woods in the eighteenth century', *Northern Scotland*, 15 (1995), pp. 19–32; Dixon, 'Forestry in Strathspey', pp. 52–3. See also below, Chapter 11.

[75] B. E. Crawford, *Earl and Mormaer: Norse-Pictish Relationships in Northern Scotland* (Groam House Museum Trust, Rosemarkie, 1995), pp. 12–16.

[76] Mitchell (ed.), *Geographical Collections*, 2, p. 447.

[77] NAS: RD 12/1059; RD 14/1662, 1651; R. G. Albion, *Forests and Seapower: the Timber Problem of the Royal Navy, 1652–1862* (Cambridge, MA, 1926), p. 207.

[78] P. Hume Brown (ed.), *Tours in Scotland, 1677 and 1681, by Thomas Kirk and Ralph Thoresby* (Edinburgh, 1892), pp. 35–6.

[79] Steven and Carlisle, *Native Pinewoods*, p. 209.

century, timber (presumably birch) was exchanged for grain from Caithness: the former was 'weel stoored with wood', and the latter had 'aboundance of cornis but indigent of wood'.[80] The overall picture in the Highlands is of a region, at least from the seventeenth century and probably before, well geared to marketing wood over short distances, supplying a lively demand within ten or twenty miles of the main production areas without much problem, and quite capable of sending it further where the flow of rivers to the open sea allowed it.

In the scattered broadleaf woods of the Lowlands the tradition was rather different, although individual woods are sometimes mentioned as supplying the country round about, just as in the Highlands. Thus an early eighteenth-century account of Ayrshire speaks of 'no country... better provyded of wood, for alongst the banks of Dun, Girvan and Stincher there be great woods, but especially on Girvan whereby they serve the neighbourhood both in Kyle and Cunninghame' for building country people's houses, 'and for all the uses of husbandrie as cart, harrow, plough and barrow at very easie rates, and the sorts are birch, elder, saugh, poplar, ash, oak and hazell'. Similarly in Galloway in the 1680s the oakwoods near Creetown furnished 'the greater part of the shire of Vigtoun... or building of houses and other uses'.[81]

In the Lowlands, however, little is heard of timber offered for open sale at the burgh markets, but a great deal about woods sold by the landowner *in situ* to small partnerships of two or three people, or to individuals who undertook to cut the wood in a prescribed way, often to enclose it after felling, and who then presumably sold much of it on either to the end-user or to another retailing intermediary. The main source of information for these sales is the register of deeds, and the sort of woodland management that these imply is described in the next chapter. The actual sale arrangements were usually straightforward enough. A contract was drawn up after a private agreement or (increasingly often after the middle of the eighteenth century) after a public auction or roup, often held in a local hostelry and sometimes advertised in the press. The buyers usually purchased the right to fell by a down-payment, followed by further predetermined payments in instalments as they began to get a return from earlier sales. Usually they were local men, sometimes with a professional interest in timber products such as wrights or tanners, but generally there is no indication that they were anything other than enterprising countrymen,

---

[80] Mitchell (ed.), *Geographical Collections*, 1, p. 203; 2, p. 559.
[81] *Ibid.*, 2, pp. 3, 70.

tenants or occasionally millers, with enough stock and skill to try their hand at a bargain.

It was a great school for small business, and in due course some of them no doubt became, in effect, professionals. For example in 1698 John Staffing, skipper burgess of Kirkcudbright, bought eight tons of timber for transport and sale to England from the wood of Compstone, about two miles north of the burgh, from Alexander and Hugh Sinklers, described as 'woodhages' – that is, woodcutters.[82] John Mackie, 'portioner' (or small freeholder) of Larbert in Stirlingshire was an interesting man. His name, or that of his son Andrew, occurred in several deeds between 1671 and 1683: in one, he and a partner bought trees on Inchcailliach in Loch Lomond on a four-year contract, and got the laird to provide sixty horse-trips ('carriages') a year to take the bark from Loch Lomond to a barn outside Stirling (a distance of over twenty miles); in another, he bought the right to cut over three years part of the forest of Alloa at Forest Mill in Clackmannanshire; in a third, his son purchased bark from Torwoodhead in Stirlingshire to be delivered to a quay on the Forth. The family were probably making a profession from dealing in tanning materials.[83] Carpenters, like tanners, also sometimes spread their net wide. William Johnston, wright in Roslin, and William Noble, wright in Lasswade parish, joined with Robert Ker, portioner of Gilmerton, to form a partnership that appears cutting woods at Humbie in 1706 and then again at Pressmennan in 1708.[84] Even more explicitly professional capitalists were Alexander Cockburne and George Carstaires from Ormiston, who described themselves as 'woodbuyers'. In 1665 they offered (along with John Gilbert) 34,000 merks Scots to be paid in ten instalments to work over Pencaitland wood for nine years, in 1668 they paid 3,500 merks Scots to work on timber at Spott near Dunbar, and in 1670 they agreed to pay 21,000 merks Scots in five instalments to work in the wood of Roslin for thirteen years.[85]

The great majority of recorded wood bargains entered in the Register of Deeds are on the financial scale of the purchase at Spott, and only eight out of sixty-eight registered between 1668 and 1683 were worth more than 10,000 merks. Three out of the four most expensive were in the hinterland of Edinburgh – aforementioned examples at Pencaitland and Roslin, one at Humbie in 1675 for 20,000 merks over nine years undertaken by an Edinburgh merchant and a clerk of session.[86] These were considerable

---

[82] C. M. Armet (ed.), *Kirkcudbright Sheriff Court Deeds, 1676–1700* (Edinburgh, 1953), 2, p. 680.
[83] NAS: RD 13/1675, 330; 14/1682, 418; 12/1683, 538.
[84] NAS: RD 13/1710, 171; 13/1710, 570.
[85] NAS: RD 13/1670, 388; 14/1670, 1390; 13/1670, 280.

sums for the age, albeit paid out over a period of years. To put them in perspective, when in 1699 the prominent Edinburgh merchant Alexander Pyper valued his stock ventured in shipping and cargoes ranging from New York, Barbados, the Canary Islands, France and Norway, it came to about 24,000 merks Scots.[87] The high values of the Lothian woods clearly reflect the strong demand for tanning materials and for timber from Edinburgh, and their proximity to that market. Roslin, Humbie and Pencaitland woods are of course still there, reflecting the fact that their heavy exploitation led to their preservation, not the reverse.

Finally, let us consider the question of how an ordinary farmer might obtain the timber he needed for his house and agricultural operations if he was not allowed to take it without payment. He might indeed buy it from the local market or from the contractors at the woods. For example, when the timber on the Arnprior estate, Kippen, was to be sold by public auction in 1782, the Annexed Estates Commissioners were advised not to reserve uncut the 'natural ash, elms, geans, plains and a gray willow or birch', amounting to about a third of the whole wood, 'as these will be very usefull to the country for many purposes of husbandry'.[88] But very often the tenants were able to buy it straight from the estate. This was regular practice, for example, at the MacDonald estate on Skye in the later eighteenth and nineteenth centuries when quantities were sold to tenants and sub-tenants (including tanning materials for cottar shoemakers) and much also directly to local craftsmen, such as wrights and boat-builders.[89] Occasionally it was bought for labour, as when the tenants who worked the pinewoods of Rannoch in place of rent were allowed a proportion of the poorer wood for their own use.[90]

A good example of the tensions that naturally arose on shifting from a system of free provision to one of payment comes from the forfeited estates of the Duke of Perth. Before the forfeiture in 1746 and for a time afterwards, the estate had apparently operated a policy by which the tenants agreed with the wood officer which trees they were to be allowed to cut, marked them and felled them at their leisure without charge. This was changed, in line with the latest thinking, to make the tenants pay for what they needed, but the factor in 1771 emphasised the absolute necessity for them to obtain

---

[86] NAS: RD 12/1680, 947.
[87] Smout, *Scottish Trade*, p. 290.
[88] NAS: E 738/59/2.
[89] Clan Donald Papers, Skye: MSS 5899, 59011–4, 3841/2, 5635; NLS: Delvine Papers, MS 1306, fos 46–7.
[90] Lindsay, 'Some aspects of timber supply', p. 45.

wood cheaply to improve their steadings, 'and I cannot help thinking that such timber should be given at reasonable rates, especially as it was the old use and wont of the estate, to give ash and other barren timber to the tenants for nothing, for the reparation of their houses, keeping up their ploughgrath, etc.' In fact the Commissioners frequently acceded to requests for free timber for mills, churches, bridges and other buildings on the estate. There was a problem of 'finding a balance between the profits gained from commercially sold wood and the general improvement and well-being of the local community'.[91] A free good might be used wastefully, as critics of the older system alleged, but to charge the tenant the going rate that wrights or tanners might be prepared to pay, might cripple his farming operations.

[91] Englund, 'Woodland management', pp. 17, 44. The previous quotation is from NAS: E 777/88/11(2). See also E 777/248.

CHAPTER 7

Managing the woods before 1770

The first sustained run of records relating to woodland management in Scotland are those of the Cistercian abbey of Coupar Angus in Perthshire between 1471 and 1558.[1] They are worth looking at closely, for they probably reflect much earlier medieval practice for which the evidence has disappeared: conversely, they show a management approach that was to continue in many respects until after the middle of the eighteenth century.

Several woods on the monastic estate are mentioned, but the most important were those of Campsie, on a pleasant and profitable grange on the banks of the Tay, where the abbot had a house and a chapel. By the fifteenth century the lands were rented out to secular tenants, but the buildings were retained to act as a resort or a rest-house for the monks.[2] Their wall-footings remain on the top of wooded cliffs overlooking a beautiful stretch of the river, and ancient woodland indicator species such as herb paris can be found nearby.

As early as 1460, two individuals had been fined for destruction and sale of the woods, and in 1471 an appointed 'forester' (presumably one of the tenants) was instructed to live with his cottars at the head of the wood. Campsie wood at that time was divided into four parts, reflecting the division of the township itself into four: each tenant was told that he had a communal responsibility to keep the wood free of cattle, 'under pain of forfeiture of the cattle found there', and no-one was to burn any of the wood unless it was pulled up from the ditch separating it from the ploughed land.[3] The prohibition on animals was modified three years later, when one of the tenants with insufficient pasture elsewhere was allowed 'license in the west part of his wood for any number of his cattle, if only without damage to the wood'.[4] In 1479 tenants were given permission to cultivate

---

[1] C. Rogers (ed.), *Rental Book of the Cistercian Abbey of Coupar-Angus* (Grampian Club, London, 1879–80).
[2] Royal Commission on the Ancient and Historical Monuments of Scotland, *South-East Perth: an Archaeological Landscape* (HMSO, 1994), pp. 110–11.
[3] Rogers (ed.), *Rental Book*, 1, pp. 130–220.
[4] *Ibid.*, 1, p. 222.

'within the walls of the wood', the first explicit mention of enclosure unless the reference to the ditch is considered such. By 1483 the four tenants were ordered to build dykes round half the forest, and in 1494 a tenant leased an area beyond 'the walls of the forest' and secured permission to run six cows in the wood together with the beasts belonging to the monks.[5] In the middle of the sixteenth century, tacks show that Campsie wood was still partly enclosed and partly not, and in 1551 one of the four divisions was described as 'waistet and destroyt' for want of sufficient dykes to exclude the animals.[6] Some of the enclosures were clearly meant to be temporary fences called 'haining', devised to protect the regrowth of broadleaf coppice after it had been cut. Thus a tack of 1558 which gave tenants 'common pasture in the open woods' also reminded them that it was not permitted to pasture cattle 'in the woods enclosed or being enclosed, commonly called "haynt woodis"', and another emphasised that animals 'cum not within our haynt woodis as sal happin to be haynt for the time'.[7]

So already before the Reformation one can detect at Campsie much of the essence of later and no doubt ongoing woodland management, including foresters appointed from the tenantry, arable and especially pasture in the wood, and enclosures, some permanent round the perimeter, some temporary internal division. One can also see the main problems, in people taking wood to which they are not entitled and above all in balancing use of the wood between animals and trees, and keeping up the dykes. James Lindsay has suggested that the monastic administration may have practised coppice rotation, cutting Campsie wood in four sections at equal intervals, each being felled every twenty-eight years and protected for seven, but to our mind that reads too much into the documents.[8]

We have located nine deeds between landowners and purchasers of their woods in the second half of the sixteenth century that further extend the picture of woodland management that we glimpsed at Campsie. The earliest of these is from the Earl of Cassillis's estate in Ayrshire in 1570, whereby three individuals purchase the right to 'cutt, peill, sell, intromett with and dispone upon' trees in part of Dalrymple wood, gaining access to do so, but 'the gaites and passages throw growand cornes and meidowes being excepted'. They have six years to carry out the task. The Earl agrees to 'build the outter dyk of the said wood' and not to sell any of the remainder

---

[5] *Ibid.*, 1, pp. 237, 242.
[6] *Ibid.*, 2, p. 69.
[7] *Ibid.*, 1, p. 65; 2, pp. 69, 179.
[8] J. M. Lindsay, 'The use of woodland in Argyllshire and Perthshire between 1650 and 1850', unpublished University of Edinburgh Ph.D. thesis (1974), p. 341.

within the six years. The wood merchants agree to cut the trees 'sufficientlie as use [is]', not peeling bark away from below the cut of the axe (or the tree would not coppice), not to cut 'in the worme moneth' (i.e. from mid-July), to cut the wood 'forward' (i.e. to clear-fell systematically), to clear up every hagg (or coup) cut in the year and to fence the haggs after them, and to leave the wood tidy after six years.[9] In its details it is typical of dozens of similar deeds in the two following centuries, and in its reference to established use, it suggests it may be typical of much that went before.

Some of the remaining sixteenth-century deeds followed this pattern. In the West of Scotland, another from the Cassillis estate in the following year was similar to the first but less detailed.[10] The Duke of Lennox sold his woods at Cruixtown in 1599 on a six-year lease to a partnership of two, including his own chamberlain: there were similar regulations to those in the Cassillis leases in respect to cutting the woods in distinct haggs, and making a fence 'for preservatioun of the young rysand grouth'.[11] In 1583, Lord Fleming signed a deed relating to Lenzie in Dunbartonshire for three years to a partnership of six, which allowed houses to be built in the wood for the workers (a common provision later), with permission to 'baik, brew and sell meit and drink in the saids woiddis during the yeirs foirsaid'. It makes no reference to duties respecting enclosure, yet the woods appear already to have been within dykes. It also said nothing about cutting regimes.[12] Rather similar were two deeds signed in the East of Scotland by George Lauder of the Bass, one of 1585 for Pressmennan and one of 1592 for 'Aikesyid', both for one year relating to a single hagg and allowing limited access for 'yokkit horses and oxin'.[13]

The main concern was often merely about preserving the grass in the wood, making sure particular trees were not cut, and ensuring that the purchasers left it neat and tidy after them. That was really all that concerned John Napier of Merchiston in 1593 when he sold woods at Edinbellie and Ballochearn in the parish of Balfron, Stirlingshire, selling all the trees except for planted ones round buildings and one particular oak, or Lord Hay of Yester in the same year, when he sold extensive woods on his estate in East Lothian, only reserving fruit trees, sycamore ('plaine treis') and all other trees growing in the yards and within the dykes of his land adjacent to the woods.[14] Finally, and also in 1593 'Alexander Tarlachsone alias

[9] NAS: Ailsa Muniments, GD 25/8/589.
[10] NAS: Ailsa Muniments, GD 25/9/30/2.
[11] NAS: GD 220/6/2004/6.
[12] NLS: MS 20786, fos 63–4.
[13] NAS: CC 8/2/16 reg. 18 Nov. 1586; CC 8/2/26, reg. 18 June 1595.
[14] NAS: CC 21/31/1 fos 132–3; NLS: MS 14755, fos 7–8.

Robertsone' sold the oakwood of Tomnagrew near Dunkeld, together with another 462 trees 'of the best of the woid of Inchnacgraniche', to two Dunkeld men, giving them the right to choose the trees and have what pasture they needed for the workhorses, fenced meadows and corn rigs always 'being keipit skaythles and reservit'.[15] Agreements like this were the opposite end of the spectrum from the Cassillis deed of 1570, being virtually a licence to cut and carry, without regard to the sustainability of the wood itself.

Good management, like indifferent management, was not confined to the Lowlands even in the sixteenth century. At Breadalbane, under 'Black Duncan' Campbell of Glenorchy, laird from 1583 to 1631, the Baron Court enforced such refinements on the tenants as planting out trees from nursery areas,[16] regulating muirburn so that it did not threaten young wood, and controlling pigs within the laird's enclosed 'parks' like that at Crannich. The laird himself was 'renowned as a planter and protector of trees'.[17] That he was not alone, even in the Highlands, in enclosing woodland at an early date is shown by James VI's 'letters of charge' forbidding the inhabitants of Urquhart in Inverness-shire from damaging Grant of Freuchie's 'dykis, parking and haining' in the Forest of Clunie in order to access the grazing, hay and timber within.[18] In sum, well before the Union of the Crowns in 1603, there is evidence from several different parts of the country for established woodland management practices devised to ensure that the resource was sustained.

In the two centuries following, woodland management changed and developed. Before we describe the details of the emerging regime, however, it is useful to explore when a more careful and systematic stewardship of the woods emerged.

There were two basic approaches to good management that can be detected in the hundreds of woodland sale deeds that have survived for the seventeenth and eighteenth centuries. The first was to enclose the wood, particularly to keep animals off the coppice regrowth in the first few years after the trees had been felled. The second was felling regulation, that

---

[15] NAS: GD 132/660.

[16] As early as the fifteenth century, the tenants of Coupar Angus had been enjoined to follow the Acts of the Scottish Parliament by planting out ash, osier, 'sauch' [willow] and broom, though not in woodland. Rogers (ed.), *Rental Book*, 1, p. xxx.

[17] M. Stewart, *Loch Tay, its Woods and its People* (Scottish Native Woods, n.p., 2000), p. 9; C. Innes (ed.), *The Black Book of Taymouth* (Bannatyne Club, Edinburgh, 1855), pp. 354, 359, 361; many of the *acta* of the court printed here went back to the sixteenth century (see NAS: Breadalbane Muniments).

[18] W. Fraser, *The Chiefs of Grant* (Edinburgh, 1883), 3, pp. 153–4.

prescribed what trees to spare from the axe for growing on and how to cut in order to maximise regeneration and regrowth.

In Table 7.1 we have tried to track the development of these two features. Whether or not a woodland sale deed has clear references to dykes, parks or haining tells us about enclosure or otherwise, and if there are further references to fencing being erected immediately after felling, this reveals a clear intention to protect coppice. It needs to be borne in mind that enclosure is often more relevant to broadleaf than to conifer, which does not coppice, and that a deed may omit to mention enclosure if the felling takes place in a single year in an already enclosed area. Mention of felling regulation may take the form of what not to cut (in an oakwood, standards to grow on among coppice, in a pinewood or an oakwood, trees below a certain size) or how to treat the stump (in an oakwood, to cut it in a manner to avoid rot prejudicing regrowth). If either of these regulation processes have been mentioned, we have counted it for our table, but if the buyers were merely told to tidy up after their operations, we have not counted it. Tidying up was universal from the earliest deeds, an extra year often being allowed to achieve it. Stump management was much commoner than mention of coppice-with-standards in the seventeenth century, but both were generally mentioned together after that.

Table 7.1 Management practice indicated from woodland sale deeds, 1600–1769

	Percentages (N = 63 in each column)		
	1600–1649	1690–1707	1750–1769
I: Enclosure			
No mention of enclosure	55	41	25
Enclosure mentioned	44	59	75
Enclosure mentioned where felling involved enclosure	35	53	63
II: Felling			
No mention of felling regulation	46	22	11
Felling regulation mentioned	54	78	89

Several features emerge. Firstly, even in the earlier seventeenth century good practice was common but by no means predominant. Forty-four per cent of the woodland deeds of this period mention enclosure, and in 35 per cent this was being actively pursued to facilitate coppice regrowth. For just over half the deeds there was mention of felling regulations, almost always just to preserve regrowth. Whether these proportions were higher than they would have been in previous eras, we cannot tell: but from information at monastic Campsie and the secular estate of the Earl of Cassillis in the sixteenth century, the practices themselves were no novelty.

Secondly, in the second half of the seventeenth century, there was a substantial improvement on this baseline. By the period before the Union, almost 60 per cent of deeds mention enclosure, and more than half mention it as a positive adjunct to regrowth: the proportion mentioning felling regulations has grown to almost three-quarters. Very clearly, improving woodland management was not a product of the Act of Union, though it might be related to a new enthusiasm for planting in Scotland after the Restoration at least partly inspired by John Evelyn in London, and exemplified by the Marquess of Tweeddale at Yester, the Earl of Leven in Fife, the Earl of Argyll at Inveraray or the Marquis of Atholl at Blair.[19] When the Murrays of Atholl raided the Campbell headquarters at Inveraray in 1684–5, Atholl took his gardener with him and allegedly purloined 34,400 trees.[20] But widespread improvements in woodland management could not rest mainly on aristocratic enthusiasm or theft. It must have reflected also a better market for wood produce, perhaps especially for tanbark.

The third conclusion from Table 7.1 is that this progress continued unabated in the first half of the eighteenth century. By the 1750s and 1760s, three-quarters of the woodland deeds mentioned enclosure, and almost two-thirds related it to immediate preservation of regrowth – a doubling of the proportion in a hundred years. Similarly, almost 90 per cent mentioned felling regulations – not only was this a great improvement on a little over a half a century before, but (what is not evident from the table) most regulations now mentioned coppice-with-standard, as well as care for regrowth from the stump. The argument for English example being important at this period is clearest in the case of the ironmasters in the 1750s, who encouraged new methods of enclosure and coppice care in Argyll (see below Chapter 9).

It is also possible to look at the growth of woodland management from a regional perspective. Table 7.2 shows how good practice varied across Scotland: the proportion of 'good practice' is defined here as the references either to enclosure or to felling regulations as a proportion of all observations within the region and period concerned (there are two observations for each deed). In the table both the proportion and the percentage are

---

[19] M. L. Anderson, *A History of Scottish Forestry* (London, 1967), 1, pp. 359–85; S. House and C. Dingwall, '"A Nation of Planters": introducing the new trees, 1650–1900', in T. C. Smout (ed.), *People and Woods in Scotland: a History* (Edinburgh, 2003), pp. 130–5.

[20] Anderson, *History of Scottish Forestry*, p. 377; House and Dingwall, '"Nation of Planters"', p. 133.

given, as a warning that some of the observations in each box are not numerous.

Table 7.2 'Good management practice' by region as indicated by woodland sale deeds, 1600–1769

Region	1600–1649		1690–1707		1750–1759	
	Proportion	%	Proportion	%	Proportion	%
Lothians and Borders	3:24	13	14:16	88	18:19	94
Perthshire	18:32	56	19:26	73	22:32	69
Dunbarton and Central	10:16	62	23:26	88	13:14	92
Clyde and Solway	27:36	75	23:28	82	27:28	96
Highlands and NE Scotland	4:18	22	7:30	23	22:32	69

Note: 'Good management practice' is defined by mention of enclosure or of felling regulations.

The table provides some striking impressions. The heartland of good practice was initially in the West of Scotland, probably in an area with Ayrshire, Lanarkshire, Renfrewshire and Dunbartonshire as its heart, perhaps along with Stirlingshire, and lowland Perthshire. The Highland character of northern Perthshire tends to depress the percentages for that county overall. Lothians and Borders appear not to have been at all to the fore initially, but to have caught up quickly in the seventeenth century, apparently learning from the west.

The Highlands scored lower than other regions until the later eighteenth century, a situation only partly attributable to the fact that 'good practice' relating to enclosure and coppice management was largely irrelevant to pinewoods. Only twelve out of forty woods involved in our sample of the area were pinewoods. Most of the improvement in the Highlands was due to more widespread enclosure of the Argyll woods in the eighteenth century.

The best source for the study of woodland management is the hundreds of relevant deeds preserved in the central Register of Deeds and in the muniments of private estates. The deeds took the form of private agreements concerning the sale of timber, signed by buyer and seller before witnesses. Sometimes they were sealed by 'earnest money', or accompanied by the present of a gown or a guinea or two 'of compliment' to the laird's lady, even (for Lady Clanranald in 1763) 'twenty guineas in money or silver plate'.[21] In Dunkeld in 1690, John Stewart of Grantully demanded that the bargain include 'ane puncheon of good French claret wine of Bordeaux and

---

[21] NAS: RD 12, reg. 17 Jan. 1764; RD 13, 1706, no. 886; RD 13, 1707, no. 155; GD 1/1–67/53.

four gallons of good and sufficient wine sack'.[22] In the eighteenth century, the private deed was partly replaced by the 'roup', where the woods were auctioned in public, subject to specified conditions, normally sold to the highest bidder and the agreement signed on the spot: it was occasionally ratified by a subsequent private deed. The first roup we have found was from the estate of the Marquess of Tweeddale in 1703.[23] They became more common as the eighteenth century proceeded, reflecting a more competitive and commercial rural ethos, but they never became predominant over private deeds. We have treated such roup agreements as tantamount to deeds in the following discussion, as their function was broadly the same.

We have used all the agreements we can find for the two hundred years after 1570: they continue after 1770, of course, but we have not explored further, nor can we reckon to have uncovered every single one within our period. In these two centuries, they recorded wood sales from Sutherland to the Solway. The purpose of the documents, apart from specifying the parties to the bargain, the position of the wood and the price and timing of sales, was to record the conditions under which the wood was to be managed during the period of felling. We have already seen their value in delineating change over time, but they can also be used to build up a more detailed picture of what management meant in these centuries.

Normally the first prerequisite of good woodland management, certainly if coppice or planting was to be involved in any form, was to make an external enclosure to control the movement of people and cattle. When a coppice 'is once begun to be cut', says a note in the Campbell of Glenure papers in the eighteenth century, 'that coppies must be cut out compleatly before another is begun to be cut so as the coppies compleatly cut may be properly fenced and preserved from being eate by bestiarl or otherways destroyed.'[24] William Boutcher thought that to plant a wood without enclosing it was among the 'idlest ways of throwing away money' and was 'laying up a fund of remorse and discontent... from the devastations of cattle and sheep'.[25]

Nevertheless there were circumstances where such enclosure was not necessary or desirable. Firstly, if the wood was on the uplands, and if it was large and scattered, the odds were that its primary use was for sheltering and grazing stock. Thousands more hectares of such woodland pasture remained in the seventeenth and eighteenth centuries than have survived

---

[22] NAS: RD 14, 1691, no. 633.
[23] NLS: MS 14754 fo. 47.
[24] NAS: GD 170/438/77.
[25] W. Boutcher, *A Treatise on Forest-Trees* (Edinburgh, 1775), p. 243.

to the present day (see above, pp. 67–70).[26] In such woods, protection, should there be any, was achieved by herding the animals in such a way as to minimise damage, and the trees would be pollarded (not coppiced) just beyond the reach of stock. If the wood was mainly of pine or birch, as so many were on the uplands, an enclosure might be of limited use because these species reproduced by seed and not usually in their own shade, so that they gradually moved their positions on a moor. On the other hand, even in these circumstances, a dyke separating the general woodland area from heavy grazing might have value.

So the characteristic enclosed wood was of broadleaf, usually but not always in the Lowlands and usually containing valuable oak, exploited by periodically cutting the trees to the ground and awaiting regrowth from the coppice stools. An enclosure would of course always be made round a park where it was intended to keep deer or domestic stock, and such parks had trees that would be pollarded if a need was felt to exploit them at all. They were common round tower houses and country homes of the landed elite from the sixteenth century onwards, but most of the woodland sale deeds relate primarily to woods rather than parks, though the latter may be mentioned.

The boundary 'woodbank' of English woods have been authoritatively described by Rackham: 'typically this consists of a bank and a ditch, the bank being on the *wood* side'.[27] The evidence for ditches around the wood is sparse in Scotland, though it is probably implied even in medieval times at Coupar Angus (see above, p. 157) and they were no doubt dug where the character of the ground allowed it – not as frequently in stony Scotland as on the English clays. They are mentioned, for example, in 1722, when the Duke of Hamilton's wood at Kinneil, of oak, ash, birch, elm, alder and other trees, was described as bounded by ditches, dykes or walls, and in 1730 Alloa wood in Clackmannanshire had 'outgates, hedges and ditches' that the proprietors had to maintain. Clearest of all was the wood of Little Park in Kirkcudbrightshire, of oak, ash, birch, willow and hazel, 'bound . . . by a stone dyke and an old ditch'.[28]

The banks or dykes themselves were usually of earth, topped by a paling of 'stake and ryce' (a wattle fence), or, in the case of more wealthy landowners and increasingly so in the eighteenth century, stone walls, which might also be topped in the same way. Earthen dykes topped with wattle are

---

[26] P. R. Quelch, 'Upland pasture woodlands in Scotland', part 1, *Scottish Forestry*, 43 (2000), pp. 209–14; part 2, *ibid*., 55 (2001), pp. 85–91.
[27] O. Rackham, *Trees and Woodland in the British Landscape* (revised edn, London, 1995), pp. 114–16.
[28] NAS: RD 12, reg. 9 Apr. 1729; RD 13 reg. 10 Dec. 1731; RD 12, reg. 9 Apr. 1729.

commonplace. For instance, on Sir William Hamilton of Preston's estate in 1681 the cutters had to provide stake and ryce to Sir William 'for the top of a faill dyke'; in Inverkip parish in Renfrewshire in 1704, John Stuart of Blackhall, the owner, undertook to provide 'ane outter feall dyke round such parts of the said wood as it was in use formerly to be built', the buyers again providing the stake and ryce to top it; at Mains of Park, Glenluce, in Wigtonshire in 1731, Sir Charles Hay reserved the 'grains [branches] and ryce he needs to peall the top of his outter wood dykes', or to make up any part of them that cannot be built of turf.[29]

Stone walls were not infrequent. An elaborate contract between John Hay of Carriber and two men in Calder, West Lothian, in 1701, specified that the buyers should be paid extra to replace an old drystone dyke with a new wall of stone and lime, 'casten and harled afterwards', six quarters (a yard and a half) high plus two 'faills' for the coping, with good foundations.[30] No doubt it would also have been topped with stake and ryce, as was explicitly mentioned at Castlecary in Stirlingshire, where the Earl of Wigton in 1722 obliged the buyer to provide such material 'to putt on the top of a stone dyke built or to be built by the said Earl'.[31] Similarly, the Earl of Traquair in 1760 contracted with masons in Innerleithen to build a good drystone dyke (also six quarters high) round part of Plora wood in Peeblesshire, the Earl providing carts and handbarrows, mattocks and shovels: he was to decide later if the lower dyke was to be made completely of stone 'or an earth dyke casten up the brae and sufficiently faced with stones'.[32] (See Figs 7.1–4.)

Not all dyke systems kept to one material. The Earl of Tweeddale at Yester in 1710 had a mixture of types for his enclosed wood: 'ane new stane dyke upon the eastend', an old faill dyke, a limestone dyke and another old faill dyke.[33] Sometimes trees rooted themselves in an established earth dyke, as at Mauchline in Ayrshire in 1688, where the seller reserved for himself 'the wood dykes and timber growing there upon which will be for hedging or laying therof'. In this case the dykes effectively became topped by a hedge.[34]

Almost invariably, the construction and maintenance of the enclosing dykes, but not the provision of stake and ryce to top them, was the responsibility of the woodland owner. A rare exception was made when Richard

---

[29] NAS: RD 13, 1686, no. 743; RD 12, 1707, no. 789; RD 12, reg. 8 Jan. 1732.
[30] NAS: RD 13, 1706. no. 592.
[31] NAS: RD 12, reg. 30 Apr. 1723.
[32] Traquair House MSS: contract dated 6 June 1760.
[33] NLS: MS 14754, fos 49–50.
[34] NAS: RD 12, 1689, no. 450.

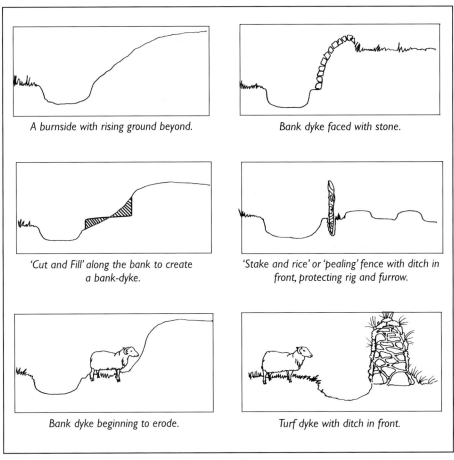

Fig. 7.1 Types of wood dyke in Sunart, Argyll. Reproduced with permission from Sunart Oakwoods Research Group, *The Sunart Oakwoods* (2001).

Ford and Company, the English proprietors of Bunawe ironworks, purchased from Sir Duncan Campbell of Lochnell in 1752 a group of Argyll woods and agreed to enclose them in the same way as Sir Duncan had done when they were last cut, 'or in any other manner of way the said Company can best devise'.[35] It was not a formula they resorted to again, for instance in dealings a month later with Captain Duncan Campbell of Inverawe, and with the Earl of Breadalbane, where they reverted to the usual strategy of laying the onus for enclosure on the owners.[36]

---

[35] NAS: RD 14, reg. 1 Feb. 1773. For another exception (Perthshire, wood of Aberdegie, 1680), see RD 13 1684, no. 1263.
[36] NAS: RD 12, reg. 5 Jan. 1762; RD 12, reg. 19 May 1755.

Fig. 7.2 Remains of a dyke with an ancient pollard oak, Firbush, Lochtayside. The oak may be five centuries old, and grew in the dyke. Woodland has since regenerated on both sides of the wall, but the croft on one side was mapped as cultivated in the eighteenth century. Anne-Marie Smout.

Fig. 7.3 The heraldic seal of the Earls of Angus displaying a 'stake and rice', or wattle, fence round the stag. Such construction was common for internal divisions in the wood, or as topping for an earthen dyke.

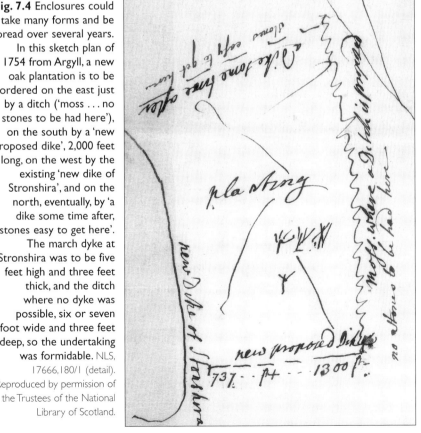

Fig. 7.4 Enclosures could take many forms and be spread over several years. In this sketch plan of 1754 from Argyll, a new oak plantation is to be bordered on the east just by a ditch ('moss ... no stones to be had here'), on the south by a 'new proposed dike', 2,000 feet long, on the west by the existing 'new dike of Stronshira', and on the north, eventually, by 'a dike some time after, stones easy to get here'. The march dyke at Stronshira was to be five feet high and three feet thick, and the ditch where no dyke was possible, six or seven foot wide and three feet deep, so the undertaking was formidable. NLS, 17666,180/1 (detail). Reproduced by permission of the Trustees of the National Library of Scotland.

Eighteenth-century Argyll provides the fullest information about woodland enclosure, perhaps because the shift from using woods primarily as pasture to using them for charcoal and tanbark was relatively new and complete. In 1754, the Duke of Argyll found he had two firms of English ironmasters competing to buy his woods, and planned to enclose a plantation and a piece of woodland at Stronshira by a dyke five foot high and three foot thick, and where no dyke was possible because of marshy ground, to dig a ditch six to seven foot wide and three foot deep.[37] Henry Kendall, partner in one of the firms which eventually built a smelting furnace on the Duke's land on Loch Fyne, gave him advice in 1756 on the 'earth fences in England about woods'. They were generally six feet high and four to six feet across at the base, tapering to ten to twenty inches: a 'tod-nosed spade' was best for making them, and he would send a pattern. Stone fences might be only three feet high and capped with a ridge of stones set on edgeways to keep out cattle, but to keep out sheep they should be four feet high with an overhanging 'cam' of three inches on either side. Some proprietors, he said, used 'Galloway dykes' full of holes, which frightened the cattle more than solid dykes. He had seen them on Lord Crawford's lands near Paisley.[38]

A recent study of Sunart adds further particulars of practice in Argyll. In 1729 the new owner of Sunart, Alexander Murray of Stanhope, asked his factor what services Argyll lairds could expect of their tenants, and was told that Campbell of Barcaldine obliged each of them every year to make 'four roods of a dyke and ditch seven feet high and seven feet wide' at his own expense, or six roods in return for a week's allowance of food. Six roods were about forty yards. In 1744 Stanhope undertook to enclose any woods he cut on the lands of a wadsetter 'with a stake and rice hedge of oak or a stone dyke for the space of four years after the cutting of the woods', showing how even some external enclosures were flimsy temporary affairs. In 1786 on the Duke of Argyll's estate on the south side of Loch Sunart, two local dyke-builders constructed round two local woods a stone dyke four foot high and three foot broad at the base, coped with two rows of turf; plus a combination of earth dyke and ditch 'seven foot high, well cop'd at the top and the green side of the sod outmost'; a stretch of paling; and two short stretches of 'paring dyke', where steep slopes were trimmed to form a vertical face of earth, perhaps faced with stone. Archaeological investigation in Sunart found remains of all these forms of construction surrounding the woods, except the stake and rice palings which had of course

---

[37] NLS: MS 17666, fos 165, 180.
[38] NLS: MS 17668, fo 219.

perished, though sometimes apparently leaving 'the merest "shadow" of a dyke, too tenuous to record'.[39]

Much of the management of the interior of the woods was specified in the deeds of sale. Usually a wood was sold to an individual or to a partnership of two to four, who would undertake to cut the timber in it over a period of years, usually between two and nine, depending on the size of the wood, with an extra year for general tidying away of any temporary buildings or rubbish left behind. Sometimes a clause was added extending the period in the event of interruption by famine, war or plague.[40] In the Highlands and the north, arrangements were sometimes made to protect the cutters from theft or lawlessness. At Dunkeld in 1690 and at Dalkeith Park in 1691, provision was made to assess damage if the wood should be damaged or destroyed by an army, a signifier of troubled times.[41]

Unless it was a pinewood where the system was different, it was normally specified that the buyers should fell a given area, known as the hagg, in the space of each year, clearing the brash after them so that regrowth was not hampered by old twigs and branches. In many cases, and increasingly so as time passed, the hagg was securely fenced ('hainit') with wattle before commencement of the next year's operations. Let there be 'ane sufficient hedg of staick and ryis, stronglie made schouldir heicht', was how an Eglinton deed of 1628 expressed it. The hagg would remain fenced for a period of years, seven to ten was normally recommended, though Robert Monteath in 1829 considered that there was value even in two to four years' protection.[42] After that the fences, also variously described as 'thorter dykes', 'char dykes', 'jarr dykes', 'run dykes', would be taken down. Their purpose was to keep animals away from the coppice shoots until they grew beyond reach, since even enclosed woods were open to stock for shelter and grazing at least for part of the year.

There is an important sense in which all woodland, and not just parks and unenclosed upland woods, formed part of a wood-pasture system, and their resource could be critical for those who relied on the grazing. 'When the wood is for pasturage', Francis Masterton in Alloa advised his son in 1699, 'keep six cows with their followers if you get grass fra ma Lord: if not, you must keep only three or four.' Lindsay says that in the well-managed

---

[39] Sunart Oakwoods Research Group, *The Sunart Oakwoods: a Report on their History and Archaeology* (n.p., 2001), pp. 45–8.
[40] e.g. NAS: GD 190/3/110/8; RD 14, 1688, no. 573; RD 13, 1688, no. 353; RD 13, reg. 10 Dec. 1731.
[41] NAS: RD 14, 1691, no. 633; RD 14, 1693, no. 220.
[42] NAS: GD 3/2/29/18; Lindsay, 'Use of woodland' pp. 352–3; R. Monteath, *Miscellaneous Reports on Woods and Plantations* (Dundee, 1827), p. 53.

and enclosed woods of the Duke of Montrose at Kincardine in Perthshire, and Mugdock, the value of the pasturage varied from 10 per cent to 30 per cent of the value of the wood produce.[43]

The responsibility for erecting and maintaining the hagg fences during woodland operations was normally the buyer's, unlike that for the outer perimeter enclosure, which was the seller's. But there were exceptions, perhaps when the seller did not think the buyer had the experience to make the fences properly. Thus at Lasswade in Midlothian in 1668 Sir John Nicholson agreed to 'fence the said haggs for preserving of the after grouth by thorter dykes', if the buyers provided the stake and ryce, and the proprietors of Alloa wood in 1730 and the laird of Invermay in Perthshire in 1703 came to the same arrangement.[44] In the last case, the fences (which were essentially temporary structures) were specified as belonging to the proprietor, but in another Perthshire instance at Leuquhatt in 1696, the buyer had the right to them 'when the samen shall be taken away'.[45] We do not know which was normal practice in this respect.

Sometimes woods had permanent internal divisions, like the 'stone dyke in the middle of Ormistoun wood' referred to in a deed of 1736, or the inner turf dykes erected around groups of haggs, and after 1758 around individual haggs, on the Duke of Montrose's wood at Kincardine in Perthshire.[46] At Ormiston there was also a line of stakes in the wood, presumably the remains of the previous fence: where hagg fences of stake and ryce were regularly erected in the same place as the wood was cut in rotation, even these may well have left some slight archaeological mark that could be traced today.

How the hagg was managed depended to some extent on the species of tree involved, but there were certain general principles. Firstly, it was to be cut 'forward', that is, systematically felled from one end to the other (excepting any trees to be preserved for growing on), to avoid the woodcutters selecting the best trees first and returning at a later date for the others. Secondly, bark must never be peeled below the cut: the tannin content of bark on or by roots was particularly high, but if all the bark was peeled off the tree could not sprout new shoots from the coppice stool. Thirdly, the trees were to be felled close to the ground, but leaving the stumps sloped so that rain could not gather and rot them.

---

[43] V. N. Paton (ed.), 'Masterton Papers 1660–1719', in *Miscellany of the Scottish History Society*, 1 (Edinburgh, 1893); Lindsay, 'Use of woodland', p. 482.
[44] NAS: GD 18/1275; RD 13, reg. 10 Dec. 1731; RD 13, 1706, no. 886.
[45] NAS: RD 12, 1706, no. 430.
[46] NAS: RD 12, reg. 1 Dec. 1738; Lindsay, 'Use of Woodland', pp. 479–80.

There were various formulations for this, and some differences of opinion. Most thought the stump (or 'stone') should be cut level with the grass, close to the ground, or 'round over by the fogg [moss] as most convenient for growing'.[47] Others specified two inches from the ground or even (from Dunkeld in Perthshire in 1690) 'skew ways half a foot or little above the ground yeirly as is of use of woods cutting'.[48] There was also room for difference about the shape of the cut. Some thought that the stump should be rounded, some ridged 'with an ascent and a descent and not even', some cut cleanly in one direction 'with a smooth sloping stone as no water may stand thereon'.[49] A deed signed in 1716 between Lord Glenorchy and James Fisher, merchant in Inveraray, specified that in woods cut on Lochawe no stocks should be left 'hollow where they are hadged'.[50] It was always assumed that the cutting would be done by an axe, but a deed signed at Langholm in 1765, dealing with Englishmen, explicitly prohibited the use of a saw.[51]

Then there was the problem of when to cut. Twice in documents from Sutherland, we encountered the archaic belief that a waxing moon encouraged regrowth. In 1642, Lord Reay stipulated that the 'Woodis of Letter Muniscarrie' be cut in February or March 'on the grouth and incres of the moone', and in 1695, George Monro of Culraine near Bonar Bridge sold his oak and birch wood of 'Meikle Altness', the buyer (a Portsoy merchant) being 'bound to cutt all the said gross timber in the samine wood onley on the grouth of moon'.[52] On Skye, Martin Martin reported the opposite belief, that a waning moon was best for cutting wood and peats, but the Baron Court of Glenorchy in 1621, instructed the inhabitants only to cut briar and thorn (a resource for fencing), 'in the waxing of the moon'.[53]

Cutting seasons everywhere depended on whether what was being felled was oak, or the other broadleaf species collectively referred to as 'barren timber' (or sometimes as 'common timber' or 'winter timber'). The season for cutting barren timber was much longer than for oak, though it varied more. Usually it extended throughout the year, often with a break in the 'flie

---

[47] e.g. NAS: GD 18/1275; GD 112/16/11/2/6; RD 14, 1688, no. 573.
[48] NAS: RD 12, 1702, no. 1279; RD 14, 1691, no. 633.
[49] e.g. NAS: RD 12, 1691, no. 357; RD 12, reg. 30 Apr. 1723; RD 12, reg. 9 Apr. 1729.
[50] NAS: GD 112/16/11/1/1.
[51] NAS: RD 14, reg. 11 Mar. 1769.
[52] M. Bangor-Jones, 'Native woodland management in Sutherland: the documentary evidence', *Scottish Woodland History Discussion Group Notes*, 7 (2002), p. 1; NAS: RD 14, 1697, no. 1571; M. Martin, *A Description of the Western Isles of Scotland, c. 1695* (edn Glasgow, 1884), p. 189. There was a similar belief in the moon in seventeenth- and eighteenth-century Denmark and here it was supposed to be best to fell on a waning moon: B. Fritzbøger, *Kulturskoven: Dansk Skovbrug fra Oldtid til Nytid* (Copenhagen, 1994), pp. 156–7.
[53] Innes (ed.), *Black Book of Taymouth*, p. 353.

month' or 'worm month', defined as fifteen days on either side of Lammas, 1 August. In the Torwood of Stirlingshire in 1687, the cutting of 'comon timber as birk, arne, eash and hizell' was allowed all year except in February. At Invermay in Perthshire in 1703, it had to stop by 1 November. At Alloa in 1730, it was allowed in every month except August. The Bunawe furnace undertaking was to cut the 'winter timber' of Sir Duncan Campbell of Lochnell in 1752 between 1 September and 1 March. In Eskdale in 1725, birch and alder might be cut all year, ash between 29 September and 1 July.[54] Ash being relatively valuable, it received special mention on other occasions, though the logic of the regulations is sometimes hard to follow: at Cranston in Midlothian in 1694 twelve ash trees were sold but they were not to be cut between 1 April and 1 August; conversely at Boturich on Lochlomondside in 1738 the felling of ash could continue until 1 August but not beyond.[55]

Oak, however, was the main crop of almost all the woodland sale deeds apart from those that dealt with Scots pine in the Highlands (see below, pp. 182–91). The main anxiety was to take advantage of the ready peeling of the bark as the sap rose in spring, while avoiding damage in the worst period of insect infection in summer.

Table 7.3 illustrates the variety in the oak-cutting season, which could begin any time between 1 March and 1 May and end between early July and early August: there was no invariable formula, though the 'flie month' was usually avoided, at least in part. If other trees were wanted for their bark, the regulations might be similar to those for oak: as a deed for Canonbie, Dumfriesshire, put it in 1739: 'the saids woods shall be cut in due seasons, that is to say the oaks, sauch and birch in summer when they peel, and the ash, elm and allar in the autumn and winter seasons'.[56]

Then there was the interesting question of what not to cut. Proprietors often excepted from the contract particular plantings in avenues, parks or orchards, and trees around their own or their tenants' houses. They might also except remarkable trees, like Wallace's Oak in the Torwood in 1687, the Sighthill Tree near Bothwell in 1701 or the 'whole old oak great trees in and about Portinhavin' at Rossdhu in Dunbartonshire in 1727.[57] They might except planted species within the woods, for example pine at Ormiston in 1736 and pine and beech at Yester in 1755,[58] or particular native species such

---

[54] NAS: RD 13, 1688, no. 353; RD 13, 1706, no. 886; RD 13, reg. 10 Dec. 1731; RD 14, reg. 1 Feb. 1773; RD 2/120², reg. 11 Oct. 1726.
[55] NAS: RD 14, 1700, no. 313; GD 47/1115.
[56] NAS: RD 13, reg. 9 Mar. 1741.
[57] NAS: RD 13, 1688, no. 353; RD 13, 1707, no. 155; RD 12, reg. 15 Feb. 1731.
[58] NAS: RD 12, reg. 1 Dec. 1738; NLS: MS 14754, fos 53–4.

Table 7.3 Examples of oak cutting dates

Year	Location	Start date	End date	References
1658	Crag Wood, Murthly, Perthshire	—	1 August	NAS: RD 12, 14 July, 1664, no. 1022
1663	Grantully and others, Perthshire	1 March	1 August	NAS: RD 14, 1671, no. 452
1678	Faskally, Perthshire	1 April	1 August	NLS: MS 1438, fos 16–17
1685	Monzievaird, Perthshire	15 April	10 August	NAS: RD 13, 1689, no. 292
1687	Torwood, Stirling	15 April	15 July	NAS: RD 13, 1688, no. 353
1701	Carriber, West Lothian	1 May	1 August	NAS: RD 13, 1706, no. 592
1703	Invermay, Perthshire	—	1 July	NAS: RD 13, 1706, no. 886
1714	Lochawe area, Argyll	1 March	15 July	NAS: SC 54/12/7 reg. 25 June 1715
1721	Kennacraig, Argyll	1 April	1 August	NAS: RD 12, reg. 2 Apr. 1723
1726	Kilfinan, Argyll	15 April	10 August	NAS: SC 54/12/13, reg. 3 Mar. 1735
1727	Aberuchill, Perthshire	—	10 July	NAS: RD 14, reg. 9 June 1731
1727	Rossdhu, Dunbarton	1 May	1 July	NAS: RD 12, reg. 15 Feb. 1731
1730	Alloa, Clackmannan	1 May	10 July	NAS: RD 13, reg. 10 Dec. 1731
1734	Kirkdale, Kirkcudbrightshire	—	15 July	NAS: RD 12, reg. 12 Feb. 1737
1736	Ormiston, East Lothian	20 April	15 July	NAS: RD 12, reg. 1 Dec. 1738
1738	Boturich, Dunbarton	1 May	7 July	NAS: GD 47/1115
1745	Calder, West Lothian	—	15 July	NAS: RD 12, reg. 18 June 1747
1753	Mugdock, Stirling	—	10 July	NAS: GD 220/6/583/12

as the ash at Yester in 1710 and at Tillydown in Banffshire in 1730, or areas of thorn and hazel for hedging, as at Bargany in Ayrshire in 1642.[59]

Sometimes exemptions of this kind were quite substantial in total. In a deed of 1731, Sir Charles Hay, selling woods at Mains of Park in Wigtonshire, reserved all the alder and saughs in the bogs, enough hazel for his own use, all the plantings about houses, yards and parks, enough stake and ryce to top the outer dykes and make up those parts of the outer enclosure that could not be built of turf; he also instructed the buyers to 'cut and dight the said wood of all bramble and brush excepting thorn, brier and broom', which gives a hint of what in the understorey was regarded as of value for hedging and other purposes. Incidentally, he also asked for trees cut in the 'overbrae', beyond the inclosure, to be 'creeled' – that is, for the stools to be individually protected by basketwork, a provision occasionally found elsewhere.[60]

An important aspect of woodland management was preserving from immediate felling some young growth as the hagg was cut. Absence of such a provision did not mean that practice was unsustainable, as new growth will spring from broadleaf stools even of a wood clear-cut. But to keep some young trees enabled a wood of variable age to become established, and it might well be associated with a full-blown system of coppice-with-standards where a defined number of trees were preserved to grow on to become more mature timber.

Already in 1628 such management was implied in the sale of Eglinton Wood in Ayrshire, reserving to the Earl, 'the just and full number of fyve scoir aike treis, quhilk the said noble lord has presentlie markit to stand still for his awin use', and also reserving to the Earl 'all thes young aike tries that ar not of the gritnes of ane stalff or ane souple [cudgel]'.[61] In 1669, on two of the upland oakwoods of Loch Etive in Argyll, without enclosure or haggs, Campbell of Glenorchy nevertheless stipulated that 'allwayes evry twentie youngest straight and smallest tries in the forsaids wods' should be spared in the clear-fell; the same proportion was to be left of those 'three inches diameter three inches from the ground' nearly a century later in another Highland felling of unenclosed woods at Arisaig and Moidart in Invernesshire.[62] In 1677, at Ardincaple on Seil Island in Argyll, the cutters had to leave in the woods 'six scoir maidines and als many mor as they please, the saids maidines being als greit neir the ground as ane ordinar mane will claspe with his mid-finger and his thumb'.[63]

[59] NLS: MS 14754, fos 49–50; NAS: SC2/59/9, no. 4344; GD 109/3011.
[60] NAS: RD 12, reg. 8 Jan. 1732.
[61] NAS: GD 3/2/29/18.
[62] NAS: RD 12, reg. 17 Jan. 1764; GD 112/16/11/2/6.

Provisions like this were repeated again and again in the later seventeenth and earlier eighteenth centuries, not just in the Lowlands. The most dramatic example was an Inverness-shire deed of 1752 relating to oak and barren timber, 'excepting ash and pine', growing between Castle Urquhart and the edge of Glen Moriston: it instructed the buyers to leave 'equally disposed through the wood as ordinally ten thousand reserves', of which 2,000 were to be twelve inches in circumference at five feet and the rest nine inches at that height. The same hint of normal practice is given in Crarae in Argyll in 1704 when 'the ordinary standers . . . not exceeding the number of two thousand' were to be left uncut, and another Argyll deed of 1701 spoke of 'sex hundreth sufficient standers or maidens' to be left in the oakwoods. None of these woods were apparently enclosed. In the smaller enclosed Lowland woods often between 100 and 300 oaks were to be left. Frequently they were to be six to eight inches in diameter at breast height, sometimes smaller, sometimes twice as big. The term 'maiden' was frequently used and implies that the trees had sprung from an acorn, not as regrowth from a coppice stool, but sometimes they were called 'standards' with the implication that they had come 'aff a stone' as singled coppice shoots.[64] Sometimes the trees were described as as big as barrel ware or as hodgit ware (barrel-staves), or big enough to take a 'wumble bore' an ell above the ground or a 'large carwumble bore' (an auger).[65] Generally oak was to be left as a standard, or ash if oak was not available: once on Cameron of Lochiel's estate in 1722 it was birch over sixteen inches round two foot from the ground that was to be left.[66] Even when, later in the eighteenth century, the prime use of the woods in Argyll was for charcoal and the preservation of trees to grow on to maturity would seem to be a lower priority, thousands of oaks were kept as standards.[67] The absence of uniformity in prescribing how coppice-with-standards should work, and the references to it throughout the period, are themselves evidence of the importance attached to growing at least a modicum of mature trees, even when the main demand was for charcoal and tanbark from young wood.

It is thus easy to show that Scottish woods long before the eighteenth century had a tradition of use that encompassed managing both the

[63] NAS: GD 47/1090.
[64] NAS: RD 12, reg. 13 Aug. 1755; SC 54/12/5, reg. 7 June 1706; RD 12, 1702, no. 470; NLS: MS 1438, fos 16–17; NAS: RD 12, 1684, no. 569; RD 14, 1685, no. 643; RD 13, 1688, no. 353; RD 12, 1705, no. 430; RD 14, 1688, no. 573.
[65] NAS: RD 12, 1707, no. 789; RD 12, 1701, no. 321; RD 12, 1706, no. 778.
[66] NAS: RD 13, 1686, no. 743; RD 12, 1707, no. 789; RD 12, reg. 17 Jan. 1729.
[67] NLS: MS 993, fos 8–15; NAS: GD 174/737; RD 12, reg. 19 May 1755; RD 12, reg. 5 Jan. 1762; RD 14, reg. 1 Feb. 1773.

coppice stools and coppice-with-standards. What is not easy to demonstrate is whether or not there was also at that period much systematic coppice rotation. Few woods or woodland estates were large enough to divide into the number of haggs – perhaps twenty or thirty – that would have been necessary for rotation on one site to continue indefinitely. Our records are too incomplete to show if a wood was revisited at absolutely regular intervals. It was always possible to have the other elements of management without rotation: landowners might use woods as a resource to raid intermittently in time of practical need or financial stringency, and yet still cut them in the manner described. 'As for the oak and the ash', Francis Masterton, a small laird near Alloa advised his son in 1699, 'cut them when needed and hen [hain] the stocks thereof by a bitt ryce dyk... as for the arns [alders]... sell them to any who will give the price... cutt wood young and it growths the better.'[68] This was not sophisticated woodmanship.

A woodland deed does not usually have occasion to mention rotation one way or another, but the balance of probability is that many landowners would have been propelled towards developing rotations, since coppice cut too early or left too long would lose value. Certainly the deeds frequently refer to the sale of a hagg cut previously, and one deed is particularly suggestive. In 1677, David Bruce of Clackmannan sold 'the wood of Clackmanane called the Birkhill... alse frielie as the samyn hes been disponed to any uther merchand or woodcutters of before', but because the wood was 'not as yet of ane full grouth and sua not for present cutting', the buyer was first to act as forester, thinning and otherwise attending to the wood, 'and sua to continew the said weiding for the space of sevin years'. After that, he was allowed to begin to cut the wood in haggs for eight years, leaving it internally enclosed as usual.[69] This suggests a rotation of at least fifteen years, and perhaps longer. His neighbour the Earl of Mar sold the wood of Alloa in five haggs in 1669 and again in 1694 – an interval of twenty-five years, which was probably not a coincidence. A note in the Duke of Atholl's papers in 1708 said that most coppice in Scotland was cut too young, and that a thirty-year rotation was the minimum desirable.[70] So there is evidence that systematic rotational coppice for a set term of years was carried out in Scotland before the Union, but not that it was widespread.

---

[68] Paton (ed.), 'Masterton Papers', p. 487. The published transcript has 'vyce dyk', clearly wrong.
[69] NAS: GD 190/3/110/1.
[70] Paton (ed.), 'Masterton Papers', pp. 468, 471; Edinburgh University Library: MS Dc. 1, 37/1/3, 12; Lindsay, 'Use of woodland', pp. 345, 352.

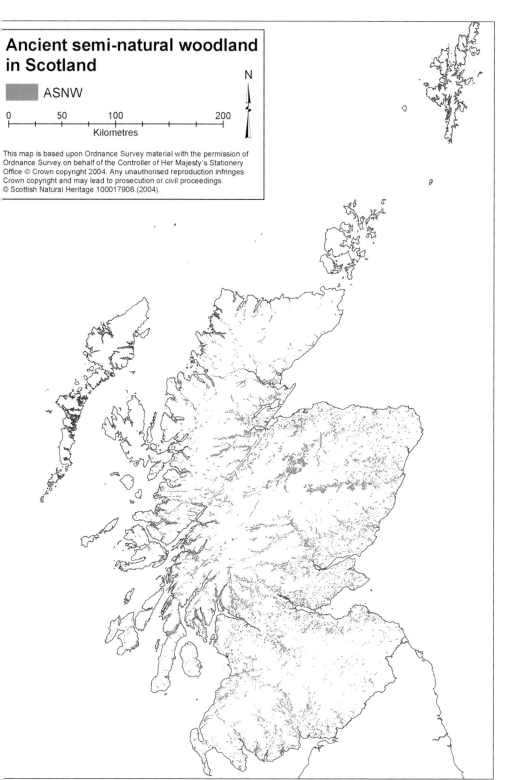

The surviving ancient semi-natural woodlands, as mapped on the Scottish Ancient Woodland Inventory.
Scottish Natural Heritage

2a. Scots pine growing on boggy ground. Beinn Eighe National Nature Reserve, Wester Ross.
Lorne Gill, SNH

2b and c. Rare pinewood flora at Loch Fleet National Nature Reserve, Sutherland:
left (b) One flowered wintergreen (*Moneses uniflora*), right (c) Twin flower (*Linnaea borealis*).
Steve Austin

3. Silver Birch (*Betula pendula*) near Rannoch, Perthshire; woods famous for their beauty.
Lorne Gill, SNH

4a. Atlantic hazel wood at Ballachuan, Argyll.
S. Coppins

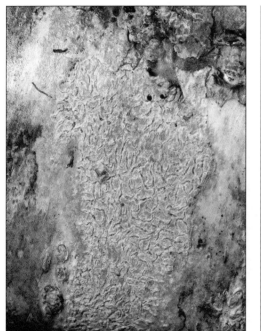

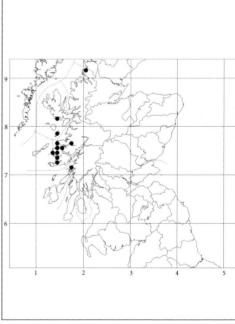

4b and c. The lichen *Graphis alboscripta* and its world distribution. It is entirely restricted to Scottish west-coast hazel woods.
4b, Lorne Gill, SNH; 4c, Map produced using DMAP

5a. Birch and aspen clothing the coastal bays of Assynt, Sutherland. The woods thrive here despite the sheep.
Robin Noble

5b. Coppiced holly in Assynt: the tree is centuries old.
Robin Noble

6a. Heather and Scots Pine at Rothiemurchus, Speyside, the Lairig Ghru pass in the Cairngorms beyond.
Lorne Gill, SNH

6b. Montane scrub: Woolly willow (*Salix lantana*) in Glen Doll, Angus.
Lorne Gill, SNH

7a. Upland wood pasture at Glen Finglas, Trossachs, veteran alder pollards in a former royal hunting forest.
Woodland Trust

7b. Wood pasture dominated by alder with a rushy group flora, at Tinnis Burn, near Newcastleton, Borders.
Mike Smith

8a. Montane scrub on limestone, Inchnadamph National Nature Reserve, Sutherland. Note how the ground beyond the fence is bitten bare by sheep and Red Deer.
Anne-Marie Smout

8b. An ancient pollard of the uplands: alder with rowan growing on it, Glen Finglas, Trossachs. See also Plate 13b.
Woodland Trust

9a. Cattle in woodland near Fungarth, Dunkeld, Perthshire: grazing was normal in Scottish woods in the past.
Lorne Gill, SNH

9b and c. A nineteenth-century outer wood dyke at Sleat, on Skye, formidably constructed to exclude stock.
Alan Macdonald

10a. A charcoal burner's stance, Sunart, Argyll, providing fuel for the Lorn Furnace at Bonawe in the eighteenth or nineteenth centuries.

Jim Kirby

10b. The site of an iron-making bloomery in Sunart, Argyll. There is slag on the ground, from a process that long predated the activities at Bonawe.

Jim Kirby

11a. A standard oak growing among singled coppice stems: it is about 250 years old. Ceol na Mara, Sunart, Argyll.
Jim Kirby

11b. Coppice stools of oak that have been allowed to develop without being singled: such trees can be identified by multi-stems. Ceol na Mara, Sunart, Argyll.
Jim Kirby

12a. A plantation of oak within Ariundle National Nature Reserve, Sunart, Argyll. It was commonplace to extend the natural woods or fill in openings when the price of charcoal and tanbark was high.
Jim Kirby

12b. The ancient rigged field within the oakwood at Ceol na Mara, Sunart, Argyll. Such cultivation was frequent within the woods, and the fields may have been protected by 'stake and ryce' fences.
Jim Kirby

13a. The late Alan Mitchell and colleagues measuring the enormous old sycamore at Birnam, by Dunkeld, Perthshire. Sycamore was introduced into Scotland probably in the late Middle Ages.
Jim Gallacher

13b. A rowan apparently 'on stilts', in fact on the ruins of an old alder, which had probably at one time been pollarded, and the seed from a bird dropping lodged in it. Tinnis Burn near Newcastleton, Borders.   Mike Smith

13c. An ancient pollarded oak growing at Fair, on the River Tweed near Innerleithen, Borders. Veteran trees such as those illustrated on this page have great importance for wildlife.
Mike Smith

14a. The pinewoods of Glen Affric regenerating freely by the lochside.
Anne-Marie Smout

14b. Birch regeneration pushing above the heather at Creag Meagaidh National Nature reserve, after the exclusion of Red Deer.
Lorne Gill, SNH

15a. Pine Marten, predator of squirrels, in a Perthshire birchwood.
Lorne Gill, SNH

15b. Red Squirrel in Speyside. Both the marten and the squirrel came close to extinction in the nineteenth century.
Neil McIntyre

16a. Nest of the wood ant (*Formica aquilonia*), Black Wood of Rannoch, Perthshire.
Anne-Marie Smout

16b. Wood ants (*Formica lugubris*), swarming over their nest, Black Wood of Rannoch.
Lorne Gill, SNH

Thereafter there are clearer indications of systematic rotations, though still little to indicate how widespread they became. Lindsay says of Kincardine Wood in Perthshire (which was very large, at 400–500 acres) that 'the whole wood was cut over in six years from 1654, and twelve years after 1681; the twenty-four-year cutting between 1705 and 1728 provides the first evidence of genuine rotational treatment'. After 1736, the rotation was lengthened to thirty-two years.[71] In Argyll, the English ironmasters after 1750 sought long leases of the woods they wished to charcoal. Richard Ford and Company at Bunawe took a ninety-six-year lease from Sir Duncan Campbell of Lochnell in 1752, undertaking to 'fell down the saids woods only four times within that space' – a twenty-four-year rotation: and twenty-three or twenty-four-year rotations were built into other leases they took from Duncan Campbell of Inverawe and the Earl of Breadalbane in the same year. The Loch Fyne iron company in 1754 envisaged cutting the Duke of Argyll's woods twice in forty-five years.[72] Such long-term thinking was still altogether exceptional, however, and exigencies of estate need probably continued over most of Scotland to decide just when a coppice would be felled, though within brackets determined by the trees being obviously too young or clearly approaching becoming too old.

Felling trees was only the start of the extractive business. Each of the three main sorts of wood product (timber, bark and charcoal) then had to be processed and taken to market. Most woods produced timber and bark, only a few produced charcoal as well. Woodcutters always needed horses within the wood as draught animals (oxen are not mentioned again in the Lowlands after the examples in Lothian in the sixteenth century) and contracts specified how they were to be grazed. Usually they were allowed only 'before the axe', that is, in those parts of the wood not yet cut and fenced into individual haggs where the trees regenerated.[73] Occasionally where grass of this sort was scarce (as in a one-year contract) the animals might be allowed elsewhere, with precautions: at Little Park in Kirkcudbrightshire in 1732 a single tethered horse was allowed, 'the said one horse doing no damnage to the young growth'.[74]

There were often strict rules to avoid damage to hay or corn inside or outside the wood, the cutters also being made responsible for harm to tenants' crops by the horses of those who came to buy bark or timber on

---

[71] Lindsay, 'Use of woodland', pp. 486–7.
[72] NAS: RD 14, reg. 1 Feb. 1773; RD 12, reg. 5 Jan. 1762; RD 12, reg. 19 May 1755; NLS: MS 993, fos 8–15; MS 17666 fos 146–55.
[73] e.g. NAS: GD 3/2/29/47; GD 190/3/110/1; RD 13, reg. 10 Dec. 1731.
[74] NAS: RD 12, reg. 6 Nov. 1733.

site, 'keeping the cornes, meidowes and haying grass free of skeath or prejudice, or als to pay the skeath done to the awners thairof', as a deed relating to Ardrostan wood in Perthshire put it in 1667.[75] The risk of harm might not all be on one side. In 1700, the Irish buyer of woods in lawless Glengarry agreed to compensate his subcontractor from Fort William for any horses stolen, just as Alexander Macdonald of Glengarry had agreed to compensate him.[76]

Frequently the cutters needed to take large amounts of bark (and occasionally timber) to a port or market at some distance, involving more horsepower than they themselves could muster. Then tenants who were obliged by their leases to provide carriage-services for their landlords found them transferred to the cutters. An early example is from Cumbernauld in 1641, where the Earl of Wigton undertook to furnish 'sex scoir carrage horss and men with them everie yier to help cary their barke out of the saidis woods to the towns of Falkairk, Linlithgow and Glasgow', with the proviso that if there was not enough bark they could be used to carry 'cutt timber or mayd worke of the said timber to the foirsaids thrie towns', or to a similar distance – that is, up to about thirteen miles.[77] Eighty-one years later, in 1722, the same estate was still offering to provide the buyer with horses to carry his bark to Glasgow or Linlithgow, though now with the proviso that the services were not to be exacted in seed time or harvest when the tenants would have other and urgent priorities for their beasts.[78] Some lairds were prepared to send their tenants even further afield with the cutter's bark – the Marquis of Montrose in 1706 offered bark carriage from Cashel and other woods on Lochlomondside to Linlithgow, about forty miles, and Alexander Robertson of Faskally in 1678 offered carriage from Pitlochry to Dundee or Forfar, also almost forty miles.[79] Sometimes the loads were specified. In 1687 Sir John Graham, selling woods at Drymen, offered fifty carriage horses yearly to carry 'ordiner secks' – the phrase was inserted in the margin and signed by both parties, suggesting a previous bad experience – and in 1685 Sir Patrick Murray of Ochtertyre, selling woods at Monzievaird in Perthshire, offered eighty horses yearly to Stirling or Perth, 'ilk horse carrying allwayes six firlots of bark'.[80] Occasionally payment, nominal or otherwise, was expected: at Duntreath in 1753 a penny

---

[75] NAS: CC 6/13/11 reg. 30 May 1672; see also e.g. NAS: GD 132/660; RD 14, 1691, no. 633; RD 12, reg. 8 Jan. 1732.
[76] NAS: RD 12, 1706, no. 735.
[77] NLS: Ch. 16445.
[78] NAS: RD 12, reg. 30 Apr. 1723. See also an example in Alloa, RD 12, reg. 8 Jan. 1832.
[79] NAS: GD 220/6/579/5; NLS: MS. 1438, fos 16–17.
[80] NAS: RD 12, 1694, no. 806; RD 13, 1689, no. 292.

a horse was extracted for the use of ninety-eight animals a year, but Cameron of Lochiel in 1722 made provision for his tenants being paid ten pence for each load not exceeding two hundredweight, carried from Achnacarry to Corpach, a distance of nineteen miles, the buyer providing the sacks.[81] Such consideration, and willingness to share the opportunity for making a little money with the tenantry, was distinctly unusual.

Cutting a wood brought people into it for most of the year, and sometimes the buyers were allowed not only to build or occupy houses, but to cultivate ground and pasture cows, as at Bargany in Ayrshire in 1705.[82] They might be allowed the use of existing building, or to put up new ones using turf, bracken and stakes, even to take them away with them, as at Bothwell in Lanarkshire in 1701 when James Machan was allowed to 'put ane little house in the said wood and to big the same with faill, divot and staicks, and at the ish heirof to transport the samyne when he pleassis'.[83] In 1715 Sir John Clerk of Penicuik sold the sycamores of Brunstane Park to a cup-turner of Restalrig, allowing him to use the vaults in the park and to 'build a convenient sitt-house and work house for himself on his oan expenses'.[84] Cup-turning was a solitary trade, but bark-peeling was labour-intensive, where many people had to be fed and refreshed. At Faskally in 1683 the facilities included a brewhouse for making ale for workers and customers.[85] One thing follows another, and at Saltoun in the 1730s the woodland management accounts mention a canvas 'bark shit', possibly to shelter a toilet for the bark-cutters, many of whom were women.[86] Bark kilns were needed to dry the bark, facilities were needed to shred it and store it, so that the whole makeshift complex might become quite elaborate: Roger Murphey in 1721, at Kennacraig in Argyll, was given leave to 'build a bark mill, kill, store house and timber yard, with cabens for workmen', though in this case he was obliged to leave all the building timber behind when he left.[87]

The production of good-quality charcoal was at least an equally skilled and specialised affair. Richard Ford and Company at Bunawe agreed with Duncan Campbell of Lochnell in 1752 to 'make sawpits, coals hearths, pitts, pittsteads and cabbines for the coaliers' and get 'floating dust and cover' (earth for the charcoal burners' heaps). All this was to be combined in Sir Duncan's woods with bark and timber extraction, for all of which they

---

[81] NAS: GD 1/1067/ 53; RD 12, reg. 17 Jan. 1729.
[82] NAS: GD 109/3499.
[83] NAS: RD 13, 1707, no. 155.
[84] NAS: GD 18/1278.
[85] NLS: MS. 1438, fos 16–17.
[86] NLS: Saltoun papers.
[87] NAS: RD 12, reg. 2 Apr. 1723.

needed to use their 'horses, carts, waggons, boats and other carriages' and to 'get stones, fearns, heather and sodds for errecting houses to put in barks and coals', but they were allowed in this case to dismantle them at the end of the lease to re-use the timber.[88] In later leases, Ford and Co. planned as far as possible to shift the cost and trouble of making the charcoal onto the landowners themselves, paying 3 shillings, or 3 shillings 6 pence, for each dozen of coals:

> The meaning of a dozen of coals is twelve bags or sacks full, the dimension being two English yards and a half in length and one English yard and breadth within, between seam and seam when empty, to be filled in the woods so as that the coal may be four foot ten inches high in the sack when fairly sett down by the carrier and stocktaker at the said furnace.[89]

It was an immensely cumbersome article to move any distance, especially as its friability could so easily turn it to valueless dust.

The management of pinewoods, or 'firrwoods' as they were invariably termed, is less well documented than the management of broadleaf woods. Fewer than 10 per cent of the contracts mention pine at all, and of those that do, only about thirty individual deeds give any hint of how they were managed, the remainder dealing merely with the sale of trees already cut. This is enough, however, to indicate the broad lines under which pine was exploited.

Compared to Lowland oakwoods, in particular, management of Highland pine often seems to have been very low key. There was, for example, no indication of enclosures before the middle of the eighteenth century because, for reasons already explained, it was not as important to enclose a pinewood to ensure regeneration (see above, p. 165). In a minority of cases, perhaps about a third, an entire wood was sold, to be followed by a clear-fell of everything except (as a rule) young trees below a specified size (see Table 7.4a). These would be expected to grow on to form the pinewood of the next generation. Alternatively, a specified number of trees was sold, which could vary from a few hundred to 100,000 (see Table 7.4b). In fact, as we shall see later (pp. 202–16), the more spectacular bargains of the mid-eighteenth century were seldom carried through, which does not negate the value of their evidence for these purposes.

---

[88] NAS: RD 14, reg. 1 Feb. 1773.
[89] NAS: RD 12, reg. 5 Jan. 1762; RD 12 reg. 19 May 1755.

Table 7.4a Pinewood sales: whole wood examples

Date	Place	Limitations	Reference (NAS)
1606	'Aldscaidrow', Lochaber	No trees below 18 in. circumference at the ground	RD 1/438, fos 229–30
1612	Glencalvie, Ross	No trees below 12 in. 'of thickness'	RD 1/206, fos 361–2
1630	Abernethy, Kincardine and Glencarrie, Speyside	—	GD 248/78/4/22
1701	Glencoe, Argyll	No trees below 12 in. circumference	RD 14, 1701, no. 406
1719	Coigach, Wester Ross	No trees under 6 in. diameter	RD 14, reg. 16 June 1724
1723	Glenorchy, Argyll	No trees below 24 in. circumference at 3 ft from the ground	RD 13, reg. 26 Mar. 1724
1730	Derry and Liubeg, Braemar, Deeside	—	GD 124/17/151/1
1762	Crathes, Deeside	No broadleaf trees	GD 345/1007

Where the mature trees were selected for felling, the buyer often had a great deal of control over their choice and therefore their extraction, which might not be in the long-term interest of the wood itself. If the best and straightest of the trees were always selected, it might be argued, the gene pool of the remainder would gradually deteriorate. There is evidence from Table 7.4b, however, that practice varied even when there was no clear-fell. The seller might have a representative forester to accompany the buyer and try to minimise abuses, or, as in Glentanar in 1708, the seller actually determined, in a wood being thinned, what the buyer could have. In the contract between the Duke of Gordon and an English buyer in Glenmore in 1715, although the number of trees was specified, the buyer was told to work forward from one corner of wood, leaving no tree uncut that could yield 'square-edged dealls ten inches broad in first cut of the root end twelve foot long', which was in effect tantamount to clear-cutting while leaving behind young trees unsuitable for timber.[90]

There was something rather happy-go-lucky about most pinewood contracts. Seldom, for example, were the buyers asked to tidy up behind them (though they were at Crathes in 1762), or to cut close to the ground or with particular tools: when they were, at Coigach in 1727, it was by a wood merchant who was contracting with others to fell what he had already bought – he wanted the trees felled with an axe as near to the ground as possible, and cut into logs with a saw.[91] On the other hand, the deeds often contain much about places for saw-pits and sawmills, about dams for storing and floating logs, and about arrangements on the rivers. Felling a pinewood sometimes required protection for the woodcutters from the locals – not an automatic expectation in the seventeenth- and earlier eighteenth-century Highlands – and sometimes other extensive facilities. When Captain John Mason signed an agreement to cut on Speyside with Sir John Grant of Freuchie in 1630 (fruitless as it turned out), he asked not only liberty to build sawmills and timber wharves, and the right for his workers with their 'wyiffs, barnes, guids and geir' to be afforded protection, but also for a hundred acres of pasture and arable land to support his workers and their animals.[92] References to oxen used with horses as draught animals in the pinewoods occur as late as 1658, from Ross-shire.[93]

In the best instances, greater care was taken over the management of pinewoods than is evident from the deeds alone. Donald Cumming, who

---

[90] NAS: GD 44/29/6/24. See also GD 44/29/6/2/10.
[91] NAS: GD 345/1007; RD 14, reg. 29 Aug. 1729.
[92] NAS: GD 248/78/4/22.
[93] NAS: RD 14/1662, no. 1651.

Table 7.4b Pinewood sales: specified quantity examples

Date	Place	Quantity	Limitations	Reference (NAS)
1653	Glen Etive, Argyll	One thousand of the best and eight score of the second best, etc.	—	RD 11, Box 312, reg. 28 June 1654
1658	'Amotnatua', Ross	Twelve hundred[a]	Buyer chooses	RD 14/1662, no. 1651
1666	Glencoe, Argyll	Twelve hundred	Buyer and seller's representative choose	RD 14/1666, no. 1474
1706	Strathcarron, Ross	Twelve thousand two hundred	Buyer chooses	RD 14, 1707, no. 145
1708	Glentanar, Deeside	Two hundred young trees	Seller chooses	GD 44/30/5/1/6
1715	Glenmore, Speyside	Three thousand three hundred	Trees must be over certain size	GD 44/29/6/24
1742	Glen Affric area, Inverness	Fifty thousand	Buyer chooses	TD 86/97 Box 3, bundle 41
1743	Abernethy and 'Glencharnick', Speyside	One hundred thousand	Buyer chooses	GD 248/23/4. no. 46
1768	Abernethy, Speyside	Twenty thousand	Trees must be over certain size	RD 12. reg. 27 June 1769
1769	Abernethy and Dulnain, Speyside	One hundred thousand	Buyer chooses with seller's representative present	GD 248/377

[a] A note in the MS says 'compting sex score to ilk hundreth', which is why the table uses words rather than numbers in this column.

had worked as overseer in Glentanar for the Earl of Aboyne, described in 1762 how good practice would cut out, or use for tenants with servitude rights, all trees that were 'crooked, deform'd, decay'd in the top and of an ugly unnatural colour', in order to favour trees that were 'straight, neat, high, fresh at the top, and green branches'. He did not rely completely on natural regeneration, but advocated sowing 'firs' thick and allowing them to thin themselves – 'weed themselves by falling down', as he put it.[94] Such management, if widespread, would have offset any tendency to allow buyers to get away with selecting the best trees at the expense of the gene pool.

Good practice was not necessarily general, however, as the Annexed Estates Commissioners found when they took over a number of pinewoods in the aftermath of the 1745 Rising. At the Black Wood of Rannoch in Perthshire in 1757, Ensign James Small, factor to the former estates of Robertson of Struan, and George Sandeman, joiner in Perth reported that:

> The whole wood has been so mangled and destroyed in the cutting, by felling trees every year in so many different places, which not only destroys a good deal of what is too small to be cutt, by the fall of the trees when cutt down, but entirely prevents the growth of any seedlings ... When trees were cutt down they only took such pieces as they pleased, leaving cutts of four, five, six, seven and eight foot long, and all the crop pieces, though quite streight and fit for a great many purposes, which made clearing of the ground like an insurmountable task.[95]

They found the brash choking the drainage ditches, birches smothering the ground where pine regeneration might have taken place and the surviving pines 'scattered and straggling', subject to windblow. The cutters had used hatchets to cut up the logs in place of saws, 'which loses one foot in length at each cutting', and had been obliged (very strangely) to tear up every tree by the root in order to maximise the tonnage of wood extracted, which made the wood full of boggy holes and a thicket of old roots that could be neither transported nor burnt.

Their two main solutions were to clear away as much brash and birch as they could, and to advocate cutting the wood in haggs. The first was a

---

[94] Quoted in I. Ross, 'A historical appraisal of the silviculture, management and economics of the Deeside forests', in J. R. Aldhous (ed.), *Our Pinewood Heritage* (Farnham, 1995), pp. 138–9. See also G. A. Dixon, 'William Lorimer on forestry in the Central Highlands in the early 1760s', *Scottish Forestry*, 29 (1975), pp. 191–210.
[95] NAS: E 783/76/2.

common prescription in conifer woodland management over the next two centuries, but extirpating birch could have implications for the long-term fertility of the ground. The second involved clear-felling a specific area, using the waste branches as a fence to protect the regeneration within.[96] The problem of hagg-felling in a pinewood was the difficulty of predicting the yield from a given cut, and though it was more than once proposed for the Black Wood, it seems that instead selective felling continued under closer supervision from the forester.[97] The Old Wood of Meggernie in Glen Lyon to the south was said in 1763 to have been 'quite exhausted' by 'extending the hagges to a greater extent than it could bear': that was an exaggeration, as the wood today covers much the same extent as it did at the time of Roy's map.[98] Lindsay comments that during the period of government administration the Black Wood was 'undoubtedly managed rather than exploited', and considers the most useful thing that the Commissioners did was to enclose the main area of the pines by a ditch and a dyke, since outliers beyond the wall were later damaged by the rise of intensive sheep-farming and muirburn.[99]

Elsewhere, the Commissioners also had an uphill task. At Loch Arkaig they struggled against a complete collapse of management in the years 1746–52 when local tenants, government officials, the military and even the minister raided the woods at will. They eventually established control and, as at the Black Wood and also at Barrisdale, they attempted an ambitious enclosure.[100] But it was not until the nineteenth century that the pinewoods, so long largely left to themselves and to the convenience of local users, underwent more radical changes in management (see below Chapter 10).

The management of a wood involved a hierarchy, at the top of which was the estate chamberlain, steward or factor, and immediately below them on a large estate the ground officers. If they had any lack of experience, the resource might suffer: when the woods of Glenorchy were devastated by the felling operations of the Irish partnership in 1725, the Earl of Breadalbane's man-on-the-spot, Campbell of Barcaldine, admitted that he had 'no more skill about firr woods than a chyld' (see below, Chapter 13). In such circumstances it was wise to appoint a specialist woodland 'overseer',

---

[96] *Ibidem*.
[97] Lindsay, 'Use of woodland', pp. 282–6.
[98] H. M. Steven and A. Carlisle, *The Native Pinewoods of Scotland* (Edinburgh, 1959), p. 145.
[99] Lindsay, 'Use of woodland', pp. 320–6.
[100] T. C. Smout, 'Cutting into the pine: Loch Arkaig and Rothiemurchus in the eighteenth century', in T. C. Smout (ed.), *Scottish Woodland History* (Edinburgh, 1997), pp. 122–4.

like Donald Cumming who advised the Earl of Aboyne at Glentanar around 1760, though we have not encountered such formally termed officials before the eighteenth century.

Everywhere and from an early date, however, there were people at the bottom of the hierarchy known as 'foresters' or 'fosters'. They were part of the local farming community, often cottars but sometimes rather more substantial tenants, given duties in respect of the wood, such as maintaining enclosures, defending it from uncontrolled grazing, marking any particular trees assigned for felling or preserving, supervising lawful timber cutters, prosecuting unlawful ones and on occasion providing the landowner with firewood or nuts.

Such was John Davidson in Burnemouth, appointed by contract with the Earl of Cassillis in 1648 to 'the custodie and keiping of his lordschippis wood of Dalrymple' in Ayrshire for a term of three years. His duties were listed as keeping the wood 'with sufficient fensible dykes', fostering the young wood from any 'cutters or destroyeris thairof, ather be nolt [cattle] or utheris quhatsomevar', and bringing to the Earl 'fyve hundreth sommer wind faggottes', as firewood, and 'ane firlot of type leimed nutes'. In return he was paid six bolls of oatmeal a year, with the right to graze one horse ('ane naig') in 'the holme' and another in the wood 'when the said nobill Earle sall have horse there'. As an additional duty, he was to herd and look after the Earl's horses in the wood, but on no account to allow his own cattle or those of other tenants to graze there, on pain of confiscation of the stock and forfeiture of the appointment.[101] Such a contract had been repeated earlier on several occasions with different foresters between 1628 and 1645, with minor variations in the conditions about nuts, firewood, grazing animals, and payment in oatmeal or a holding, as though the Earl did not fully trust any single tenant with his wood for long.[102]

A different kind of arrangement was struck between David Smith, who in 1677 had bought the timber from the laird of Clackmannan in 'the wood of Clackmanane called the Birkhill' and 'Henry Bruce, late foster of the said wood'. Smith had bought the Birkhill on condition that it was not cut for seven years to allow it to come to greater maturity, and re-appointed Bruce as its custodian, to 'weid, cutt, peall and inclosure the same' as directed. He was to act as 'haguman' as well as foster, making, keeping and preserving dykes and palings as instructed, he was not to cut or sell any wood without permission, and when it was sold he was to make 'just compt, reckoning and payment' to David Smith. He was to continue to pay to the laird four

---

[101] NAS: GD 25/9/78/3/6.
[102] NAS: GD 25/9/78/3/1–5.

bolls of meal for his house and two acres of foster land, but to receive from Smith six and a half bolls of meal and grazing in the wood for two cows (and one follower) plus a mare (and her follower), as well as 'the usuall foster fie of the timber sold'.[103]

Before the abolition of most of their powers in 1747, the Scottish system of baronial and regality courts provided judicial supervision over many aspects of rural life. Ensuring that the foresters looked after the wood and calling unruly neighbours to justice was often an important part of their activities.

In sixteenth- and seventeenth-century Glenorchy, for example, the courts exercised careful control over everything to do with woodland. The tenants were regularly prosecuted for offences against the wood that might vary from unauthorised removal of timber to allowing uncontrolled muirburn, keeping goats, breaking fences and permitting cattle to stray. If damage was found that the foresters had not prosecuted, they became liable to the fine themselves, so that they had to maintain a state of constant vigilance over their neighbours. But their duties did not necessarily make them social outcasts. In one interesting case, the tenants found one of their number pilfering wood and locked the stolen timber in the forester's barn: in this instance the forester was, exceptionally, a woman. The culprit returned, broke the lock on her barn door, made off with the goods and ended up in court.[104]

In Glenorchy, each forester was required at every sitting of the court to make a statement on oath regarding the condition of the wood for which they were responsible, and this statement was then verified by the ground officer. If a particular wood had not been 'sighted' in this way, a verdict was held over until the next sitting of the court. Similarly, in Glentanar a jury of birlaymen were appointed to inspect the condition of the woods: in 1694, for example, they reported back that the birchwoods were neither better nor worse than the last appreciation in 1691, and therefore 'finds that the forester hath done his duete'.[105]

The uses to which baronial jurisdiction was put varied sharply from place to place and from time to time. The focus might be on the revenue from the fines rather than on the management of the wood. Menzies of Weems, neighbours to the lairds of Glenorchy in Perthshire, used their foresters primarily as a device to license their tenants to take wood for their

---

[103] NAS: GD 190/3/110/2.
[104] F. Watson, 'Rights and responsibilities: wood management as seen through baron court records', in T. C. Smout (ed.), *Scottish Woodland History* (Edinburgh, 1998), pp. 101–14.
[105] Watson, 'Rights and responsibilities'; Ross, 'Historical appraisal', p. 139.

farms, the fines essentially purchasing permissions for use. This both limited exploitation and provided a tidy income that, in 1673 alone, reached over £485 Scots.[106] In Ayrshire around 1700 it was complained that the Earl of Cassillis virtually sold the offices of depute or clerk of court of the Barony of Maybole, because they were valuable and the court had jurisdiction over all the neighbouring countryside:

> For by the plenty of wood and water in this countrey which tempt men to fish and cutt scob or wattles for necessary uses, they find a way yearly to levy fines for cutting of green wood and killing fry or fish in prohibite tyme, that makes a revenue to these offices and is a constant taxe upon the people.[107]

When Baron Courts were run with an eye on the fines, offences against the wood were often a high proportion of the total cases (sometimes on the Menzies estates, 80–100 per cent) and involved many people (in 1719 on the Menzies lands at Appin of Dull, 258 men were fined, and half the offences were against the wood).[108]

At the other extreme was the regality jurisdiction of the Grants of Grant, covering Abernethy and other parts of adjacent Speyside. Here in the 1690s there were only nine cases heard against the wood, of which six were for the standard offences of cutting without permission or taking away and selling wood, the most elaborate of these, in 1692, involving two foresters who named six individuals and one ordinary tenant who accused another. The standard fine in these cases ranged from £10 to £50 Scots, so over a decade the income to the laird was inconsiderable.

The other three cases concerned fires, that usually arose from muirburn getting out of control. They varied from the absurd to the appalling. In 1692 Patrick Bayne in Inshstomach was called before the court and asked what he knew about his own and others 'burneing of the woods'. He replied that 'he heard that it was the hird of Rathiemoone that burnt the wood caled Cowhillhawne', and he confessed that he himself had lost his axe 'and for to find it he burnt ane piece of the wood besyde his owne howse'. For this he was fined £100 Scots. Very much more serious was the case in 1695 when a crofter, Alexander Gardner, and four children (a boy aged eleven, two other boys and a girl all aged thirteen) were accused of setting fire to the

---

[106] Watson, 'Rights and responsibilities', p. 103.
[107] A. Mitchell (ed.), *Geographical Collections relating to Scotland made by Walter Macfarlane* (Scottish History Society, Edinburgh, 1906), 2, p. 16.
[108] Watson, 'Rights and responsibilities', pp. 103–5.

pinewoods near Nethybridge, 'whereby the firr woods young and old in the forsaid plaices wer destroyed and brunt to ane extraordinary walwable skaith'. The crofter confessed to having started the muirburn, but on the safe side of a stream. Two of the boys confessed that they took the fire from there to 'the heather besyd the wood, wherby by the impetuous and uninteruptable fyr the adjacent woods wer kendled', the girl confessed that she was 'the person that burnt the woods', and the other boy admitted to putting the fire nearer to the woods but said that in trying to save an ox from the flames 'he did cast the fyre hither and thither but not of design to doe skaith'. The judge decided that their masters and landlords must pay for the damage once it had been calculated, and retired to consider a punishment for the guilty 'whereby others may be terriffied and feared to doe the lyk in tyme coming'.

Twelve days later the court reconvened. Alexander Gardner and the youngest child were not punished. The three remaining children were immediately taken to the gallows 'at the moor of Belintone' and their ears 'nailed to the said gallowes and to stand ther at the judge his pleasure'. Surprisingly enough, the expectation that this would act as a deterrent was foiled the following year, when another group of children, two boys and a girl, were found guilty of another muirburn that got out of control and burnt part of Glenmore forest and an adjacent part of Abernethy. Apart from fines, the boys again suffered the 'nealling of the luges to the gallowes', but the girl's punishment was referred back to the laird of Grant himself for his consideration. Perhaps even the strong stomachs of the age had revolted at the earlier treatment and hesitated to extend it again to a girl. But no-one ever again, as far as is known, nailed children to the gallows to show how much the owner cared for his woods.[109]

---

[109] University of Guelph archives, Ontario, Canada: 'Transcript of the Regality of Grant Court Book, 1690–1723'.

CHAPTER 8

Outsiders and the woods I: the pinewoods

Much of the focus of this book so far has been on local use and short-distance trade. Within the period 1600–1850, however, external demand, sometimes arising from a market hundreds of miles away, became an increasingly important factor in determining how the woods were used. This demand was often, though by no means always, mediated through individuals with no connection to the communities who had hitherto used the woods. Such entrepreneurs were English-speaking Lowland Scots, Irishmen or Englishmen, but their usual theatre of operations was the Gaelic-speaking Highlands, a circumstance which seems to give a quasi-colonial flavour to events. However, the outsiders operated often on the prompting and always with the co-operation of the local chiefs and lairds, the indigenous elite who owned the woods, and who were still the lynch-pin of Gaelic social structure. The landowners were also perfectly capable of sensing an external market on their own account, and acting as their own entrepreneurs. This was a phenomenon that went far beyond timber extraction, and has been called by Allan Macinnes the 'commercialisation of clanship'.[1] It began a century and a half before the 1745 Rising, and was as common among Jacobites as among Hanoverians. The Camerons of Lochiel were no less commercially minded than the Dukes of Argyll, though they might ultimately have been less effective.

It has often been asserted that satisfying external demand leads to particularly short-sighted and destructive exploitation, as when Fraser Darling maintained that cutting for local use would have little impact, but that 'real danger' comes when a product is seen as having value for export – a process which in the case of the Scottish woods he dated to around the period of the Union of the Crowns.[2] The modern world is replete with

---

[1] A. I. Macinnes, *Clanship, Commerce and the House of Stuart, 1603–1788* (East Linton, 1996), p. 172.
[2] F. F. Darling, *Natural History in the Highlands and Islands* (London, 1947), p. 59. He actually relates it to 'the beginning of the 16th century', but the context of his discussion suggests that this is a slip for '17th century'.

examples of logging companies moving into one area, extracting resources, and moving on to another, indifferent to the damage caused.[3] On the other hand, it has been argued that by giving a commercial value to a renewable resource like timber, external demand encourages better care of it, as the owners will wish not to destroy a source of income renewable in the medium to long term.[4] Much may depend on the degree of local control still maintained over both the resource itself and over the processes of exploitation. A main aim of this chapter and the next will be to examine the degree of impact by the outsider and the external market on the Scottish woods, and to judge whether its direction was for good or ill.

Three types of external demand were most in evidence in these years – coming from shipbuilders and timber merchants, from ironmasters, and from tanbarkers. The first were primarily interested in pine, the last two in oak. In this chapter we shall examine only the pinewoods (Map 8.1). In some cases, more detailed accounts of the outsiders' activities are available in the four chapters of case studies, summarised and cross-referenced here.

After the Union of the Crowns in 1603, there was a heightened sense of awareness, even excitement, about the timber resources of the Highlands, in the context particularly of providing naval stores to protect the newly united realm. In 1608, James VI wrote to his Scottish Privy Council, linking the 'great quantitie of growing timber and woddis of all sortis discoverit', to the need to provide shipping 'necessarye for saiftie, as being one of the most principall bulwarkis and defenses of the Iyland againe all forrane invasioun'. He ordered Council to take measures to prevent the export of Scottish timber (there was perhaps already a market for it in Ireland) and to ban the erection of new ironworks in the Highlands because of their destructive effects on the woods there. Three days later, the co-signatory of the royal letter, Sir Alexander Hay, the Scottish secretary, was given a warrant for £500 sterling to go to Scotland to buy 'masts and scantlings of Scotch fir to be delivered within the Thames for the Navy'.[5]

The response of Privy Council to the king's letter was equivocal. Unusually, they demurred, writing back that timber was not exported from Scotland, but had to be imported, and that a royal ban might elicit similar responses from foreign countries on which Scotland relied for its wood: in place of a general prohibition, would it not be better that the restraint be

---

[3] For a survey of contemporary world forestry, see A. S. Mather, *Global Forest Resources* (London, 1990).
[4] e.g. T. C. Smout, 'Woodland history before 1850', in T. C. Smout (ed.), *Scotland since Prehistory: Natural Change and Human Impact* (Aberdeen, 1993), p. 43.
[5] *RPCS*, 8, pp. 775–6; *CSP Dom, Series James I, 1603–1616*, p. 473; *DNB*, 9, p. 250.

The Native Woodlands of Scotland, 1500–1920

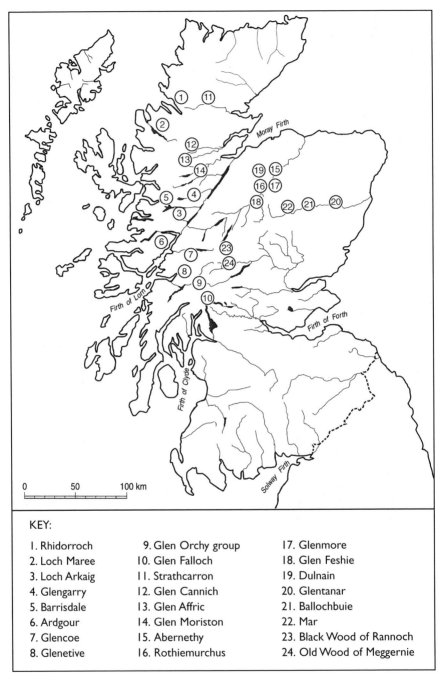

KEY:

1. Rhidorroch
2. Loch Maree
3. Loch Arkaig
4. Glengarry
5. Barrisdale
6. Ardgour
7. Glencoe
8. Glenetive
9. Glen Orchy group
10. Glen Falloch
11. Strathcarron
12. Glen Cannich
13. Glen Affric
14. Glen Moriston
15. Abernethy
16. Rothiemurchus
17. Glenmore
18. Glen Feshie
19. Dulnain
20. Glentanar
21. Ballochbuie
22. Mar
23. Black Wood of Rannoch
24. Old Wood of Meggernie

Map 8.1 Main native pinewoods known to have been subjected to commercial exploitation, 1600–1830. Based on H. M. Steven and A. Carlisle, *The Native Pinewoods of Scotland* (1959), with modifications and additions.

'onlie astricked to suche particular wodis as your Majestie intendis to mak use of for your awne schipping?' As for 'irne mylnis', there were none in Scotland, only a little quantity of iron made (presumably in bloomeries), from the 'scroggie boughis brancheis and auld stokis and cuttingis of tymmer which can serve for no other vse': besides, it might seem to impinge on 'your maiestries subiectis who by the lawes and custome of this kingdome are free to mak vse of thair awne commodeteis within the cuntrey to thair best advantage'. The wording suggests a real debate within the Council, perhaps an awareness of vested interests at risk from a blanket ban.[6]

Next year, however, after a further letter from the king, Privy Council banned the export of timber from Perthshire, Argyll, Inverness-shire and Caithness (was the omission of Ross-shire significant?) and the Convention of Estates banned ironworks. The Act referred to the discovery, by the grace of God, of 'vaynes of rich mettall', and of 'certain wodis in the Heylandis, whilkis wodis by reasoune of the savagnes of the inhabitants thairabout wer aither unknawin or at leist vnproffitable and vnvsed'. Ironworks should be prohibited to save the woods 'for mony better useis'.[7] It is highly likely that the debate hinged on the resources of Loch Maree, in Wester Ross, the property of the Mackenzies, Lords of Kintail from whom the well-connected and ambitious laird and courtier, Sir George Hay of Netherliff, (earlier busy in the western Highlands with the Fife Adventurers) obtained a lease of the woods in 1611. In the previous year he had secured a 'Commission and licence [to] mak yrne and glass' within the kingdom of Scotland from James VI, who had presumably either abandoned his previous train of thought or was too unwary to put two-and-two together.[8]

It is hard to believe that those involved in this tangle were unaware of the near-contemporary description of Loch Maree ascribed to Timothy Pont, though not of absolute certainty by him:

> compas'd about with many fair and tall woods . . . in sum places with fair and beautifull fyrrs of 60, 70, 80 foot of good and serviceable timmer for masts and raes, in other places are excellent great oakes whair may be sawin out planks of 4 sumtyms 5 foot broad.[9]

---

[6] *RPCS*, 8, pp. 543–4.
[7] *RPCS*, 8, pp. 234, 552–4; *APS*, 4, p. 408.
[8] NAS: RD 1/213, fos 354–67; *APS*, 4, p. 515.
[9] A. Mitchell (ed.), *Geographical Collections relating to Scotland made by Walter Macfarlane* (Scottish History Society, Edinburgh, 1906), 2, p. 540.

In other words, this was timber ideal for shipbuilding. As for Alexander Hay and his £500, we know nothing more apart from a chance remark in 1611 by John Speed, the English cartographer, that Crown surveyors had gone north and found:

> firr-trees for masts in north-west Scotland upon the banks of Lough Argicke [Arkaig] of such great height and thickness that at the root they bear 28 handfuls about and the bodies mounted to 90 foot of height they beare at that length 20 inches in diametre: and at this present growing upon the lands of the right worthy knight Sir Alexander Hayes, his Maiesties principal Secretay for Scotland.[10]

It is most unlikely that Alexander Hay had bought the land, but he may have used some of the royal cash to buy the timber. There is no indication that it was ever felled and brought to the Thames.

After 1630, there was a chance that a similar scheme might be revived, but on Speyside. The seller of the woods was Sir John Grant of Freuchie who had long had an active interest in exploiting Highland forests from Glen Urquhart to Morar, Loch Arkaig and Glen Loy, as well as in Speyside.[11] The buyer was Captain John Mason, who in 1612 had been commissioned by the Privy Council to collect the 'assize herring of the Northern Isles' from Dutch fishermen and who two years later sold his ship to the Scottish Crown for use in maritime policing.[12] Mason received a forty-one-year lease of the woods of Abernethy (along with neighbouring Kincardine and Glencairn) for £20,000 Scots, and shortly after reported to the Commissioners of the Navy that the timber was suitable for naval purposes.[13] There is, however, no evidence that any naval timber was extracted as a result, and a few years later Captain Mason sold part of his interest in the lease to the Earl of Tullibardine.[14] The tenuous connections that had been established in the first half of the seventeenth century with the Royal Navy thus all seem to have evaporated without evidence of any wood extracted.

In the second half of the century, wood was indeed shipped south for the

---

[10] Quotation in H. M. Steven and A. Carlisle, *The Native Pinewoods of Scotland* (Edinburgh, 1959), pp. 164–5.

[11] W. Fraser, *The Chiefs of Grant* (Edinburgh, 1883), 1, p. 220; 3, pp. 444–6, 424–8, 431–4.

[12] J. Goodare, *State and Society in Early Modern Scotland* (Oxford, 1999), pp. 165, 250.

[13] NAS: GD 248/78/4/22; J. Munro, 'The golden groves of Abernethy: the cutting and extraction of timber before the Union', in G. Cruickshank (ed.), *A Sense of Place: Studies in Scottish Local History* (Edinburgh, 1998), pp. 157–8; E. Grant, *Abernethy Forest: its People and its Past* (Nethy Bridge, 1994), p. 6.

[14] Fraser, *Chiefs of Grant*, 1, p. 277.

use of the navy in two episodes, probably in 1652–3 and certainly in 1665–9, in each case from Strathcarron in Ross-shire on the estate of Ross of Balnagown. The story is told in detail in Chapter 12 below, and is only summarised in what follows.

Renewed interest in Scotland as a possible source for naval stores came with the outbreak of the First Dutch War in the summer of 1652. The Cromwellian government had already received reports on the possibility of extracting timber, tar and masts from Speyside, Deeside and Glen Moriston, including the comment that Abernethy was the best wood in Scotland for tar, 'a red fir and full of sap'. However, it was clear that the military situation and the lack of local co-operation would make extraction difficult. When war broke out, the navy found its Baltic supply line under threat, but the government discovered that it had in its prisons following the Battle of Worcester, Colonel David Ross of Balnagown, owner of some of the most promising pinewoods in Scotland. It switched its attention to the north, sent a ship to collect the masts, and ordered soldiers to protect the workers in the woods if necessary: possibly one boat-load was extracted, before a rising in the Highlands in 1653 and the end of the Dutch war brought the experiment to an end.[15]

The episode, however, was not forgotten after the Restoration, and renewed hostilities against the Dutch in 1665 brought Captain Phineas Pett, 'on[e] of his majesties shippe buylders', north. He called on David Ross, younger, of Balnagown, whose father had died in prison in England, in order to strike a bargain to sell 1,000 masts to the Royal Navy. When they began to be delivered, however, despite the personal supervision of Pett who bought an impressive property in Inverness, they were found to be expensive and unsuitable as masts, so the enterprise came to an end in 1669.[16] It is doubtful if the felling of a few hundred pines for this purpose would have made any impact on the woods. David Ross between 1667 and 1678 then turned his attention to providing deals for the reconstruction of Holyrood Palace as well as running a considerable trade elsewhere in the Firth of Forth, the Firth of Tay and the north: this seems to have involved much heavier exploitation of the woods than his involvement with the navy, and was undertaken with little obvious care for the future by a laird heavily burdened with debt. How far this damaged the woods or altered their character, however, is not clear.[17]

---

[15] See below, pp. 323–4.
[16] See below, pp. 325–31.
[17] See below, pp. 331–4.

Activity initiated by private interests was always much more extensive than anything undertaken by the navy. Wood merchants were at least intermittently busy in the north-west from immediately after the Union of the Crowns. In a contract dated 8 February 1606 and signed in Inverness, George, Earl of Enzie, Lord Gordon and Badenoch, sold to Robert Dumbar 'of Newtoun' and Edward Jones, 'Inglischman', the 'haill firr wods of Aldscaidrow, by and within the lordschip of Lochaber and sherefdome of Innerness', for four years, the wood to be delivered at Inverness. 'Aldscaidrow' is possibly the same wood as 'Slisgarrow': nineteen years later, in 1625, Robert Dumbar, 'with consent of Edward Jones, principal tacksman of the woods of Slisgarrow... within the lordshipe of Lochabbir', bought from 'Richard Besand, Englishman, and William Bonthron, smith in Inverness', their rights in these 'fir wods' for the next four and a half years. Here at least is evidence of interest in pine extraction in Lochaber over a period of a quarter of a century by a partnership of two, one of them English.[18]

Equally interesting are two documents of 1622 relating to Sir John Grant of Freuchie, the same who sold Abernethy to Captain Mason eight years later. In one, he purchased from the Macdonald lairds of Lundie in Glengarry the rights to their woods in Morar, which had hitherto been unprofitable to them because they 'lye in the far and barbarous hielandis, circuite and invironit about with evill neyctbouris, quha continewallie cuttis, destroyis and takis away the samen in great quantitie', involving the lairds of Lundie in deadly feud and discouraging merchants who would not come near them 'for fweir of thair lyveis'. Sir John Grant took a thirty-one-year lease of this intimidating property in return for half the profits that might be made by exploiting the woods, with a provision to sublet to any interested merchants who might be persuaded to operate – presumably after Sir John had dealt with the neighbours.[19] The other document was a contract of some complexity: in effect, the penniless new inheritor of lands in the west, Sir William Mackintosh of Torchastell, leased the actual property of Glen Loy and Loch Arkaig to the Camerons of Lochiel, but sold the standing wood on that land to Sir John Grant of Freuchie, to help cover his debts. To get the Camerons to co-operate and protect his operations, Grant then became the conjoint seller of the woods with Lochiel and his son, who in turn promised the help of the whole clan to keep any merchants operating in the woods secure from harm, in the same way 'as Lord Lovat or Kintail are bound to their merchants that bought their woods'.[20] There

---

[18] NAS: RD 1/438, fos 229–30; RD 11/175, reg. 21 July 1626.
[19] NAS: GD 248/8, no. 342; Fraser, *Chiefs of Grant*, 3, pp. 424–8.
[20] Fraser, *Chiefs of Grant*, 3, pp. 431–4.

is no evidence in either document that the merchants would be either Lowlanders or Englishmen, nor, indeed of any exploitation consequent on either agreement. But they are redolent of an atmosphere of commercial interest in the woods, held at bay and hampered by the endemic disorder of the region.

Clearer indications of regular trade with outsiders in the earlier seventeenth century comes from the lochs of the west coast. Although there is little doubt that Sir George Hay's main interest in Loch Maree was centred on plans for an iron furnace there (see below, pp. 229–36), the deed of 1611 by which he bought woods on the estate of Mackenzie of Kintail involved much more. It names not only the timber of Loch Maree, but also that of Loch Alsh, Loch Carron and Loch Duich, and four miles inland, from the shores of all the named lochs, a resource that (in the words of the deed) 'hes heirtofoir importit littill or na commoditie to the said nobill Lord', though it did mention an existing sawmill at Kinlochewe and also reserved to the Earl the right to cut 'fir tymber necessar for bigging of his galeyis abone the water'. Hay bought 'the aik, firre, esche, elme, esp [aspen], birk, hasill, holene and all uther kyndis of wood treis and tymber, with the bark and bewch [boughs] theirof, alsweill standard and growand, as cuttit down and fallin'.[21] From this there appears to have developed a trade in timber produce of enough substance to have recognisable 'Lochew' measures. In 1617, three Orcadians bought from Robert Symson in Aberdeen twenty-five score bolls of bark 'of Lochewis mett in the Ilis', Hay having sold Symson the privilege 'to peill and wyn barkis in Lochew'.[22] And in 1621 a contract between George Hay of Kirkland and Patrick Bell of Glasgow spoke of the former selling his 'schippis laidning, callit the *George of Lochew*', of 'Lochew daills', to be brought to the Clyde: the deals had to be 'of the ordinar lenth of Lochew daills' and nine inches broad at the narrowest end, suggesting local sawmilling.[23] The term 'deal' implies softwood.

There is other evidence of significant trade down the west coast in the seventeenth century. In 1620 Duncan Stewart of Appin contracted with John Sempill of Aitkinbar, burgess of Dumbarton, to supply from his woods on Loch Leven (probably in Glencoe), enough timber of pine and oak to fill four ships of fifty to sixty tons a year for five years, of specified types and sizes. Half was to be made up equally of joists and 'double ruiff sparris' of pine ('reidwod'), their length to be twenty-two feet, the spars to be at least four to six inches square, and the joists eight to ten inches square. The other

[21] NAS: RD 11/85, reg. 27 Feb. 1612.
[22] NAS: RD 11/119, reg. 8 Aug. 1618.
[23] NAS: RD 11/138, reg. 22 Nov. 1621; RD 1/313, fo. 127.

half was to be great timber logs of pine and oak, the shortest to be about seven feet long and a foot and a half 'in the squair' and of pine: but 'onie cruikit aiken timber or ony uther sort of timber... weill cuttit' was also acceptable. It was a substantial order, and presupposed that Stewart of Appin would make the capital investment to process the wood and also to make a building to shelter Sempill's crews while their boats lay loading.[24]

An alternative market for pine from Loch Leven appears in 1666, when the Earl of Argyll sold twelve hundred pines from his woods in Glencoe (they marched with those of Stewart of Appin) to John Parry, carpenter in Dublin – an early testimony to Irish enterprise in the west which was to be a marked feature of the next seventy years. The trees were to be eight to twelve inches square measured six foot from the ground, to be felled by Parry after agreement with the Earl's representative as to which were suitable, and carried to the shore at the Earl's expense.[25] Several decades later, in 1701, Argyll sold 'the present standing cropt and grouth of the firrwoods of Caolisnacon, lying in Glencoe', to Mungo Cochrane, merchant in Glasgow, allowing the felling of all trees over twelve inches circumference (but strictly reserving all those below that size), for £40 sterling – a small sum which suggests only a modest wood remaining. It was still sufficiently substantial an asset fifty years later, however, for Argyll and Stewart of Appin in 1750 to agree a formal division of the Glencoe firrwoods between them, roughly two-thirds to Argyll and one-third to Appin.[26]

Further south, Campbell of Glenorchy was also selling pine to outsiders. In 1653 he sold to William Robertson, merchant burgess of Glasgow, a thousand of the best trees from 'his firr woods of Glenettive under the lands of Buchiletive', plus eight score of the second-best trees thrown in free ('the woods of Altossibruicke being exceptit'), and also two hundred of the best trees growing in his woods of Glenorchy at 'Innergawnane and Creichastillan', with forty old trees not above 'eight insh of the square of reid wood' as free additions. Robertson was to cut the wood and Campbell to have it carried down Glenorchy to Loch Awe, and the bargain made provision for amendment in case the work was interrupted by 'troubles or incursions of any sortes of broken men'.[27]

Although we have found no seventeenth-century deeds relating to sales in the area, Ardgour was another west-coast locality from which extensive early exploitation of pine took place, as expressed in an account of Conaglen

[24] NAS: RD 11/132, reg. 20 Feb. 1621; RD 1/305, fos 92–4.
[25] NAS: RD 14/1666, no. 1474.
[26] NAS: RD 14, 1701, no. 406; NLS: MS 17664, fos 192, 194.
[27] NAS: RD 11, Box 312, reg. 28 June 1654.

and Glen Scaddle, full of pines and cattle, and in the latter the trees easily transported to the shore: 'there uses manie shipps to come to that Countrie of Ardgoure, and to be loadned with firr jests, masts and cutts. This glen is verie profitable to the Lord.'[28] Irish customs books, surviving for 1683–6, specify about 200 tons of timber a year imported from Scotland at that juncture of the seventeenth century. The wood could well have come from Ardgour, but did not necessarily do so.[29] Possibly there was not much left in Ardgour after this period, for a deed of 1713 sold the firwood of 'Conglen or Inverscaffidell' to Coll McAlister of Tarbert in Kintyre, along with some neighbouring oakwoods, all for £40 sterling, reserving all young firs 'below six inch square timber, to be measured five foot from the butt'.[30]

Irish interests in Scots pine also reached above Ardgour into the Great Glen at the close of the seventeenth century. In 1700, Captain John Fleming, residenter in Dublin, was described as the sole tacksman 'of the whole standing firr trees growing in the wood of Sleighgarre in Glengarry'. He had a tack for fourteen years, and was arranging for William McBean of Maryburgh to extract the wood and bring it to the head of Loch Linnhe.[31]

We can complete the survey of the seventeenth century by mentioning that there was at least a limited interest in the Speyside pinewoods by outside merchants in the second half of the century. Abernethy and Rothiemurchus attracted the interest of Benjamin Parsons, surely an English name, successively described as a merchant of Aberdeen and then of Leith. He obtained a lease of Rothiemurchus in 1658 and between that year and 1666 he had also bought up all the rights over Abernethy that Grant of Freuchie had originally assigned to Captain Mason for forty-one years in 1630. Parsons, financed by Samuel Collins of London, set up a sawmill at Garmouth at the mouth of the Spey to make shipbuilding timber, and in 1666 his partner Francis Arneill of Edinburgh, sold to two Forres merchants, 360 pines, twenty-five feet long, twelve inches broad and ten inches thick, to be delivered at Garmouth. By 1672, however, Parsons abandoned his efforts to make money in the north, and Collins was allowed to claim from the assets, including sawmills, cut and uncut timber, along with a new ship on the stocks at Garmouth, 'with haill new rigging belonging to her'.[32] It seems unlikely that these activities made much impression on the forests of this region. In 1704 William Batt of the Royal Dockyard at Deptford, on the search for masts, reported that the woods of Abernethy

---

[28] Mitchell (ed.), *Geographical Collections*, 2, p. 165.
[29] E. McCracken, *The Irish Woods since Tudor Times* (Newton Abbot, 1971), p. 110.
[30] NAS: SC 54/12/7, reg. 2 Oct. 1714.
[31] NAS: RD 12, 1706, no. 735.
[32] Munro, 'The golden groves of Abernethy', pp. 100–1.

were the 'likeliest to serve Her Majesty and Government', and in 1706 James Spreull, Glasgow merchant and pamphleteer, identified Strathspey as capable of providing 400 masts 'for the greatest ships in England', as well as being suitable for producing tar, pitch and resin. But comments like these were not followed by decisive actions, any more than those of one hundred years earlier had been. They testify more to the eternal optimism of half-informed outsiders than to the realities of commercial opportunity.[33]

The one area of Scotland where the seventeenth-century timber merchants probably made an impression on the pinewoods, was on the western seaboard, firstly because exploitation was probably more intense there than elsewhere, taking advantage of easy access by boat, secondly because in the wet and mossy west, conditions for natural regeneration were much less favourable than elsewhere. Some pinewoods exploited then, in Glen Etive, Glencoe and Glenorchy, are no longer there, and though they did not all vanish before 1700 they may have been so eroded that their final demise was seriously hastened. Others, as in Ardgour, have only a comparatively small fragment of their former forests surviving. It is hard to detect much strategy for sustainable use, though the marking by the Earl of Argyll's agent of the trees to be felled in 1666 could be interpreted as such. The responsibility for this failure (if failure there was) was the lairds' rather than the merchants', since it was the seller who set the terms for the exploitation and who was presumably in a position to make conditions stick had he wished to do so.

In the eighteenth century, the pattern of outside interest gradually changed, though both the impact and the rhythm of exploitation differed from one region to another. Down the west coast, the earlier pattern of fellings by the Irish and the Lowlanders continued, though interest in logging is sometimes hard to tell apart from interest in tanbark and charcoal, which increasingly also involved the English. The Irish, drawn to the area in the second half of the seventeenth century, remained active until the 1730s. However the only operation that we know of in this area which involved them extracting logged timber as well as bark were the ambitious activities of Murphey, Galbraith and partners, dealt with in detail in Chapter 13 below. To summarise, in 1721 and 1722 a Dublin tanner, Roger Murphey, became heavily involved in contracts for cutting oak with a range of Scottish landowners along the coast from Argyll to Morar: he subsequently joined with four of his countrymen to establish an iron furnace at Glen Kinglass on Loch Etive, and in 1723 to set up the Glenorchy Firwood Company. The latter can be presumed to have been primarily a

---

[33] Grant, *Abernethy Forest*, p. 9.

lumber concern, though as the company cut oak as well as pine in Glenorchy the partners were presumably, even here, also feeding their tanbark and charcoal interests.

Within two years the result of their operations apparently proved devastating. The Irish were alleged to have felled the oak without any regard to its regrowth, and the pine (of which little or no natural regeneration seems to have been taking place at the time) was, by them, largely cleared from the area. The oak may have recovered (at least in part) after the landowner enclosed the cut stumps, but the pine was clearly badly damaged. The Irish were traditionally blamed thereafter for the deforestation of Glenorchy, but the Earl of Breadalbane himself can hardly escape censure as his agents had failed to supervise the felling or to mark the trees to be cut as stipulated in the contract between the estate and the company.

In any case, the pinewoods of Glenorchy did not completely vanish at this point, as in 1809 'the fir woods of Cronich, Glenfua, Corryriggar and Corrycherry in Glenorchy, containing about 50,000 feet of old natural trees' were advertised for sale.[34] Fragments still remain, especially near Loch Tulla. Between them, the eighteenth- and early nineteenth-century episodes of exploitation caused very extensive and irretrievable damage to the pinewoods of the area. The Irish partnership in the 1730s moved north to cut timber and take tanbark from the woods on the north side of Loch Maree, but it is unclear if they did any long-term damage there, or whether they felled pine.

In the course of the eighteenth century there were other operations down the west coast by outsiders, many of which will have left little trace. One that has been studied is the exploitation of Cameron of Lochiel's pinewoods at Loch Arkaig which had already attracted the navy at the start of the seventeenth century and now again attracted the English.[35] In 1739, a company of Whitehaven merchants had 'large dealings' with Lochiel, and extracted what one source described as 'the best part' of the woods, but then quarrelled over the payment of some £1,300 and abandoned operations. There is a strong suggestion that the timber, either in quantity or in quality, failed to live up to their expectations. After the 1745 Rising and the consequent forfeiture of the estate, severe depredations against the wood were carried out by insiders, not outsiders: the Commissioners for the Annexed Estates effectively lost control of local pilfering, and discovered

---

[34] M. L. Anderson, *A History of Scottish Forestry* (London, 1967), 2, p. 62.
[35] For a fuller account see T. C. Smout, 'Cutting into the pine: Loch Arkaig and Rothiemurchus in the eighteenth century', in T. C. Smout (ed.), *Scottish Woodland History* (Edinburgh, 1997), pp. 115–25, which gives full details of sources.

after enquiry that not only the local tenants and tacksmen had been helping themselves, but also the minister, the customs officer, several of the military establishment from Fort William, and even the Sheriff Depute of Inverness himself. It was not until after 1755 that the Commissioners assumed effective control, and only in 1764 were they able again to interest outside capital in the form of a Glasgow partnership including Provost George Murdoch. But they, like the Whitehaven men, found the bargain not worth persisting in because the wood was difficult to extract and the size of the timber too small. When they made it over to a local partnership headed by Ewen Cameron of Glenevis, the latter also failed to find markets either locally, on the Clyde, at Campbeltown or in Ireland, and also complained of the size of the timber. Nothing seems to have been cut again in the eighteenth century after 1772, and there is no indication that the total number of trees felled for distant sale exceeded 12,000 in the whole century – a trivial number compared to the estimated size of the wood (162,200 trees) at mid-century.

The woods of Loch Arkaig are, of course, still substantial and magnificent. Less happy was the fate of the extensive pinewoods on the south shore of Loch Hourn in Barrisdale and Knoydart. Barrisdale estate also fell into the hands of the Commissioners for the Annexed Estates in the eighteenth century. Despite their attempts in 1774 to enclose the wood between Skiary and the main body of Loch Hourn, and to compensate the tenant for his loss of grazing, nothing remains now except for a few scattered pines. How or why this wood disappeared is not known, though in nearby Glen Barrisdale some pinewoods do survive. Local tradition relates that fellings and a sawmill operated in this area during the Napoleonic Wars and possibly the Loch Hourn woods went then, or their remnants may have been the victim of nineteenth-century sheep farming.[36]

In the same general area south of Loch Hourn but in Knoydart, Thomas Pennant in 1772 reported seeing:

> Vast numbers of pines scattered among the other trees and multitudes of young ones springing up. A conflagration had many years ago destroyed a fine forest: a loss which in a little time, it is to be hoped, will be repaired.[37]

Fire does not normally extirpate a pinewood, but often encourages its regeneration. Nothing of the forest that Pennant saw now remains, apart

---

[36] Steven and Carlisle, *Native Pinewoods*, p. 161.
[37] T. Pennant, *A Tour in Scotland and Voyage to the Hebrides, 1772* (London, 1776), 1, p. 396.

from the local place-names Inverguseran and Gleann na Guiserein in very bare territory, but how it disappeared is also not clear.

Better documented is the history of the pinewoods of Rhidorroch in the hinterland of Ullapool, the northernmost on the west coast of Scotland, which have also declined since the eighteenth century, though not to extinction. Gordon of Straloch provided a map from the 'many imperfect papers' of Timothy Pont showing extensive woodland on both banks of Loch Achall and extending some way up the strath to east and west, with a note that described the 'said river pleasant in woods and plentifull in salmond'. In 1676 the Earl of Cromartie appointed the local tacksman 'to look carefully to the woods of Ridorach', as in the role of forester, but no attempt seems to have been made to interest outsiders until, in 1719, the second Earl sold 'the fir woods of Coigach to John Innes, an Edinburgh lawyer to whom he was in debt, from which it passed to William Innes his son.[38]

The relationship turned acrimonious when the Earl complained that his neighbour, General Ross of Balnagown was getting better prices than he, to which Innes replied that had the trees of Rhidorroch been as well cared for as the General's, and as 'narrow' he would have got better prices: 'but they have not, for they have been made havok of at pleasure, not only by all your own people in this country but also by the neighbourhood without much control'. He went on to explain that none of the trees large enough for naval masts, which would have earned a 20 shilling bounty paid by the government after the Union, could be transported because of the 'linn', a torrential narrowing of the river below Loch Achall, obstructed by a rock that the merchants had wanted blown up, but which proved too much for gunpowder technology. The trees therefore had to be cut into logs twelve to fourteen foot in length, and even then could only be moved with great difficulty. He further complained that there were not as many available for cutting as Cromartie had said.[39]

Further discussion was pre-empted when the Earl's oldest son, George, Lord Tarbat, discovered that his father had been selling the trees illegally: the estate was entailed and the first Earl, his grandfather, had left the trees to him, and not to his father, the second Earl. A complicated lawsuit followed, ending with Lord Tarbat conveying the woods again to William Innis, who subcontracted their working to Norman Macleod, the chamberlain of

---

[38] M. Clough, 'Early fishery and forestry development on the Cromartie estate of Coigach, 1660–1746', in J. Baldwin (ed.), *Peoples and Settlement in North-west Ross* (Edinburgh, 1994), pp. 235–7; NAS: RD 14, reg. 16 June 1724.
[39] Clough, 'Early fishery and forestry', pp. 237–9.

Coigach, and John Brander, a 'squarewright' of Fochabers in Morayshire, who were to fell them, take them down to sawmills constructed on the river, and manufacture them into building timber. The contract envisaged 5,000 logs (equal to about 2,500 trees) to be cut yearly for four years, which was some 10,000 trees: all trees under six inches in diameter were to be left, which should have ensured sustainability.[40] As, judging from the contemporary deeds and maps, the woodland probably stretched for some ten kilometres and was over a kilometre wide, it must have largely consisted of very sparse, open pinewood, mixed with other trees: birch is mentioned, and several local place-names suggest hazel. William Innes's description of the General's pines as 'narrow', in contrast to the Earl's, implies the spreading, 'granny pine' character of the trees at Rhidorroch. Perhaps the wood was another 'mechtie parck of nature', such as Timothy Pont described the environs of Loch na Sealga twenty kilometres to the south, which was yet another west-coast pinewood that disappeared for no known reason.[41]

The success and extent of Innes's exploitation of Rhidorroch was unclear, but enough of the woods were left for the Royal Navy to take the trouble to burn them as part of the campaign of retribution after the Jacobite Rising in 1745, in which the Earl had taken part. This, however, would have done no more permanent damage than the fire that Pennant commented upon in Knoydart. A report ten years later spoke of the pine and birch woods 'greatly destroyed after the rebellion', but also of 'young wood' and the need to assess it for sale. According to the historians of the Cromartie estate, timber sales from the area 'continued through the eighteenth century', which confirms recovery.[42] The decline to their present condition, covering less than half the area suggested by eighteenth-century documents, was perhaps due mainly to neglect, muirburn and overgrazing when large-scale sheep farming was introduced in the early nineteenth century.[43]

Probably a good deal of wood was extracted from west-coast sites for markets on the Clyde and elsewhere without the actual enterprise coming from outside, especially when timber prices were high in the Napoleonic Wars. In 1806, General Campbell of Monzie sold 1,200 trees from Coinletter

---

[40] Clough, 'Early fishery and forestry', pp. 239–42; NAS: RD 14, reg. 23 June 1726; RD 13, reg. 5 Sept. 1728; RD 14, reg. 29 Aug. 1729.

[41] We have received great help in understanding Rhidorroch, especially in locating places and the meaning of placenames in Rhidorroch, and in estimating the extent of the wood, from Mr Bill Gilmore of Ullapool. We also received valuable bibliographical help from Alex Eaton, Curator of the Ullapool Museum and Visitor Centre.

[42] Clough, 'Early fishery and forestry', p. 242; J. Dunlop, *The British Fisheries Society, 1786–1893* (Edinburgh, 1978), p. 37; E. Richards and M. Clough, *Cromartie: Highland Life 1650–1914* (Aberdeen, 1989), p. 42.

[43] Richards and Clough, *Cromartie*, p. 144.

in Glen Etive; this wood no longer exists. Similarly in 1808 about 5,000 pine were sold from the Glenfalloch area, where only about 200 'thinly scattered trees' remained in 1967.[44]

To turn from the west coast to the north-east, the story of the woods on the Balnagown estate is related at length in Chapter 12 below. To summarise, the woods of Strathcarron were the focus of interest from the Royal Navy in the second half of the seventeenth century, but little sustained exploitation resulted, and it was the lairds themselves who were to the fore thereafter, selling pine for the reconstruction of Holyrood after the Restoration, and in the eighteenth century exploring opportunities to make tar and to sell wood to merchants from the Firth of Forth and Dublin. The trees in Strathcarron were of a quality that made them suitable (at least in theory) for masts and shipbuilding timber.

In 1723 the woods were being worked directly for General Charles Ross by Joseph Avery, an Englishman with interests in military surveying and in forests: he declared himself frustrated by the character of the workforce, 'a stiff-necked and turbulent people', and he soon moved on. Thereafter little is heard until the 1790s, when the woods of the parish of Kincardine, that included Strathcarron, were described as 'not one half, either in respect of size or extent of ground, to what some living remember'.[45] There could, however, have been some recovery later.

Strathglass has a different history, for here woods that had been little touched by outsiders in the seventeenth century aroused interest in the eighteenth (Fig. 8.1). At first, it was low key. In 1704, Thomas Fraser of Struy sold forty 'great ridd fir loggs', twenty-four to twenty-six feet long and fourteen to eighteen inches 'of squair measur in breadth' (big trees) to be delivered at Beauly to a partnership of six, five from Inverness (including Thomas Jaffray, master of the grammar school) and Robert Bell from Burntisland.[46] Perhaps the timber was also to be used for shipbuilding there. A new chapter opened, however, a quarter of a century later when the London-based land speculators, the York Buildings Company became involved with the Chisholms in Glen Affric and Glen Cannich. Between 1725 and 1730, Joseph Avery undertook on behalf of the company a survey of the Moray Firth and its hinterland, in which he showed in detail these richly wooded glens.[47] In 1730, the company promised to pay £2,000 for the

---

[44] Anderson, *History of Scottish Forestry*, 2, p. 62.
[45] W. Macgill, *Old Ross-shire and Scotland as seen in the Tain and Balnagown Documents* (Inverness, 1909), 1, p. 284; *OSA*, 3, p. 508.
[46] NAS: RD 12, 1705, no. 1080.
[47] Inverness Museum: Joseph Avery, 'Plan of the Murray Firth'.

Fig. 8.1 Glen Affric, photographed by R. M. Adam in 1929 before the modern hydro-electric scheme. The woods are similar in extent to when they were first mapped by Joseph Avery, around 1725. St Andrews University Library.

Chisholm woods in four instalments but, characteristically, only paid the first £500: they were said in 1735 to have cut 30,000 great trees and left them to lie on the ground.[48] They may well have found it impossible to get the timber out, just as Innes did at Rhidorroch. In 1742, a partnership of English merchants from Liverpool and Preston agreed an even more ambitious bargain, subject to proof that the timber could be moved. They agreed to pay Chisholm £2,500 for 50,000 trees, and to pay Simon, Lord Lovat the same sum over ten years for way-leave to float the wood down the river through Lovat lands, with power to build floodgates and dams and blow up rocks. A trial was made with 300 trees that ended up 'torn, broken, scattered and rendered useless', Chisholm arguing apparently in vain that the attempt at floating had been incompetent and that the English merchants should not be allowed to wriggle out of their bargain.[49] It had been set up in the

---

[48] J. Munro (ed.), *The Inventory of Chisholm Writs 1456–1810* (Scottish Record Society, Edinburgh, 1992), pp. 124–6.

[49] Munro, *Chisholm Writs*, pp. 130, 132–3; NAS: TD 86/97, Box 3, Bundle 41.

first place by John Lummis, a shady character from Shropshire who operated as a timber agent and go-between for other estates around 1740 – certainly for Rothiemurchus and probably for Loch Arkaig. Lummis had told the company that the Chisholm glens contained 100,000 trees measuring four feet to twelve feet in circumference and forty to eighty feet in length, but to judge by what happened in similar circumstances when he intervened (see below, pp. 295–6) this was probably a wild exaggeration.[50]

The landowner in 1742 placed no restriction on the size of the trees to be cut, and there is clear evidence that before the end of the century that the economic quality of the woods was deteriorating. After their unfortunate experiences with English companies, in 1765 the Chisholms turned to an Inverness merchant, Hugh Falconar, and sold him 10,000 trees over ten years for £1,250 sterling. When this was renewed for a further five years in 1775, a new contract provided for 1,300 trees a year in place of 1,000, Alexander Chisholm writing 'I am satisfied that the trees are now smaller in size than what you cutt down for those ten years past': at this level, the contract was renewed again until at least 1790.[51] At the end of the eighteenth century, logs were still floated down the Rivers Cannich and Glass (Fig. 8.2), and there was a sawmill above the Kilmorack Falls with four wheels and seven saws.[52] Yet all this activity resulted in little or no permanent reduction in the area under pine. If one compares Joseph Avery's map of 1725–30 with Steven and Carlisle's map of approximately 1950, the native woods of Glen Affric and Glen Cannich seem still to be in the same place and of approximately the same extent.[53] These eastern woods often had a power of regeneration and recovery that those on the west apparently had not.

Two woods higher up the Great Glen, at Glen Loyne and on the former Macdonell estates of Glengarry, are intermediate between east and west, though on roughly the same longitude as the forest in Glen Affric. Glen Loyne has the distinction of holding the three oldest pine trees known in Scotland (500–550 years old), but the wood has almost disappeared, only eighty-five living trees remaining in 1994 in an area that both Avery's map of 1725–30 and visitors' accounts of 1769 and 1777 describe as wooded with natural pine.[54] Glengarry's native pinewoods immediately to the south

---

[50] NAS: RD 12, reg. 21 Apr. 1742.
[51] Munro, *Chisholm Writs*, pp. 143–4, 1014, 1023.
[52] Steven and Carlisle, *Native Pinewoods*, pp. 177–8.
[53] Inverness Museum: Avery, 'Plan of the Murray Firth'; Steven and Carlisle, *Native Pine woods*, pp. 179, 182.
[54] A. Bartholomew, D. C. Malcolm and C. J. Nixon, 'The Scots pine population at Glen Loyne, Inverness-shire: present conditions and regenerative capacity', *Scottish Forestry*, 55 (2001), pp. 141–8; Steven and Carlisle, *Native Pinewoods*, p. 155.

Fig. 8.2 The problems of floating wood down Strathglass: 'When the river is most swelled with rains, they take advantage of the encreased torrent, to float down the timber cut in the woods, on its way to the sawmill. On these occasions, vast logs of wood, hurled against the rocks, and dashing on each other as they float along the impetuous waves, increase the horrid roar of the waters.' Charles Cordiner, *Remarkable Ruins and Romantic Prospects of North Britain*, 1795. St Andrews University Library.

survived well until the twentieth century, when they were seriously damaged by a combination of heavy felling and replanting with exotic species, by fire, and by excessive grazing by sheep and deer. They had, however, evidently no problem in the nineteenth century with regeneration, since a survey of the area in 1908 described a class of trees 15–20,000 strong aged 80–100 years, and a further class still younger.[55] The difference between the two has to do with soils and altitude, Glen Loyne being 'located in an extreme environment for the species in Scotland', with heavy rainfall and poor minerals, while Glengarry is relatively sheltered and mineral rich. Both were exposed to excessive grazing by sheep and deer, and even Glen Loyne shows some slight signs of recovery and regeneration in recently

[55] Steven and Carlisle, *Native Pinewoods*, p. 157.

fenced areas.[56] In the great antiquity of its trees and its gradual collapse, it seems to have much in common with the vanished woods of Little Loch Broom or Loch na Sealga in Wester Ross (see pp. 53, 58–9 above). Whatever the cause of its near demise, heavy exploitation by outsiders is not involved.

The most dramatic examples of English external interest in the Scottish pinewoods occurred not in areas where the pinewoods have since declined, but in areas where they still flourish, pre-eminently in Speyside. Of the three great forests of Abernethy, Rothiemurchus and Glenmore, it was Abernethy that first attracted major attention. As early as 1704, the Master Mast-maker at the Royal Dockyard at Deptford had identified Abernethy as the Scottish forest 'likeliest to serve her Majesty and Government' and the lairds of Grant began casting around for English purchasers soon after the Union, attempting to sell masts from Abernethy in London in 1714 (they failed to meet stipulations regarding quality), and signing contracts in 1721 with John Smith, shipwright and William Francis, both Quakers from Co. Durham, to sell 60,000 'sound deals', with uncertain outcome.[57] All this was far eclipsed, however, by the spectacular adventure of the York Buildings Company. Initially a London-based water-supply company, it first became heavily involved in Scottish landholding when it bought up, as a speculation, a series of estates forfeited from Jacobite rebels after 1715. Already over-extended, and in a precarious financial position, it fell under the spell of Aaron Hill, poet and erstwhile manager of Drury Lane Theatre, who persuaded the governor, Colonel Horsey, that Abernethy forest, despite previous failures, could indeed provide a rich store of masts for the Navy. The company in 1728 accordingly purchased from Sir James Grant of Grant 60,000 pines for £7,000 sterling, to be felled over fifteen years.[58]

Too late, however, it was discovered from the Master Mast-maker that, although the wood was of excellent quality, there was not a tree suitable for a main mast in any ship of His Majesty's service. Undeterred, the company pressed on, embarking on a programme of heavy investment in sawmilling and in the removal of obstructions on the Spey by explosives in an attempt to reduce the high costs of extraction. Within four years they had felled 20,000 trees but made a net loss of £28,000 sterling, and Colonel Horsey, on a visit to Scotland, found himself arrested for debt on the initiative of the laird, who was owed £1,000 for an unpaid instalment of the contract

---

[56] Bartholomew *et al.* 'Scots pine population', pp. 145, 147.
[57] Grant, *Abernethy Forest*, pp. 9–10; J. Skelton, *Speybuilt: the Story of a Forgotten Industry* (Garmouth, 2nd edn, 1995), p. 18.
[58] D. Murray, *The York Buildings Company: a Chapter in Scotch History* (edn Edinburgh, 1973), pp. 55–7.

money. 'But in order to show that there was no ill feeling on account of his little *contretemps*' as the nineteenth-century historian of the company put it, the Governor entered into another contract with Sir James Grant to supply charring wood for ironworks. It is not clear that pine would have been felled for this purpose, rather than birch, but apparently 'great quantities of wood' were cut and 'large shipments' of charcoal allegedly made to England and Holland. But the venture was short-lived. By Christmas 1732 the debts of the ironworks were £7,000, and in 1734 and 1735 the laird's factor raised actions against the company and seized, for debt, twenty sawmills, a smith's bellows and anvil, and fifty-two tons of locally made pig iron. That put an end to their activities.[59]

What damage did the York Buildings Company actually do? They felled 20,000 pines for which they paid 2s. 4d. sterling each: according to one expert in 1733, there were another 10,000 of that quality left when they stopped operations, but the remaining 30,000 specified in the contract were not worth 6d. a tree.[60] This can be taken as indicating that they felled two-thirds of what proved, after all, to be not a very large resource of fairly big trees. As for the charcoaling, the brevity of operations limits the damage that might have been done, and because pine is reckoned to make poor charcoal, the operators may well have cut out the birch within what was probably a much more mixed forest than the present one. Besides, woods regrow.

Fatuous though its business sense was, the York Buildings Company achieved two positive things for Speyside. The first was to alert the outside world to its potential, though whether that was real or imaginary had still to be properly tested. Up to that time, commented William Lorimer in a report to the laird thirty years later, 'your woods had never been touch'd almost from the flood'. The second was to introduce the art of floating the wood in rafts, done before by a man who attended the loose logs in the river from currachs, 'wicker baskets covered with hides', as Lorimer described them.[61]

With the failure of the company, much of the subsequent exploitation of Abernethy, like that of Rothiemurchus, was again directed to local markets, or at least to markets within the Highlands that could be reached by overland transport to Inverness or river transport to Garmouth on the Moray Firth. But from time to time the Grants of Grant tried again to reach

[59] *Ibid.*, pp. 63–5.
[60] *Ibid.*, p. 58, footnote.
[61] G. A. Dixon, 'William Lorimer on forestry in the Central Highlands in the early 1760s', *Scottish Forestry*, 29 (1975), p. 209.

more distant markets. In 1743, they signed a contract with Alexander Grant and George Steevens, merchants of London, to sell 100,000 pines at three shillings each, over a period of twenty years, the partners having the right to withdraw at the end of each year if they were losing money.[62] The contract cannot have lasted, for in 1746 Sir Ludovic Grant was providing wood for Ralph Carr of Newcastle, and for three years (until the War of Jenkins's Ear came to an end) he struggled to sell spars and deals in the North of England. Problems of trying to compete with Norway, however, proved insuperable in peace time. There were difficulties with shipping. Garmouth was a small harbour with which skippers were unfamiliar, and the freight-charge on the wood was high because it was hard to find a suitable return cargo from Newcastle – attempts to sell coal, glass and tallow locally were not very successful. There were even worse problems over quality. Ralph Carr wrote:

> I am surprised William Grant should have shipped such miserable [boards] as these last by Abernethy, some hundreds of these are only six and seven inches broad – they will not pass here for deals under nine inches and to sell these for the price of battens is to sustain a loss.[63]

Grant was told at the time that 'the most convenient part of your woods is cut already', and a sense of the loss of pristine quality was repeated by William Lorimer in 1763 when he told the laird that those felled by the York Buildings Company and for manufacturing for Newcastle, as well as many cut for local use, 'have not been succeeded by an equal number of new ones in other parts of your woods'.[64] It is possible, though, that small trees (as at Rothiemurchus) were preferred by local customers who had neither the means to transport nor any use for very large trees.

James Grant of Grant looked again for external buyers in 1769, having failed the previous year to sell 20,000 trees for cutting and manufacture to a ship-master and a wright in Banff, who could not find guarantors for their part of the bargain. This time he sold 100,000 'of the choice trees of the fir-woods of Abernethy and Dulnan' over a period of fifteen years, for nearly £8,000 to be paid at a rate of 19d. per tree, to a London partnership of Alexander and James Cumming, William Allan and Alexander Grant. The Scottish and clan associations are obvious. Alexander Cumming, the

---

[62] NAS: GD 248/23/4. no. 46.
[63] A. Thomson, 'The Scottish timber trade, 1680 to 1800', unpublished St Andrews University Ph.D. thesis (1990), pp. 144–6.
[64] Dixon, 'Lorimer on forestry', p. 209.

principal partner, was the son of the inn-keeper of Aviemore, and had made a substantial reputation as a watch-maker in London – he had become clock-maker to George III and was the author of a standard work on the subject.[65] The main intention was to operate a boring mill to supply water pipes for the London market, a project first mooted in 1766. Less than three years after signing, the partnership withdrew from the bargain on a penalty of £200 because:

> A prejudice has taken place against Scots firr-pipes that may not be easily removed, and that all the companys here have now got into the way of boring their own pipes, and the prices which they allow for timber comes far short of the expence of bringing them to market.[66]

Once again, extraction and transport costs foiled the interest of the outsider in the Scottish woods. But optimism did not falter. In 1773 Abernethy was again sold to 'certain Englishmen' for nineteen years, paying £1,000 a year for the right to cut 12,000 trees annually. At 1s. 7d. a tree, this low price was connected with the small size of the timber. Only one tree was noted as large, thirteen feet in girth and fifty feet in height.[67]

In the nineteenth century, the estate took over all aspects of management, becoming heavily involved in planting as well as in conserving and exploiting the native wood. It is unclear just how much was sold to distant markets in the first quarter of the century, though when the railway came to Strathspey in 1863 it facilitated access to the south. Active forestry was, however, abandoned in 1878. Abernethy had become a deer forest following the clearing, in the previous decade, of at least 104 people from twenty crofts: five employees sufficed to serve the sporting interest.[68]

Studies of the extent of Abernethy Forest, based on maps of ca. 1750 (Roy), of 1812 (George Brown) and of 1830 (William Johnson) show a progressive and substantial contraction of woodland over the period, though subsequent maps show how it revived thereafter, to cover, in 1955, an area of 'surprising similarity' in outline to what it had been two centuries before, albeit truncated in the upper forest and with some of the replenished wood being plantation (Map 8.2). On the other hand, what map-makers meant by

---

[65] NAS: RD 12, reg. 27 June 1769; NAS: GD 248/377; G. A. Dixon, 'Forestry in Strathspey in the 1760s', *Scottish Forestry*, 30 (1976), p. 56.
[66] Dixon, 'Forestry in Strathspey', p. 56.
[67] Grant, *Abernethy Forest*, p. 16.
[68] *Ibid.*, pp. 16–19. The author's claims for substantial shipments of wood from Abernethy via Speymouth in the first part of the century are not substantiated by the sources cited, yet are likely to be true.

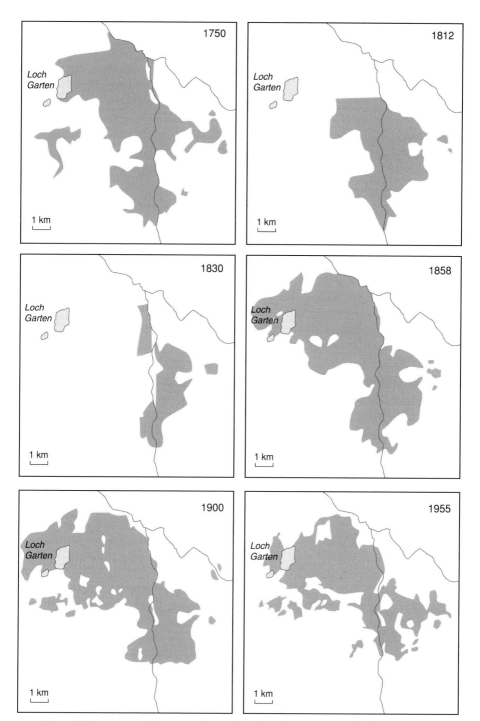

Map 8.2 Outlines of Abernethy forest, showing changes in woodland cover since 1750. From R. Summers (1999), p. 10 with permission from Royal Society for the Protection of Birds, original sources O'Sullivan (1973) and Ordnance Survey.

their symbols is often ambiguous, and some areas shown as treeless in Abernethy in the nineteenth century still contain pines over two hundred years old, so the land on which they stood was probably lightly wooded or regenerating at the time of Brown's and Johnson's maps, rather than devoid of cover. What the maps do show, however, is that exploitation in the century after 1750 was heavy and, indeed, that only a very small proportion of Abernethy is shown as continuously wooded over the last two hundred years.[69]

Yet, contrary to common assumptions, there is no indication that the main cause even of this temporary decline was the activities of the outsider. On the contrary, external enterprise was usually distinguished by its inefficient and short-lived character, though it probably brought about the destruction of the largest trees. The activities of local people probably had a much larger overall effect. More difficult to assess is the relative importance of local and external demand, but the inroads into the extent of the wood in the early nineteenth century are very likely to have been encouraged by the high price for timber on the national market, particularly when supplies from the Baltic and Scandinavia were severed by war or hampered by tariff policy. This was certainly the impetus for heavy exploitation elsewhere in Speyside and Deeside.

It is, however, important to emphasise that as a Caledonian pine ecosystem Abernethy is still there, and of a similar size to what it was in the earlier eighteenth century. Felling in a pinewood mimics the effects of fire or windthrow, and if the ground beneath is not converted to arable or its regeneration grazed away by beasts, the forest should return of its own accord. Only if something had happened to the ground cover to prevent regeneration – by becoming wetter and more mossy or peaty due to climate change – would this not occur. On the dry, well-draining glacial soils of Speyside a geomorphological change was less likely than in the west. Abernethy was not destroyed in the eighteenth and nineteenth centuries.

Of other forests of the area, Rothiemurchus is the subject of Chapter 11 below, so only a summary of its history is necessary here. In many ways it closely resembled that of Abernethy. There were early eighteenth-century expressions of interest from short-lived partnerships involving men from Banff, Edinburgh and London, but they did not survive the 1715 Rising. In 1738, John Lummis (whom we have met before in Strathglass), acted as go-between for the laird and a Hull merchant partnership who were persuaded to take a contract to fell 40,000 trees, half of them at least ten to

---

[69] P. E. O'Sullivan, 'Land-use change in the Forest of Abernethy, Inverness-shire (1750–1900 AD)', *Scottish Geographical Magazine*, 89 (1973), pp. 95–106; R. Summers, *The History and Ecology of Abernethy Forest, Strathspey* (RSPB Internal Report, Inverness, 1999), pp. 10–11.

twelve inches square: they resigned the contract within two years, saying they had not been able to find 500 trees of that dimension in the whole wood – a familiar tale by now on Speyside. In 1771 the laird's brother William, medical practitioner in London, set up a boring mill to manufacture water pipes for the capital. After manufacturing 4,000 pipes, the enterprise collapsed in 1774 due to competition from English elm and, again, to the small size of the trees: the parallel with Abernethy is close again.

These were the only interventions in the eighteenth century occasioned by outsiders or demand from distant markets, and they can have had no material impact. On the other hand there was continuous exploitation of the Rothiemurchus woods by local merchants who sent the wood down the river or away by horseback to serve markets within fifty miles at most. For their needs, it was an advantage that the woods were kept full of smaller timber.

In the early part of the nineteenth century, a combination of the penury of Sir John Peter Grant of Rothiemurchus and a favourable price for pine resulted in something akin to the clear-felling of the forest by 1834, an episode repeated again before 1869 by his successor. This superficially devastating exploitation, however, was carried out on the initiative of the lairds themselves. The markets appear to have stretched from Inverness to Crieff (the latter after the coming of the railway), but with most of the produce being rafted downriver to Garmouth where, as we shall see, there was definitely outsider involvement in the marketing. Following each of these episodes of heavy exploitation, there was ready regeneration throughout the forest, so that modern Rothiemurchus, like Abernethy, covers much the same area as it did in the eighteenth century. The strict exclusion of sheep from the forest from as early as 1824, and the relatively modest number of red deer in it until the end of the century, undoubtedly helped to ensure its regeneration and ecological continuity.

The third great forest of Speyside, the Duke of Gordon's Glenmore, was less directly accessible to the Spey than either Abernethy or Rothiemurchus, which may have enabled it to keep its larger trees for longer. There was some interest in the early eighteenth century when in 1715 John Baxter, timber merchant from Durham and, after his death three years later, John Smith, shipwright from Stockton and William Francis of Oumby in Co. Durham, yeoman, signed contracts to extract wood. Baxter was to take 3,300 trees a year for seven years, and Smith and Francis (who were also busy at Abernethy) were to cut as many trees as would make 20,000 deals a year for three years, ten feet to fourteen feet long.[70] But the contract, if

---

[70] NAS: GD 44/29/6/24; GD 44/29/6/2/10.

fulfilled, does not seem to have been renewed, and apart from a fruitless offer for 5,000 trees by a Francis Grant and Company in 1755, nothing further is heard of outside interest in the Duke's forest for sixty years. In 1782, however, the wood was advertised for sale as containing 'a hundred thousand trees full grown and fit for the Royal Navy', and subsequently sold to Ralph Dodsworth, former mayor of York, and William Osborne of Kingston upon Hull.[71]

Glenmore was described in the *Statistical Account* of 1791 as 'the oldest, the largest and the best quality firwood in Scotland', which was probably true at that time. What followed was the most significant direct intervention by outsiders in the history of the natural pinewoods, at least before the twentieth century. Dodsworth and Osborne bought the whole wood for £10,000 payable in instalments within fifteen years. They had twenty-six years in which to extract the wood, and were restricted to cutting trees over eighteen inches girth at five feet from the ground – clearly larger than those in contemporary Rothiemurchus. They had the right to assemble timber on Loch Morlich and to erect dams and sluices that would take it down the subsidiary burns and rivers into the Spey.[72] They erected on the Duke's land at Garmouth (alias Speymouth) a timber-holding and shipbuilding facility that they called Kingston Port. Operations commenced in 1784, and the first vessel, a brig of 110 tons named the *Glenmore*, was launched before the end of 1785. By 1791, some nineteen vessels, some over 500 tons, had already been built at Kingston, and in the space of twenty-two years (1784–1806) the company built and launched forty-seven sail of ships (the largest of 1,050 tons), totalling 19,000 tons, as well as exporting quantities of timber. The *Statistical Account* reported that in one year ending October 1782, eighty-two boats had sailed from the Spey laden with wood.[73]

The success of Dodsworth and Osborne, in stark contrast to all their fellow countrymen who had gone before, must have been at least in part due to their identification of the good quality of Glenmore trees for their purposes. In the spring of 1783 Dodsworth wrote to Osborne, after visiting the forest, that though there were not 100,000 trees of first quality, 'we observed many coarse crooked trees standing in the forest which we understood were of little or no value, but as we intend to build ships they will be useful to us'. At the conclusion of the lease, having effectively cut out all the

---

[71] Skelton, *Speybuilt*, p. 22.
[72] *Ibid.*, p. 23.
[73] *Ibid.*, pp. 7–8, 28–32. The largest ship was the *Lady Madolina Sinclair* of 1797, not the *Glenmore*, as stated by J. Atterson and I. Ross, 'Man and woodlands', in C. Gimingham (ed.), *The Ecology, Land Use and Conservation of the Cairngorms* (Chichester, 2002), p. 122.

Fig. 8.3 The commemorative plaque on the cross-section of a 260-year-old pine, the largest cut in Glenmore forest by Osborne and Dodsworth, and presented to the Duke of Gordon on completion of their operations in 1805. Courtesy of Professor Alexander Mather and the University of Aberdeen.

suitable wood two years before its end, they presented to the Duke of Gordon a plank about six feet high and four feet wide, cut from a great tree calculated to be nineteen feet in girth and 260 years old, named locally as the Maighdean Coire-chun-claich, or Maiden of Glenmore (Fig. 8.3).[74] Some of the local sense of outrage at their brash and aggressive enterprise is conveyed in the words of a local bard: 'our ears are stunned by the crash of falling trees and the clamours of the Sassanachs'.[75] Around 1830 Thomas Dick Lauder visited Glenmore and was impressed by the skeletons of huge trees, twenty feet and more in girth, that had been too decayed to justify felling. He was also struck by the vigorous regeneration which bore witness

---

[74] *Ibid.*, p. 24.
[75] Steven and Carlisle, *Native Pinewoods*, pp. 126–7.

to the fact that Glenmore, like Abernethy and Rothiemurchus, had not been erased by its heavy exploitation. It later became a sheep run, and then a deer forest, but in the absence of any nineteenth-century felling the regenerated trees survived to become the Glenmore of the twentieth century.[76] It is still a magnificent forest, though it had to endure felling in the world wars and an episode of heavy planting of Norway spruce and Sitka spruce fifty years ago: most of those non-native trees have now been at least temporarily cleared. There are no traces now, however, of the very old and large trees that once characterised this forest, and its invertebrate fauna resembles that of Rothiemurchus and Abernethy.

The success of Dodsworth and Osborne bred imitation, and other shipbuilders began to operate at Garmouth from as early as 1787 and continued for more than a century, eventually constructing over 500 ships and over 80,000 tons of shipping, though as the nineteenth century proceeded, less Speyside wood was used and more material was imported from the Baltic or North America. According to M. L. Anderson (who gives no sources but was usually very well informed) Dodsworth and Osbourne and no doubt some of their successors, also sent timber to Hull and to the Royal Dockyard at Deptford, as well as along the north coast from Aberdeen to the Isle of Skye, 'the wood being of the very best quality, highly prized and highly priced'.[77] Skelton also speaks of a number of ships built in other ports 'including Peterhead, Hastings and Shoreham', where they were 'described as being built of "Spey fir" or "Spey timber"'.[78] For Rothiemurchus, and probably for Abernethy, the trading and shipbuilding outlets at the mouth of the Spey must have formed an important market in the phase of exploitation that lasted from early in the nineteenth century to 1830. Its influence spread at least as far as Glen Feshie, where, for example, in 1819 Sir Aeneas Mackintosh of Mackintosh sold 10,000 trees to Alexander Duncan, wood merchant of Garmouth. Many of these seem to have been rather small – 4,000 were 'spar trees', twenty-four feet long but only six or seven inches in diameter.[79]

It is interesting to contrast the history of Speyside with that of Deeside. To the former, external interests were repeatedly attracted, albeit generally to their cost rather than to their profit. In the latter, the outsider was seldom seen. Deeside ran an increasingly important timber trade downriver to Aberdeen and also a less dynamic, traditional trade over the hills to

---

[76] Ibid., p. 127.
[77] Anderson, *History of Scottish Forestry*, 1, p. 478.
[78] Skelton, *Speybuilt*, p. 29.
[79] Ibid., p. 26. The interest of the owners of Glen Feshie in selling wood from Garmouth appears to go back to the middle of the eighteenth century: Skelton, *Speybuilt*, p. 18.

Angus and Moray, and the effects on the forests by the middle of the nineteenth century were as dramatic as those on Speyside, but the enterprise and initiatives were driven by local landowners and mainly involved local people. Only one exception has been found, when in 1762 Sir Thomas Burnett of Leyes sold to Samuel Bewley, merchant in Newcastle upon Tyne, the pinewood, 'commonly called the Westwood', west of Crathes Castle, paying a sum for each tree extracted and having six and a half years to clear-fell the wood.[80]

The most important names on Deeside were those of the Earl of Fife, owner of the Forest of Mar, Farquharson of Invercauld, owner of Ballochbuie, and the Earl of Aboyne, owner of Glentanar. In each forest, cutting was well organised locally, and the enterprise admired by visitors. Thomas Pennant in 1771, for example, described the pines of Farquharson of Invercauld as a 'magnificent forest . . . of many miles extent'. He measured trees of ten to twelve feet in circumference and sixty feet in height, 'having, as is supposed, seen two centuries', and admired Farquharson's enterprise: 'by sawing them and retailing them, he has got for eight hundred trees five-and-twenty shillings each'. The planks were ten feet long, eleven inches broad and three inches thick.[81] William Lorimer in 1762 contrasted Deeside with Speyside:

> In the knowledge and care of woods the people of Strathspey seem to be as far behind those of Braemar and Glentanner as the Spaniards are behind the rest of Europe in knowledge in general.[82]

Even in the eighteenth century, natural regeneration of the pine forests was being supplemented by extensive planting of Scots pine and larch in the parishes of Aboyne and Birse, and on the Invercauld estate.[83]

If the niche for entrepreneurship was already filled by experienced and established locals, this could well have been a factor in making Deeside less attractive to outsider speculation. Yet the final result was much the same in Deeside and in Speyside, with much heavier exploitation in the first half of the nineteenth century associated with external demand leaving the main forests largely felled, though capable of regeneration.

The increase in the scale of exploitation was brought about by the rising price of wood during the Napoleonic Wars: and by the coincidence of a storm in 1808 that flattened a large number of trees in Finzean and Glentanar,

---

[80] NAS: GD 345/1007.
[81] Quoted in Steven and Carlisle, *Native Pinewoods*, p. 101.
[82] Dixon, 'Lorimer on forestry', p. 192.
[83] Atterson and Ross, 'Man and woodlands', pp. 121–2.

and 'triggered extensive felling programmes on both estates that were to continue for some thirty-five years'.[84] In 1808 there was a sale of '15,000 fir trees of great age and size in the forest of Crathie', and further sales in 1809 from Glentanar and the forest of Mar totalling 24–30,000 trees. At Glentanar, where in the 1760s six or seven axe-men had comprised the Earl of Aboyne's workforce, there was in 1809 a woodcutter's village of fifty huts, 'and during the 25 years occupancy of this hamlet 150 children were born there'. Irvine Ross has argued that one of the basic differences between the eighteenth century and the early nineteenth was that the timber was no longer manufactured locally but floated down to Aberdeen for shipbuilding or onward shipment to (for instance) Leith and Berwick, depriving the local economy of the added value.[85] So intensive was the floating that in 1812 the bridge over the Dee at Potarch was brought down by logs, and the Bridges (Scotland) Act was passed the following year to regulate the traffic.[86] Detailed work has not been done to show how far the Deeside forests occupy the same areas today as in the seventeenth or eighteenth century, though there is evidence of shrinkage at Ballochbuie and disappearance of pines from Creag Ghiubhais and Pannanich, as well as between Braemar and Ballater on the north side of the Dee.[87]

What conclusions can we draw from this study of the pinewoods over two and a half centuries? Firstly, there was a general difference between east and west. In the west, exploitation was heavier than in the east in the seventeenth and earlier eighteenth centuries, and often associated with outsiders. It was less marked than in the east after about 1750 because by then several of the woods had been extirpated or had declined in size. Climatic reasons and a failure to protect the woods from grazing, as well as continued use by the local population, are factors to be considered alongside the intervention of outsiders. In the east, there were larger forests and, ultimately, heavier exploitation, the most successful of which was generally undertaken by the estates themselves, albeit increasingly in response to external market forces. These culminated in the late eighteenth and early nineteenth centuries in something close to clear-felling of all the marketable

---

[84] *Ibid.*, p. 122.
[85] I. Ross, 'A historical appraisal of the silviculture, management and economics of the Deeside forests', in J. R. Aldhous (ed.), *Our Pinewood Heritage* (Farnham, 1995), p. 141. See also Anderson, *History of Scottish Forestry*, 2, pp. 61–4.
[86] Anderson, *History of Scottish Forestry*, 2, pp. 62–4; N. A. Mackenzie, 'The native woodlands: history, decline and present status', in C. Gimingham (ed.), *The Ecology, Land Use and Conservation of the Cairngorms* (Chichester, 2002), p. 110.
[87] Steven and Carlisle, *Native Pinewoods*, pp. 92–3, 95, 101–2; Ross, 'Historical appraisal', pp. 136–44.

trees on all the larger forests. The eastern pinewoods, however, unlike those of the west, generally recovered much of their former extent and status, due to their greater powers of regeneration on dry mineral soils. The woods of the north were intermediate between those of the west and east, Strathcarron's history being more like that of the western woods, and that of Strathglass more like that of the eastern ones.

Outsiders suffered a very high rate of business failure for their rash interventions on the ground, of which the troubles of the York Buildings Company were among the most dramatic. On the other hand, the influence of the external market itself became increasingly apparent, notably in the big fellings on Speyside and Deeside between 1780 and 1830 and the associated opportunity to sell downriver to Garmouth and Aberdeen for shipbuilding, construction and onward sale. After around mid-century though, cheap wood from abroad again limited the sale of Scottish pine. By 1841, it was reported that the cost of taking the timber from Glentanar to Aberdeen was barely met by the price obtained for it in the city.[88] This simultaneously stopped felling in its tracks and removed any rational incentive to preserve the woods as an economic resource, though there were times and places when the market revived for short periods sufficiently to justify another onslaught on the resource.

At the start of the Victorian period, indeed, it seemed to many as if the native pinewoods had been fatally damaged. As P. J. Selby put it in 1842:

> The indigenous forests of Scotland, which formerly occupied so large an extent of its territory, have, within the last sixty years, been greatly reduced, in consequence of the demand for pine timber, occasioned by the difficulty of obtaining wood from the Baltic during the late wars: some, indeed, are nearly obliterated.[89]

In fact, heavy felling did not in any way render regeneration impossible, especially if the smaller trees were deliberately left, as at Glenmore and also at Mar. Without anything being specified in contracts, too, timber merchants would generally leave decayed, twisted or excessively crooked trees that would act both as seed sources and as biodiversity refuges for lichens and invertebrates. Few of the natural pinewoods would have been dense or close-grown like a modern conifer plantation, particularly those that had been traditionally used for wood pasture, or cattle shelter in winter, so the

---

[88] Steven and Carlisle, *Native Pinewoods*, p. 95.
[89] P. J. Selby, *A History of British Forest Trees, Indigenous and Introduced* (London, 1842), quoted in Ross, 'Historical appraisal', p. 142.

extraction process by horse would not have been particularly damaging to regrowth. The felled woods freely regenerated where grazing pressures allowed. Deer had not yet reached their modern numbers on Speyside, though they were certainly already enough to inhibit regeneration in the Forest of Mar and around Braemar.[90] Sheep pressure on Speyside and Deeside was not as great as it was on the north and west. Furthermore, many Victorian landowners in the area remained keen on forestry as a hobby or for potential long-term investment, as at Rothiemurchus or Abernethy, or found the woods a romantic asset to their estates, as Queen Victoria did when she bought the woods of Ballochbuie in 1870 to prevent them being felled by an Aberdeen timber merchant.[91]

What is remarkable, perhaps, is not that so many of our most famous and beautiful woods of the eastern Highlands were levelled by the axe two hundred years ago, but that they have so often reappeared, of similar extent and magnificence to what they were before. That the same cannot unfortunately be said of the west probably has more to do with different climate, soil and grazing pressures than either with different levels of pressure from the external market or superior forestry practices.

---

[90] A. Watson, 'Eighteenth century deer numbers and pine regeneration near Braemar, Scotland', *Biological Conservation*, 25 (1983), pp. 289–305.
[91] Steven and Carlisle, *Native Pinewoods*, p. 102.

CHAPTER 9

Outsiders and the woods II: charcoal and tanbark

The native species of broadleaf trees, as a group, have always been far more numerous and widely distributed than the conifers, despite the attention given to Scots pine as quintessentially the Scottish tree and undisputed queen of the mythical Caledonian forest.[1] However, whereas the attraction of Scots pine to the outside world was as building timber, broadleaf trees were mainly influenced through external markets by the demand for charcoal for smelting iron, and the demand for bark for tanning leather and hides. Oak beams and boards were certainly carried significant distances for building, especially of royal palaces and warships in the sixteenth century, (see pp. 42, 45, 81, 97 above) but after 1600 exploitation of this kind, though not unknown, was normally a by-product of other considerations. Oak was also the preferred species for charcoaling and tanning, but a wide range of other species could be used if supplies of local oak were insufficient or unavailable.

External interests followed local use. Primitive ways of making iron, using mainly bog ore from the upland mosses, were widespread since before Roman times: the iron ore was smelted with charcoal in a bloomery hearth to produce a paste that then had to be hammered in a smith's forge to produce wrought iron.[2] Traces of such bloomeries are widely scattered through the Highlands, with few elsewhere. As long ago as 1886, the Victorian metallurgist and antiquarian W. I. Macadam, identified ninety-four sites where he considered bloomery iron had been made.[3] The National Monument Record of Scotland has pinpointed eighty-one sites as bloomeries, by no means all coincident with Macadam's; but the pattern is 'heavily influenced by the small number of field workers who actually

---

[1] See, for example, H. Miles and B. Jackman, *The Great Wood of Caledon* (Lanark, 1991); R. Mabey, *Flora Britannica* (London, 1996), p. 21.

[2] For an account, see H. Hodges, *Artifacts: an Introduction to Early Materials and Technology* (London, 1964), pp. 80–90.

[3] W. I. Macadam, 'Notes on the ancient iron industry of Scotland', *Proceedings of the Society of Antiquaries of Scotland*, new series 9 (1886–7), pp. 89–131.

recognised them and the picture of the overall distribution pattern may be misleading'.[4]

Table 9.1 Bloomery sites in Scotland

	Macadam, 1886	NMRS, 2002
Highland	33	16
Moray and Nairn	17	2
Banffshire	3	0
Aberdeenshire	2	2
Argyll, Bute and Arran	24	33
Perth and Kinross	12	6
Stirling	1	14
Dunbarton	1	0
Ayrshire	0	1
Dumfries and Galloway	1	7
Totals	94	81

Sources: See text, above.
Note: These lists are indicative but not exhaustive. See also, for example, J. Williams, 'A mediaeval iron smelting site at Millhill, New Abbey', *Transactions of the Dumfriesshire and Galloway Natural History and Archaeological Society*, 44 (1967), pp. 126–32, and J. Wordsworth, 'Report on the investigation of various iron-working mounds in the Ben Wyvis area in November, 1992', report to Scottish Natural Heritage, northern area (Inverness).

There are, however, very substantial problems about dating such sites. Opinions are divided on how common bloomeries were after 1600. Lindsay thought that by then such peasant iron-making was restricted to the far north, but even in Sutherland, where 'in the late sixteenth and early seventeenth centuries there was a fair amount of iron working' (as in the wood of Skail and the woods of Creich), it had 'come to an end long before the end of the seventeenth century'.[5] Other recent work, however, suggests that medieval associations between bloomery iron and elite weapon-making may have continued in places until the decline of the Highland military machine in the eighteenth century.[6] Tittensor also believed that it persisted

---

[4] Sites reported on www.rcahms.gov.uk/canmore. The comment is from Jack Stevenson of RCAHMS, *in litt.*

[5] J. M. Lindsay, 'Charcoal iron smelting and its fuel supply; the example of Lorne furnace, Argyllshire, 1753–1876', *Journal of Historical Geography*, 1 (1975), p. 284; M. Bangor-Jones, 'Native woodland management in Sutherland: the documentary evidence', *Scottish Woodland History Discussion Group Notes*, 7 (2002), p. 2.

[6] E. Photos-Jones, J. A. Atkinson, A. J. Hall and I. Banks, 'The bloomery mounds of the Scottish Highlands, part 1: the archaeological background', *Journal of the Historical Metallurgy Society*, 32 (1998), pp. 15–32; E. Photos-Jones and J. A. Atkinson, 'Iron-making in medieval Perth: a case of town and country?', *Proceedings of the Society of Antiquaries of Scotland*, 128 (1998), pp. 887–904.

as far south as Stirlingshire until the eighteenth century.[7] In any case, in the seventeenth and eighteenth centuries, outsiders played an increasingly dominant role in Scottish iron-making, as technology turned away from bloomeries and towards new water-powered blast furnaces and forge hammers. It is relatively simple to trace the impact on the woods from the new ways of smelting because the outsiders (Lowland, Irish and English) formed a grouping of entrepreneurs easy to identify (Map 9.1).[8]

With the tanbark industry, the position is rather different. As we have already seen, cutting tanbark was for centuries a characteristic activity of local people operating generally within a ten to twenty mile radius of their own homes, often building up small businesses to supply the nearby towns with bark (see above, pp. 153–4). This pattern scarcely changed: the technology did not alter, capital requirements per unit of production did not grow, external firms did not become involved and the activity was sustainable.

But where the external world did decisively impinge was through the market price for tanning materials. When the demand for leather increased at home and in the colonies, and when bark became in short supply through wartime checks on imports and for other causes, it increased in value in the Scottish woods. As we shall see, this inspired the same attention to good coppice practice and led to the same preference for oak over other broadleaves that the iron industry had itself inspired a little earlier, but on an altogether wider scale. Whereas the impact of the ironworks was restricted mainly to areas within a few miles of the shore in Argyll, and to a lesser extent in Solway and Wester Ross, the impact of tanbarking was felt in all those areas, but also strongly across inland Dunbartonshire, Stirlingshire and Perthshire, in a belt along the southern Highlands quite beyond the reach of the sea but readily accessible by land carriage to the main Lowland centres of leather manufacture such as Glasgow, Stirling, Linlithgow, Perth and Edinburgh. Thus, although enterprise in tanbark may have remained local, the forces that drove and rewarded it were no less external than in iron smelting.

In this chapter we shall look first at iron smelting with reference to the outside partnerships and firms involved in charcoal-fired blast furnaces from 1610 onwards, but consider tanbarking mainly in the eighteenth and early nineteenth centuries when external market forces were most evident.

---

[7] R. M. Tittensor, 'History of the Loch Lomond oakwoods', *Scottish Forestry*, 24 (1970), pp. 103–4.
[8] J. M. Lindsay, 'The iron industry in the Highlands: charcoal blast furnaces', *Scottish Historical Review*, 56 (1977), pp. 49–63.

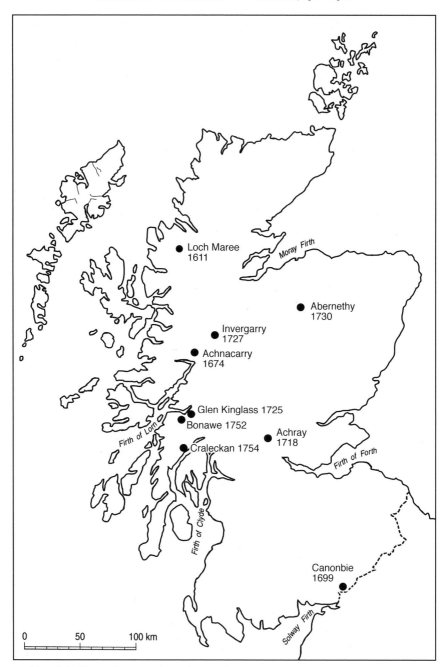

Map 9.1 Charcoal ironworks with outsider interests. *Note:* Dates are those when agreements with landowners were reached, not necessarily when work began, which was often a year or two later. Based especially on J. H. Lewis (1984) in *Proceedings of the Society of Antiquaries of Scotland*, 114, pp. 445–63, with modifications.

The first blast furnaces in Scotland were established around Loch Maree in the early seventeenth century, and, as noted in the last chapter, originated in the interest that Sir George Hay had in the north-west Highlands through his involvement from 1605 with the Fife Adventurers (Fig. 9.1). It is not clear exactly when his attention was caught by the possibilities of using sophisticated iron technology in the area, but that something was afoot in the Highlands is indicated by an Act of Parliament of Scotland of 1609 that tried to prevent 'some personis' taking advantage of the general peace in the area, who 'wald erect yrne milnis in the same pairtis, to the vtter waisting and consumeing of the saidis wodis'.[9] The phrasing suggests that nothing had happened yet, and Parliament's prohibition on such activity was perhaps an attempt at a pre-emptive strike, given their long-standing concern that Scotland was short of good quality woods except in the Highlands. It did not, however, deter Sir George Hay one whit. On Christmas eve 1610, he obtained from the Crown a 'Commission and licence [to] make yrne and glass' within the kingdom of Scotland for thirty-one years, a patent confirmed by Parliament two years later.[10]

In June 1611, he signed an important contract with the Mackenzie laird of Kintail.[11] In it, the debts owing to him by Mackenzie's late father were liquidated in exchange for another 7,000 merks a year paid to Mackenzie and the sale to Hay of all the woods within four miles of the shores of Loch Maree, Loch Alsh, Loch Carron and Loch Duich, with certain exclusions: in this area, Hay was to have for thirty-one years all the oak, pine, ash, elm and aspen, but only half the birch, hazel, holly and other trees, and also to leave enough pine for Mackenzie to build his galleys and only to cut the oak once. Mackenzie also sold all the iron ore and other metal ores (except gold and silver) and granted liberty to Hay to erect sawmills and 'mylnis for making and fyning of iroun, kills, fornaces and all utheris airt work and ingyne' necessary, including the right to make mill leads, construct dams and houses, sail ships, pasture horses, make roads, and so on. Finally, Mackenzie bound himself to defend Hay's enterprise and his workmen 'fra the injurie, violence and oppressioun of all uther inland hielandmen, illismen and also the said John McKenzie of Gairloch'. The deed made clear that hitherto 'the woods, mettallis and utheris eftirspecifeit hes heirtofoir importit littill or na commoditie to the said nobill Lord', and made no mention of pre-existing bloomeries, though it did refer to a sawmill already at Kinlochewe. Although 1607 has been mentioned as a date that saw the

---

[9] *APS*, 4, p. 408.
[10] *APS*, 4, p. 515; *RPCS*, 11, pp. 138–9.
[11] NAS: RD 11/85 reg. 27 Feb. 1612.

Fig. 9.1 Sir George Hay, first Earl of Kinnoull, painted by Daniel Mytens the elder in 1633. The wild scene in the top right must refer to his speculations in Wester Ross.
National Galleries of Scotland.

commencement of industrial activities round Loch Maree,[12] it does seem from this document that nothing substantial had happened before 1611. It has also been suggested that Hay may have set up a glassworks in the area at the same time, but there is no mention of one in the document.[13] Thereafter things moved relatively quickly. Nothing much could happen with the new blast-furnace technology (which reduced the iron to a molten liquid, not, as in a bloomery, merely to a workable paste) without help from the south. Hay involved English partners from the Sussex weald, assigning (in August 1611) one-fifth equally to John Middleton of Horsham and Henney Scheillis of Worminghurst, and one-fifteenth to Anthony Fowll of 'Ratherfeild'.[14] English workmen were also employed, and in 1612 Sir George Hay obtained from Privy Council the rights to exercise justice over the 'grite nomber of strangeris' as well as Scots in his workforce, whereby he had 'interprysit and undertane . . . the arte and practize of making of irne'.[15] A little English colony developed on Loch Maree, accorded rights to bear 'haglibuts and pistolats' against the danger of 'the injurie and malice of the disordinat persouns nixt adjacent to them'.[16] A location beside Loch Maree was known in the late nineteenth century as 'Cladh nan Sasunnach', or the burying-ground of the English, and a pool in a marsh called Lochan-Cul-na-Cathrach, was supposed to be where they had flung their tools on leaving the district: nevertheless there were still families in the area named Kemp and Cross, traditionally held to be descendants of the original Englishmen.[17]

For a decade or more, Sir George Hay's ironworks on Loch Maree operated actively: as well as using the local bog iron specified in the contract of 1611, he shipped clayband iron ore from Fife, encountering opposition from Lord Sinclair at Dysart in 1613 and problems from fishermen at St Monance in 1620. In the latter case, his local English expert who had been stockpiling the ore found it all pilfered and loaded as ballast aboard boats going to the herring drave, 'to his grite hurte and inconvenient'. In each case Privy Council backed his lawsuits, perhaps not surprisingly, as by 1616 Hay had become Lord Clerk Register and returned to Edinburgh to pursue a profitable career in public service.[18] In 1613 Privy Council also protected Hay's supplies by forbidding the export of iron ore from Scotland (not that any known exports ever took place), in consideration that certain people in the

---

[12] J. H. Dixon, *Gairloch in North-west Ross-shire* (Edinburgh, 1886), pp. 72–96.
[13] J. Turnbull, *The Scottish Glass Industry, 1610–1750* (Edinburgh, 2001), pp. 64–70.
[14] NAS: RD 1/213. fos 347–67.
[15] *RPCS*, 9, p. 351.
[16] *RPCS*, 14, p. 567; Turnbull, *Scottish Glass Industry*, p. 67.
[17] Macadam, 'Ancient iron industry', pp. 108–9, 116; Dixon, *Gairloch*.
[18] *RPCS*, 10, p. 160; 12, p. 187; Lindsay, 'Iron industry in the Highlands', p. 51.

kingdom had brought the manufacture of iron 'to ane reassounable good perfectioun'.[19] The iron manufactured was apparently good enough to sell to the Master of Works at Edinburgh Castle in 1617, when the substantial quantity of 1,194 stone was sold for 26s. 8d. a stone, compared to Swedish iron selling at 28s. a stone.[20] In 1621, Hay obtained the right to send his iron to any port or harbour of any burgh, notwithstanding the existing privileges of royal burghs.[21]

Shortly thereafter, however, operations in the north-west were seriously interrupted, perhaps for reasons that had more to do with Hay's own priorities than with any insuperable weakness in the business. By 1622 he had become Lord High Chancellor of Scotland, a prelude to his elevation in 1627 to a peerage as Viscount Dupplin and Lord Hay of Kinfauns and then to securing in 1629 the office of Collector General of Taxes for himself and his son.[22] There were manifestly more profitable and easy things in life than making iron in Wester Ross.

In a letter apparently dated July 1624, Mackenzie of Kintail, now Earl of Seaforth, wrote to the Lord Chancellor.[23] He informed him that the furnace at Loch Maree was blown out and the stock of raw materials exhausted. An unnamed Englishman recommended by the Lord Chancellor was considering whether to take a new lease of woods in the area (perhaps in partnership with George Hay, the latter's nephew), but meanwhile Seaforth had dismissed all the workers until a new stock was made, except for those who worked at the sawmill. He hinted that charcoal wood was at least temporarily in short supply at Loch Maree – the Englishman saw 'sick woods as was thair' – but that those on Loch Carron would suit him better – 'they may best do his turne, and may mak me maist benefeit'. The Earl was now actively managing affairs locally, perhaps having acquired a controlling interest in the works following non-payment of rent.

No immediate action seems to have followed this English show of interest, but in 1626 James Galloway, Master of Requests (who had been mentioned in passing in Seaforth's letter as a person through whom the unnamed Englishman could be contacted), and Nathaniel Udwart, well known as a risk-taking entrepreneur and adventurer from Leith, sought a

---

[19] *RPCS*, 10, p. 24.
[20] J. Imrie and J. G. Dunbar (eds), *Accounts of the Masters of Work, for Building and Repairing Royal Palaces and Castles 1616–1649* (Edinburgh, 1982), pp. 65, 67.
[21] *RPCS*, 13, p. 13.
[22] Turnbull, *Scottish Glass Industry*, p. 64.
[23] J. Maidment (ed.), *Letters and State Papers during the Reign of King James VI, chiefly from the MS Collection of Sir James Balfour of Denmyln* (Abbotsford Club, Edinburgh, 1836), p. 365.

patent of monopoly for twenty-one years to make ordinance and shot, ultimately confirmed in 1628 by Privy Council.[24] The words of the grant suggest that casting cannon was then a novelty in Scotland, and there is, indeed, no evidence that Hay had attempted it in his own venture.

In order to finance the undertaking, Galloway and Udwart originally sought a grant of £2,000 from the Crown, but when this was not forthcoming they entered into partnership with the Earl of Seaforth, 'without whose helpe and concurrence thay could not undergoe so weightie a charge', in the words of the Privy Council Register.[25] Other documents show that there were several other individuals involved in complex financial and logistical agreements that are difficult fully to understand. In one agreement of 1628, the Earl was to receive 46,665 merks Scots (about £2,592 sterling) from a partnership consisting of the Lord Chancellor, his kinsman Lord Hay of Yester, Galloway and Udwart, for the lease of the woods and ironworks of Loch Maree, and the partners were further to provide money for stocking the works for a three-month trial to see if cannon could satisfactorily be made there.[26] Presumably the Hays had to be involved, even as sleeping partners, because the original 1610 iron-making contract was still valid.

There was a second and subsequent agreement between the Earl on one part, Galloway and Udwart on the second part, and four English gentlemen on the third part – Sir Robert Vernon of Fordham in Cambridgeshire, Richard Bathhurst of Bromley in Kent, Dennis Fleming of Camberwell in Surrey and Alexander Thomas of Lamberhurst in Kent.[27] The Earl was to provide on Loch Maree 'one sufficient furnace and one sufficient forge', together with the outbuildings, casting vaults, sawmills, water courses, dams, tools, boats and so forth, necessary for making iron ordinance, shot and 'other engynes of warre'. He was further to provide two years stock of iron ore, that is, 2,000 tons of 'the best stone myne and bogg myne well washed and clensed', and an equivalent stock of 2,000 loads of charcoal at twelve dozen per load, at specified prices. The option of shifting the concern to Loch Carron after two years was left open. After two years, the two sets of partners were to have full possession of the works and to enjoy all the privileges that George Hay had had in his earlier agreement over the lands of Loch Maree and Loch Carron in relation to cutting, digging and transporting timber and ore, and otherwise carrying on their business. The ordinance and shot made was to be jointly owned by the Earl and the partners, who agreed a common interest in the patent of monopoly and

---

[24] *RPCS*, 2nd series, 1, p. 290, pp. 338–9, p. 377. p. 433, p. 449. p. 482; *RMS*, 8, no. 1272.
[25] *RPCS*, 2nd series, 2, p. 64.
[26] NLS: MS 14476/19. See also NLS: MS 14476/18.
[27] NLS: Ch. 10779.

that it should only be exercised on the Earl's lands. The English partners agreed to provide 'good skillfull and expert able workmen' for the works, that is to say a founder, a mounter, a hollower, a borer and a cutter (or as many of these as the appointed founder considered essential). Every year, the net profits after deduction of the running expenses were to be divided between the three interests. Provision was made for royal pre-emption of any of the material, and for turning broken guns into bar iron. It was agreed that if the three-month trial of the furnace showed that ordinance could not be made, the contract was to be void.

What happened thereafter is difficult to determine, but it is unlikely that the concern continued for very long. In 1629 the patent holders petitioned Privy Council to say that the works were 'being now begun and in a good way to continew' but also that they were hindered for want of sufficient bog ore. They were granted the right to take it wherever they could find it, paying compensation to the owners.[28] They do not seem to appear again in the record, and when Blaeu's *Atlas* was published in 1654 with a text compiled from notes made by Robert Gordon of Straloch a decade or two earlier, the author seems doubtful if the works were still going.[29] Though oral history reported much later suggested they had struggled on until the 1660s, there is no documentary evidence for this.[30]

Archaeologists have uncovered the remains of three different ironworks around Loch Maree. Those at Fasagh on the north-east shore, four miles from Kinlochewe, revealed only bog ore, and though described as 'a complex industrial processing site' there is no evidence of a blast furnace rather than a bloomery. Indeed, archaeologists have recently concluded after excavation that the site was 'an iron-working as opposed to an iron-making installation'.[31] The second was probably constructed at Letterewe higher up the north-east shore, described as probably 'the site of a high bloomery later converted into a blast furnace': the ores around it include both bog ore and mined ore, some of it characteristic of ores from the Scottish central belt, including Fife, but some of it apparently haematite of the kind found at Ulverston in Lancashire, suggesting an additional connection with

---

[28] *RPCS*, 2nd series, 3, p. 151.
[29] J. Blaeu, *Theatrum Orbis Terrarum Sine Atlas Novus*, 5 (Amsterdam, 1654), p. 98. The Latin runs: *Lacus Ew, undique densis silvis obseptus, ubi superioribus annis ferrariae exercitae sunt, nescio an adhuc desitum sit.*
[30] Macadam, 'Ancient iron industry', p. 123; Dixon, *Gairloch*.
[31] Turnbull, *Scottish Glass Industry*, p. 66; J. H. Lewis, 'The charcoal-fired blast furnaces of Scotland: a review', *Proceedings of the Society of Antiquaries of Scotland*, 114 (1984), pp. 436, 443–4; GUARD, 'Scottish Bloomeries Project: Interim Report' (1996), p. 16.

north-west England in the search for raw materials.[32] The third is at Red Smiddy on the east bank of the short stream that connects Loch Maree with the sea at Loch Ewe: this is unmistakably a blast furnace and associated forge, using the same mix of ores as were found at Letterewe.[33] The Red Smiddy was the only site from which a fragment of cannon has ever been recovered.[34] It would be well sited for the import of materials both by sea and by fresh water from the shores of Loch Maree, but whether it predates Letterewe or the final works made by Seaforth for the ordinance monopolists (as is perhaps more likely) is not yet established.

What was the impact of these early ironworks on the woods of Loch Maree and the adjacent areas? They should be considered in conjunction with the rest of Hay's timber business in the area, which was described in the last chapter. Taken together, these operations must have cut down the wildwood described so vividly in Timothy Pont's account – 'many fair and tall woods as any in all the west of Scotland' (see p. 15 above). On the other hand, this would not of itself have led to deforestation. The ironworks would have left the pine alone (it was regarded in the 1628 contract as unsuitable for making charcoal) but, as we have seen, the pines were the subject of a considerable trade in deals. The broadleaf woods, especially the oak, were the charcoal burners' target, and the Earl of Seaforth's lukewarm phrase about 'sick woodis as was thair' in 1624 suggests a certain impact from Hay's furnaces on Loch Maree. In 1628 the Earl undertook to provide (not necessarily from Loch Maree alone), 2,000 loads of charcoal over two years, which, (following Lindsay's calculations of the consumption at Bonawe) equates to the yield of about 427 acres of Highland coppice each year: to have kept this up on a twenty-year rotation would have demanded 8,540 acres set aside for the purpose.[35] On the other hand, this can be considered as evidence that the Earl thought there was plenty of wood left on his estates to meet this demand, and there is also no sign that such demand ever actually materialised in the years after 1628. The effect of all this exploitation in the first half of the seventeenth century was probably to convert old growth woodland into secondary growth, and then perhaps expose it to further loss and erosion as a result of uncontrolled grazing of

---

[32] Lewis, 'Charcoal-fired blast furnaces', p. 445; Macadam, 'Ancient iron industry', pp. 109–19.
[33] Lewis, 'Charcoal-fired blast furnaces', p. 440–3; Macadam, 'Ancient iron industry', pp. 119–23.
[34] Macadam, 'Ancient iron industry', p. 122.
[35] NLS, Ch. 10779. The agreement was to provide over two years, one thousand load of charcoal, defined as 12 quarters English measure (presumably the English 'dozen') each year. A single acre might, according to Lindsay, yield 2.34 dozens, so 427 acres would produce the thousand loads required. Over a 20-year period, this would require 8,450 acres on a managed rotation. See Lindsay, 'Charcoal iron smelting', pp. 288–9.

regenerating trees. There is no sign that the Earl enclosed any wood, though he was careful not to allow Hay or his successors to return to any oak once they had cut it. Substantial woods certainly still remained on Loch Maree in the early eighteenth century, sufficient to attract Irish timber-merchants (see below, pp. 345–6), and of course important woods still exist there today. The works themselves were also so transient that the effect they had, except in converting old growth wood to secondary growth, must have been rather limited.

The smelting of iron on an industrial scale was apparently not seriously attempted beyond Loch Maree in the first half of the seventeenth century. Robert Edward in 1678 wrote of a project by Sir David Lindsay to build a smelting house in the Wood of Dalbog near Edzell in Angus, ca. 1600, but 'there is no record of any iron produced'.[36] Sir George Hay in 1611 received permission from Inverness Burgh Council to divert water to serve a lead mill and ironworks that he intended to build nearby, but no evidence survives that he did so.[37] Hay did construct a manufactory at Limekilns near Dunfermline with a substantial forge that operated for many years probably using Swedish bar, for it had no smelting capacity.[38] Archibald Primrose was granted privileges to make iron in the sheriffdom of Perth in 1612, but there is no sign he ever used them, and Sir John Grant of Grant was in the 1630s similarly considering erecting ironworks in Strathspey or at Urquhart on Loch Ness, but did not pursue the idea further.[39]

In the second half of the century, however, there were two ventures which, while adding up to little in themselves, were forerunners, in different ways, of things to come. The first was a venture of Irish capital on Cameron of Lochiel's land north of Fort William. The Irish were also becoming interested in Scottish tanbark around this date, and their own involvement in ironworks had been both extensive and destructive of Irish woods in the seventeenth century, though what they now did in Scotland was on too small a scale to have much impact. In 1674, Cameron of Lochiel made a contract with John Davidson and William Munro (their nationality is not stated) 'for building and erecting of an iron miln and other works' at Achnacarry on Loch Arkaig. In 1681 Davidson and Thomas Rickaby of Lambay in Co. Antrim bought all Lochiel's woods, except the pinewoods at Glen Loy and Loch Arkaig (pine was widely regarded as unsuitable for charcoaling). In 1688 'one half of the iron forge at Achnacarry' was assigned

---

[36] Lindsay, 'Iron industry in the Highlands', p. 54; R. Edward, *The County of Angus, 1678* (ed. J. MacNair, Edinburgh, 1883), pp. 24–5.
[37] Lindsay, 'Iron industry in the Highlands', pp. 50–1.
[38] J. Shaw, *Water Power in Scotland 1550–1870* (Edinburgh, 1984), pp. 89–91.
[39] Lindsay, 'Iron industry in the Highlands', p. 54.

to Lochiel for non-payment of debt by Mathew Riccaby and James Stammie (or Tammie) of Lorn in Ireland. Nothing else is known about the concern, except that in the same year the Earl of Breadalbane bought a small quantity of 'excellent' iron from the works, still managed by John Davidson, for window bars and hammers.[40] It foreshadowed the interest of more ambitious Irishmen in the woods and iron-making possibilities in the west coast in the first part of the eighteenth century. (See Chapter 13.)

Equally a precursor of things to come was the involvement in 1699 of three Englishmen from Cumbria, in this case in a site at Canonbie near the Solway coast in Dumfriesshire.[41] Richard Patrickson 'of Calbrodie', Thomas Fawcett, clock-maker and Charles Russell 'hammerman', contracted with the Duchess of Buccleuch to buy woods to make charcoal and to 'erect one or two forges or furnaces in . . . convenient places'. The sums offered were large: £30 sterling annual rent for the industrial site for nineteen years, and £150 sterling a year for the woods for ten years. Nevertheless, five years later the partners were bankrupt and the estate seized the 'forge and all materials'. Yet it limped on, acquiring first in 1712 three new owners from London (William Hall, ironmonger, Thomas Davis and Robert Child) and then in 1715 two further new owners, also from London, John Henry Boock and Thomas Dod. These two continued the works in production, after a fashion, until 1729. According to a Swedish visitor to Scotland, Henry Kalmeter there were in 1719 'no ironworks in Scotland' except those at Canonbie 'where there is a smelting house and four forge-hammers'. He explained that it was 'not going so strongly' and used remelted scrap iron rather than local or imported ore.[42] A letter of 1732, after closure, however, suggests that attempts had at least been made to mine local iron ore, and also to use pit coal as a fuel alongside charcoal. The concern had evidently not compromised the charcoal supply in its thirty years of operations, as the correspondent wrote that 'the wood may last to maintain a forge with addition of pit coals to make iron for generations'.[43] Indeed, in 1737 the

---

[40] J. Munro (ed.), *The Lochiel Inventory, 1472–1744* (Scottish Record Society, 2000), pp. 55, 61–2; Lindsay, 'Iron industry in the Highlands', pp. 54–5.

[41] For the full account see A. R. MacDonald, 'The first English ironworks in Scotland? The "forge" at Canonbie, Dumfriesshire', *Transactions of the Dumfriesshire and Galloway Natural History and Antiquarian Society*, 73 (1999), pp. 209–21.

[42] T. C. Smout (ed.), 'Journal of Henry Kalmeter's Travels in Scotland 1719–20', in *Scottish Industrial History, a Miscellany* (Scottish History Society, 1978), p. 19.

[43] MacDonald, 'The first English ironworks', p. 220. The writer of this letter was a John Davidson, who had been involved in the management of the works since 1704. The coincidence of the name with the builder and operator of Lochiel's iron works at Achnacarry from 1674 to 1688 is interesting. Although the length of time involved makes it very unlikely they were the same individual, they may have been father and son.

woods of Canonbie were sold to a tanner from Hexham in Northumberland to make charcoal and process bark, so they were still in sustainable use.[44]

In the early decades of the eighteenth century there were a few other concerns which showed some ambition, if not much success. The ill-fated venture by the York Buildings Company at Abernethy in the early 1730s has been discussed in the previous chapter (see above, p. 212), and that of an Irish partnership at Glen Kinglass between 1725 and the later 1730s was also short-lived (see below, p. 345). Another Irish partnership was involved in 1718, when John Smith of Castlefinnan in Donegal and John Irvine of Newtonwood in Co. Tyrone paid £2,000 sterling for the rights to cut oakwood on the Montrose estates in Buchanan and Menteith. This was probably primarily for tanbarking, but they also secured the rights to make charcoal, to prospect for iron ore and to build an iron mill. In the event, the ironworks (just a large bloomery and forge) was financed by Glasgow and Dumfriesshire Scots and built at Achray, east of Loch Katrine. It used local ore and Dutch scrap, its fuel supply was charcoal from the local birchwoods, of which John Smith had also secured a ten-year lease (for £222 sterling, vastly cheaper than the oak). This concern may not have survived beyond 1730 and was always small-scale.[45] Still less is known of a small blast furnace built by the Earls of Cathcart around 1732 at Terrioch, west of Muirkirk in Ayrshire, of which archaeological but scarcely any documentary evidence survives.[46] Apart from Canonbie, it was the only charcoal-fired blast furnace constructed in the Lowlands. It cannot have amounted to much to leave so little trace.

Much better documented was the Highland concern of Thomas Rawlinson and partners, owners of an existing ironworks at Furness in Lancashire, who set up operations at Invergarry on the Great Glen following an agreement in 1727 by John Macdonell of Invergarry to sell his woods for thirty-one years for the very modest price of £320 sterling. At considerable expense a furnace was built with skilled English labour, and, in 1729, production began using English ore that had to be transhipped overland from Corpach by Fort William. Scots cut the wood, Irishmen turned

[44] MacDonald, 'The first English ironworks', p. 220.
[45] Lindsay, 'Iron industry in the Highlands', pp. 55–6. Lindsay doubted whether it had furnace capacity. The phrase in A. Mitchell (ed.), *Geographical Collections relating to Scotland made by Walter Macfarlane* (Scottish History Society, Edinburgh, 1906), 1, p. 343 that suggests the iron was made from local 'tar' (and Dutch scrap) is obviously a mistranscription for ore (or 'oar'). Robin Maclean of Brig O'Turk has drawn our attention to papers relating to this concern in NAS: RH 15/120/91, which confirm that it was a forge and bloomery, not a blast furnace, and also that it used ore and scrap.
[46] Lewis, 'Charcoal-fired blast furnaces', pp. 464–5.

it into charcoal and English smelters used it in the furnace: 295 tons of pig iron were produced in the first six months of operations and marketed in England and (to a smaller degree) elsewhere in Scotland. In 1731 a second furnace was started but experienced operational difficulties. Chronic problems of high cost for the transport of ore and the manufactured pig iron overwhelmed the advantage of cheap wood supply, the works began to operate intermittently, ran into debt and closed in 1736.[47] It was not quite the end of the story, as the indefatigable Thomas Rawlinson in 1737 began negotiating with John Champion of Bristol (the main purchaser of the Invergarry pig-iron output), and his partners, for further finance. This was to continue mining ventures in the western Highlands, but based now on mineral concessions from John Mackenzie of Applecross and existing 'considerable and (as is conceived) profitable' leases of wood in Moidart and Arisaig from Ranald Macdonald of Clanranald and Donald Macdonald of Kinlochmoidart.[48] Nothing came of this because the minerals (iron, copper and lead) were not there, but it further serves to illustrate the irrational optimism of many English entrepreneurs of the period about wood-based opportunities in the Scottish Highlands, a phenomenon that appeared again in the activities of the York Buildings Company and of John Lummis and his associates (see above pp. 207–12, 295–6).

Although, as at Loch Maree earlier, Thomas Rawlinson and partners can have had little long-term effect on afforestation, if only because their onslaughts were so brief, they may well have provided significant short-term experience of paid employment for the local population. According to Lindsay, output from Invergarry, which was very irregular, sometimes achieved as much as 583 tons of pig iron within a limited period, which may have necessitated the felling, in a year, of as many acres of wood to supply the charcoal.[49]

Just as Invergarry and similar concerns such as the York Buildings Company in Strathspey began to get into difficulty, the Trustees for Establishing the Colony of Georgia in America started to look for Highland recruits to establish on the frontier a settlement they were eventually to call Darien. Many of those who took up the Trustees' offer came from the Great Glen and the northern or north-central Highlands – Mackintoshes, Macdonalds, Munros, Mackays, Grants and many more. Apart from the clansmen's

---

[47] NAS: GD 1/168. There are good accounts in H. Hamilton, *An Economic History of Scotland in the Eighteenth Century* (Oxford, 1963), pp. 189–91, and Lindsay, 'Iron industry in the Highlands', pp. 58–9. See also A. Fell, *The Early Iron Industry of Furness and District* (Ulverston, 1908), pp. 343–89.
[48] NAS: RD 12, reg. 9 June 1737.
[49] Lindsay, 'Iron industry in the Highlands', p. 58.

Fig. 9.2 Bonawe, Loch Etive, 1836. The big timber transported in the foreground is too large to be charcoaled, but the hill in the middle distance shows clear evidence of coppice felling. Thomas Allom, in William Beattie, *Scotland Illustrated*. St Andrews University Library.

martial prowess, their main attraction to the organisers of the colony was stated to be their skill in managing cattle and in cutting trees.[50] The nascent capitalism of the 1730s in various parts of the Highlands, with the provision of wage payments for cutting timber that suddenly came to an end, may well have shaken loose suitable recruits for just such an adventure.

A new and much more significant start was made in the charcoal iron industry in Scotland when, in 1752, Richard Ford and associates of the Newland Company from Furness in Lancashire set up what was to become known as the Lorn Furnace Company, at Bonawe on Loch Etive (Figs 9.2 and 9.3). This concern was to last for a century and a quarter, though in the closing decades its operation was very spasmodic. Two years later, Henry Kendal, William Latham and partners of the Duddon iron works, also in Furness, planned to build a furnace of comparable capacity at Craleckan on Loch Fyneside. It became known as the Argyll Furnace Company, and

---

[50] A. W. Parker, *Scottish Highlanders in Colonial Georgia: the Recruitment, Emigration and Settlement at Darien, 1735–1748* (Athens, GA, 1997), esp. pp. 31–2, 56–7.

remained in operation until around 1813. Both produced pig iron from imported Cumbrian ore, and exported it back to the parent concerns for finishing, though Craleckan later had limited forge capacity.[51]

The business climate of the early 1750s was very different from what it had been in the 1730s. The Highlands were settled: banditry had still been regarded as a problem in 1729 when Invergarry was started. Landowners were anxious to be seen as improvers encouraging industry in their domains. Demand for iron south and north of the Border was growing rapidly. True, the use of coal in iron smelting was a recognised new technology, but an uncertain one. When Carron Iron Works opened near Falkirk in 1759 to utilise coal, the owners were so unsure of its future that in 1760 they paid £900 to Patrick Grant for a twenty-year lease of his woods in Glen Moriston to secure a charcoal supply, although they lay eight miles on bad roads from the navigable waters of Loch Ness.[52] Despite the increasing success of coal-based iron smelting in subsequent years, there long remained a limited niche for charcoal iron for certain qualities. When Bonawe eventually closed, the Newland Company were operating the last three charcoal blast furnaces in Britain, both of the others being south of the Border.[53]

Bonawe and Craleckan opened in 1753 and 1755 respectively, and were substantial undertakings of comparable size. Production at Bonawe in the later eighteenth century was of the order of 700 tons per annum, falling to around 400 tons in 1839 and becoming intermittent after 1850.[54] Lindsay has suggested that 10,000 acres would have been the necessary area to sustain such an output on a twenty-year rotation, though planned rotations by the company were sometimes as long as twenty-four years, which would increase the need to 12,000 acres.[55] Assuming that, at its peak Craleckan was no less productive, the two would have called upon the resources of at least 20–24,000 acres when their output was greatest. In 1798, the extent of the natural woods of Argyll was estimated at 20–30,000 acres, though other and later estimates put it much higher, one as high as 98,000 acres which probably included much scrub pasture. The demands of the two furnaces clearly had the potential to affect a substantial proportion of the mature oak resource of the region.[56]

---

[51] Lindsay, 'Iron industry in the Highlands', pp. 60–1; Lewis, 'Charcoal-fired blast furnaces', pp. 473–6.
[52] NAS: RD 12, reg. 10 Nov. 1760; J. Walker, *An Economical History of the Hebrides and Highlands of Scotland* (Edinburgh, 1808) 2, p. 209.
[53] Fell, *Early Iron Industry*, p. 414.
[54] Lindsay, 'Charcoal iron smelting', pp. 283–98.
[55] Ibid., p. 289.
[56] J. Smith, *General View of the Agriculture of the County of Argyll* (Edinburgh, 1798), p. 131.

Fig. 9.3 The buildings of the Lorn Furnace, Bonawe, photographed in the 1960s before restoration: (above) the furnace and (opposite) the charcoal store. Despite the immense size of the charcoal store, there was at least as much land under oakwood in Argyll when they ceased operations in 1876 as when they commenced in 1753. Royal Commission on the Ancient and Historical Monuments of Scotland. Crown copyright.

Contemporaries believed this to be the case, but not in any destructive or inimical sense. James Robson in 1794 commented on the 'frequency of returns to the proprietors from keeping the oak in coppice wood', which was due to the high price of tanbark and the demand for charcoal 'in that quarter where it is used in smelting iron': these two factors 'naturally induces these proprietors to keep their woods in coppice'.[57] James Smith in 1798 was still more explicit. After citing the sale of the Glenorchy pinewoods to Murphey and Galbraith 'for a mere trifle' (see below Chapter 13), he went on:

Some time after that, however, the remaining deciduous woods in the

---

[57] J. Robson, *General View of the Agriculture of the County of Argyll* (London, 1794).

country were brought into greater estimation, by means of two English companies who set up iron forges, the one near Inveraray, and the other at Bunaw. Ever since, our natural woods are in general tolerably cared for; and though the long leases granted to those companies, of some of the woods, and the want of a sufficient compensation for the rest, has hitherto kept some of them low; yet they are always of more value to the proprietors than any other equal extent of ground, arable land excepted.[58]

Good silviculture in such woods consisted of enclosure, division into haggs, a systematic coppice-with-standards regime, exclusion of all grazing animals for six or seven years after felling, thinning and weeding the coppice, the eradication of inferior wood such as birch and hazel and the planting up of gaps with acorns to provide a more uniform oakwood.[59] Such treatment would, of course, fundamentally modify the old open wood pastures where

---

[58] Smith, *Agriculture of Argyll*, p. 129. See also J. Williams, 'Plans for a Royal forest of oak in the Highlands of Scotland', *Archaeologica Scotica*, 1 (1784), p. 29.

[59] Smith, *Agriculture of Argyll*, pp. 130–1; R. Monteath, *Miscellaneous Reports on Woods and Plantations* (Dundee, 1827), p. 15.

peasant animals had grazed freely and trees had been pollarded for local uses, or the thickets of oak, thorn, hazel, birch, ash and other trees described on Loch Fyneside by Dugald Clerk of Braleckan in 1751 (see above, p. 73). The modern oakwoods of Argyll, such as those in the National Nature Reserves at Taynish and Ariundle, have been profoundly affected by these forestry regimes in the past and bear obvious traces to this day.[60] To find a wood which bears little or no trace of this kind of industrial management one may have to visit some of the remote Hebridean localities, such as the woods of Loch na Keal on Mull, or near Cluanach in the middle of Islay.

The two Argyll furnaces were plainly stretched to find wood sufficient for their needs, and outsiders feared the worst. Pennant in 1769 considered that the 'considerable iron foundry' at Bonawe would 'soon devour the beautiful woods of the country', and Saint-Fond believed in 1784, after talking to the company agent, that the works would soon be forced to close because the available woods were too small and too slow growing.[61] The companies avoided their nemesis by promoting good practice among landowners (thus Henry Kendall advised the Argyll estate on the best methods of wood enclosure[62]) and by searching out new wood supplies. As Map 9.2 shows, Bonawe came to be supplied from sources up to about seventy miles by sea from the furnace, ranging from Jura and Loch Tarbert in the south to Arisaig in the north and probably beyond.

None of this led to deforestation. Lindsay, in a very careful examination of woodland cover in the Muckairn parish, where the Lorn Furnace Company had its largest long-term contract with Campbell of Lochnell, showed that there appeared to be considerably more wood in 1876 than in 1750. Part of this was apparent, not real, as Roy's surveyors omitted woods in the west of the parish which were actually mentioned by name in the contract of 1752, but unless they recorded 'less than half of the true woodland area of 1750 ... the extent of woodland in Muckairn increased or at least remained stable during the company's management'.[63] This very cautious conclusion reinforces what contemporaries said about the ironworks assisting the preservation of the woods.

When the companies were first established, they aimed at long-term contracts at fixed prices which put as much of the onus as possible on woodland owners. Campbell of Lochnell, who had already had experience of selling his woods to the Irish ironmasters at Glen Kinglass (see below

---

[60] See, for example, Sunart Oakwoods Research Group, *The Sunart Oakwoods: a Report on their History and Archaeology* (n.p., 2001).
[61] Lindsay, 'Charcoal iron smelting', p. 293.
[62] NLS: MS 17668 fos 219, 240.
[63] Lindsay, 'Charcoal iron smelting', pp. 294–6.

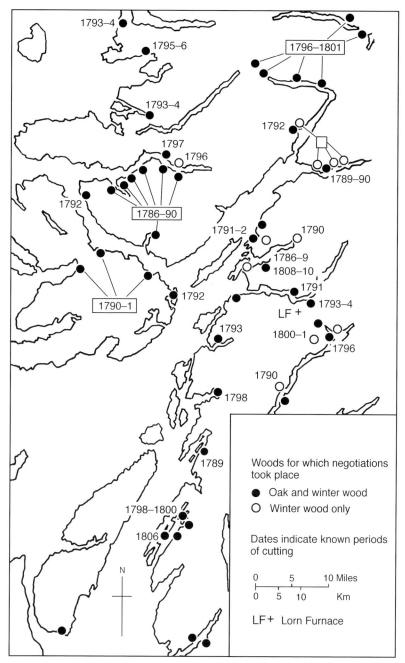

Map 9.2 Supplementary sources of charcoal for the Lorn furnace at Bonawe, 1786–1810 (other than the main contracts with Campbell of Lochnell and the Earl of Breadalbane in the close vicinity). From J. Lindsay (1975), *Journal of Historical Geography*, 1, pp. 295, 297, with permission.

p. 345), agreed to sell to the Lorn Furnace Company the wood on nineteen farms, to be felled in the first instance over the space of fourteen years, from 1754 to 1767, for the sum of £1,500 sterling. In this period the woods were to be cut in the accepted manner and in due season, the company leaving appropriate numbers of standards within the coppice and not felling pines, or certain areas of birch, hazel and other trees reserved for tenants' use, and undertaking 'to inclose and preserve sufficiently' the woods they felled in the same way as the estate had enclosed them 'when last cut' (presumably for the Glen Kinglass ironmasters), 'or in any other manner of way the said Company can best devise'. After 1767, these woods were to be let for a further ninety-six years to the company to make charcoal on four rotations of twenty-four years each, paying Campbell a fixed price of 3 shillings per dozen of coals for the first two cuttings and 3 shillings 6 pence per dozen for the last two. This might be thought a considerable gamble on price trends, as indeed it proved to be, though in the previous century general price levels in Scotland had not varied much (we have no specific information on charcoal prices). In the first period of fourteen years, the company also had the right to sell and strip the oak bark: in the second, of ninety-six years, this was reserved to the landowner. In both periods, the oak standards were reserved. The company, but not the landowner, secured the right to withdraw from the contract at two break points, in 1767 or 1815, options they did not take up as the agreed price of charcoal proved highly favourable to the buyers at a time of general price inflation, especially during the Napoleonic Wars.[64]

In the same year the company signed similar contracts with the Earl of Breadalbane for woods on or near Loch Awe. Again there was an initial lease of the woods, in this case for ten years, for £1,500; again oak standards and pine timber were reserved, as well as certain trees to be used for mill axles and wheels; again the haggs were to be enclosed, but now by the landowner. This was followed, the company explaining that they could not proceed 'without contracting for farther quantities of charcoal for keeping their works agoing', by a second contract, whereby the estate itself undertook to make charcoal from two further systematic rotations of the wood, each of twenty-four years, ending in 1810. The product was to be delivered at Bonawe not for 3 shillings a dozen but for 18 shillings a dozen, the large difference between the prices reflecting the costs of manufacture and transport, which in the case of the Lochnell contract were borne by the company and in this case by the landowner.[65] In yet another lease of the

[64] NAS: RD 14, reg. 1 Feb. 1773.
[65] NLS: MS 993, fos 8–15; NAS: RD 12, reg. 19 May 1755.

same year signed between the Lorn Furnace Company and Captain Duncan Campbell of Inverawe, £275 was paid for an eight-year lease of woods that the landowner was to enclose, and thereafter to sell charcoal for one rotation ending in 1784 at 18 shillings a dozen delivered at Bonawe.[66] As early as 1756 the company was looking as far afield as Mull, buying an eight-year lease of the woods of Lochbuie for £500, but with no further stipulation about making charcoal.[67] All this generated a great deal of employment. Although the works themselves were run by a handful of men, probably under a score, even as late as 1845 it was said that almost 600 people were employed at some seasons in the woods, including women and children.[68]

Arrangements at Craleckan seem to have mirrored those at Bonawe, though less evidence has emerged of so wide a search for fuel among different proprietors. In 1754, the Duddon Company signed a lease with the Duke of Argyll to rent land on which to erect their ironworks, at a rent of £35. 10s. 0d. yearly and to buy an interest in the woods of thirty-one farms in seven parishes, including one as far away as Dunoon on the Firth of Clyde. The initial lease was to be for twelve years for a consideration of £3,000 sterling (later raised to £3,120), and included rights both to make charcoal and to cut tanbark, with the usual reservations of standards and mill timber. Thereafter there were to be another two cuttings of the wood at twenty-four-year intervals, taking the entire lease to 1812: in these the Duke reserved the right to utilise the tanbark, and either to sell the charcoal at 3 shillings and 6 pence a dozen, if the company manufactured and transported it, or at 18 shillings a dozen if the estate or its subcontractors did so.[69]

Both concerns had bid for the Argyll estate woods in 1754, the Lorn Furnace Company offering £3,840 as opposed to their rival's bid of £3,437, calculated as including the rent of the farms on which the furnace was to be built.[70] The lower figure was nevertheless accepted, no doubt because the Duke preferred to have a furnace on his own territory: indeed, it seems likely that the Argyll estate actively promoted a second ironworks in the area in order to introduce an element of competition.

There is no sign that at the time local estates thought they had made a bad bargain. Indeed, the price of 3 shillings 6 pence a dozen fell exactly in

---

[67] NAS: GD 174/737.
[68] Lindsay, 'Charcoal iron smelting', pp. 286–7.
[69] NLS: MS 17667, fos 88–9.
[70] NLS: MS 17666, fo. 165. John Fisher, a local wood merchant, also bid unsuccessfully for some of the woodland in 1754, though he had no interest in the charcoal: fo. 156. And the Lorn Furnace Company had earlier bid for woods on Argyll land on Loch Awe, apparently without success: NLS: MS 17665, fo. 124.

the centre of the range of 2 shillings to 5 shillings that Argyll had been independently advised was the value of charcoal in 1751. But in due course as the price of charcoal elsewhere increased, landowners came to feel resentful of the fixed price to which they had agreed. The Craleckan furnace ceased to exist around 1811, possibly because of competition from coal-smelted iron elsewhere, possibly because the Duke declined to renew the woodland leases. The Lorn Furnace Company went on for much longer, albeit at a reduced output, able to rely for at least an important proportion of its fuel on cheap charcoal from the Lochnell contract. As late as 1829, the beneficial effect on the woods was held up for admiration:

> Travellers are delighted in passing through any district in which regular attention is paid to the forests, even when reared for the purpose of charcoal. No one can traverse the lands, held under long lease by the Lorn Furnace Company from General Campbell of Lochnell, between Oban and Inveraray, without a feeling of this kind; when the former waste state of that large tract of country is also contemplated, and contrasted with the marks of industry which appear, and the beautiful appearance of the regularly preserved and rising trees.[71]

It was a transformation in the opposite direction to that which has so long and so wrongly been assumed to have been the environmental impact of the charcoal iron industry.

The iron industry was only one of the determinants of the fate of the oakwoods. The other, which had a wider and longer impact, was the tanning industry. In this case, it was movements in price that determined and reflected the intensity of outsiders' interest. Table 9.2 shows that there were two periods when tanbark prices were generally inflated above previous levels – the early eighteenth century, and the period from around 1790 to 1820. In the first period, when rather sparse data suggest the price may have doubled, the most obvious sign of external interest came from the Irish. In the second, much more significant rise, when prices were up to five or six times the original level, it was the local entrepreneur, backed by knowledge of a secure outlet in the Lowlands, who made the running.

Of course, there is a sense in which in all other periods the influence of the external market was also critical, as even in the seventeenth century, bark from Perthshire was sent to Edinburgh, bark from Argyll sent to Glasgow, or bark from Loch Lomond and the Trossachs sent to Stirling,

---

[71] W. Singer, 'An essay on converting to economical uses trees usually treated as brushwood', *TRHAS*, 7 (1829), p. 139, quoted in Lindsay, 'Charcoal iron smelting', p. 204.

Glasgow or Linlithgow, usually involving local wood purchasers who cut and chopped the bark and sold it on. But it is useful to concentrate our attention on the two periods of particularly high price.

Table 9.2 Price of oak tanbark delivered at Glasgow ca. 1690 – ca. 1825

	Per Stone	Per Ton
Late 17th century	4d. – 5d.	—
1735	9d. – 10d.	—
1753	—	£3. 10s. 0d. – £3. 11s. 0d.
1754	6½d.	£3. 15s. 0d.
1757	6d. – 7d.	—
1787	12d.	£6. 8s. 0d.
1790	—	£5. 0s. 0d. – £6. 0s. 0d.
1795	—	£12. 0s. 0d.
1797	—	£8. 5s. 0d.
1809	—	£17. 0s. 0d. – £18. 0s. 0d.
1810–11	—	£12. 0s. 0d. – £12. 12s. 0d.
1812–14	—	£18. 0s. 0d. – £20. 0s. 0d.
1817–19	—	£14. 15s. 4d.
1821–3	—	£12. 17s. 4d.
1825–7	—	£12. 12s. 0d.

Source: J. Lindsay, 'The Use of Woodland in Argyllshire and Perthshire between 1650 and 1850', unpublished Edinburgh University Ph.D. thesis, 1974, pp. 401–6; NLS: MS 17665 fo. 248; 17666, fo. 160; W. Nicol, *Planter's Kalendar* (Edinburgh, 1812); R. Monteath, *Forester's Guide and Profitable Planter* (3rd edn, London, 1836), pp. 253–5.

The involvement of the Irish originated in their particular circumstances at the end of the seventeenth century, when wood became in short supply due to over-exploitation and clearance, but the tanning trade increased due to British prohibition on the export of live cattle and the concomitant need to process meat and hides at home.[72] The Irish tanners publicly deplored that 'the planting of trees [and] the enclosing and fencing up of copses ... for many years past had been greatly neglected' in their country, and that they had become dependent on imports of bark.[73] Their main supply was across the Irish Sea from the oakwoods of Wales and western England.[74] Certain communities like Troutbeck in the Lake District came to

---

[72] L. Cullen, *Anglo-Irish Trade 1660–1800* (Manchester, 1968), pp. 5, 11; E. McCracken, *The Irish Woods since Tudor Times* (Newton Abbot, 1971).

[73] National Library of Ireland: P. 4034, *The Case of the Tanners of Ireland Briefly Represented* (n.d., but early 18th century). See also L. E. Cochran, *Scottish Trade with Ireland in the Eighteenth Century* (Edinburgh, 1985), pp. 12, 22, 72–3n.

[74] L. A. Clarkson *The Pre-Industrial Economy in England, 1500–1750* (London, 1971), pp. 179, 199. English tanners, 1711–19, petitioned in vain to stop the Irish buyers.

depend on their purchases. Their interest in Scotland was not so intense or long-lasting, but it was part of a wider involvement in the timber trade and in ironworks (see Chapter 13).[75]

John Spreull, one of the best-informed commentators on Scottish trade on the eve of the Union of 1707, was the first to mention the interest of the Irish in the Scottish bark trade along the west coast. He said that when a wood was cut 'most of the bark is carried into Ireland, without which they could not tan their leather, their own woods being worn out'.[76] These comments were in general terms, and the Irish were directly responsible for only a small proportion of all known bark contracts of the first quarter of the eighteenth century, though these were spread over a wide area. For instance, in 1702 and 1703 Thomas Sawyer, merchant in Dublin, signed contracts with Sir James Douglas of Kelhead and his son to buy woods of oak and birch near Annan in Dumfriesshire for £1,050 (later reduced after independent valuation to £783). Sawyer was also to supply 1,200 deals and enough slate to cover the mansion house at no additional charge. From the quantities of peeled bark lying in the local storehouse in 1703 it was clear where his main interest lay.[77] We have already noted (p. 238 above) how, further north, John Smith of Castlefinnan, Co. Donegal and John Irving of Newtownwood, Co. Tyrone, contracted in 1718 to pay the Duke of Montrose £2,000 for oakwoods on Inchcailliach in Loch Lomond, and in Menteith, together with the right to establish an ironworks. Later the same year John Smith offered a further 14,000 merks Scots (£889 sterling) for additional woods on the estate, though he later sold most of his rights to a local man.[78] It is quite possible that Irishmen working from a distance found it too difficult to organise and supervise in the woods a workforce that consisted largely of women and children who peeled the trees after men had felled them, and that such outsiders preferred ultimately to operate through those better placed to manipulate the local labour markets.

In the Highlands, too, the Irish appeared, possibly in larger numbers. The first record of any commercial exploitation of the Sunart woodlands in Argyll was in 1706, when John McLauchlane, the principal tenant in Resipole, was collaborating with his uncle McNeill of Barra, to make 'a bargain of barks to be sent to Ireland'.[79] Cameron of Lochiel, never a family to miss a

---

[75] In 1700, Henry Sandwich of Dublin, merchant, sold to Captain William Richardson 'thirty tuns of good and mercatable oak timber' to be delivered at the head of Loch Creran. It must have been local Scottish timber that he was selling on. NAS: RD 14, 1700, no. 1219.
[76] J. W. Burns (ed.), *Miscellaneous Writings of John Spreull* (Glasgow, 1882), pp. 63–4.
[77] NAS: RD 12, 1702, no. 1279; RD 12, 1703, no. 385.
[78] NAS: GD 220/6/580/5–6; RD 13, reg. 26 Dec. 1727.
[79] *Sunart Oakwoods*, p. 45.

commercial opportunity, and with previous experience of the Irish both in selling pine deals and at his ironworks at Achnacarry, signed three contracts between 1701 and 1721 for the sale of oak and oak bark around Loch Arkaig and elsewhere, to merchants from Co. Down, Co. Meath and Drogheda, before falling in 1722 to the blandishments of Roger Murphey and his associates.[80] By then, Murphey and Galbraith had become by far the best-known Irishmen in the west of Scotland, and the logical conclusion of all the commercial interest and probing that had gone before, but their story is left until later (see Chapter 13). In 1724, when Alexander McDonald of Aikbrechlan in Argyll sold Felix O'Neill of Drogheda certain woods at Kinlochbeg at the head of Loch Leven, with power to 'peill the bark, coall, cord and carry the said bark and timber away', for the modest sum of £47 in two instalments, it was stipulated that when the first was paid, the cautioners for the second were to be 'Captain Arthur Galbraith of Dubline and Roger Morphy, tanner, of Dubline'.[81] Not all reputations for probity are deserved.

After around 1730, direct Irish involvement in the bark trade disappeared, though the Irish remained very active in the trade from England until the end of the century, and also used quantities of German bark.[82] They must have continued to buy from Scottish suppliers indirectly, at least to some degree: for example, an advertisement for an oakwood for sale in Glenelg, Inverness-shire, in 1797 spoke of its suitability to supply 'the smelting furnaces on the west coast and sale of bark to the Irish market'.[83]

The second notable period of price rise, that began in the 1780s and lasted until about 1815, is in some ways less easy to study, as the contracts after 1780 have not been analysed in the same detail (an opportunity for further research). On the other hand, there is much more contemporary discussion of it in the literature, both by agricultural reporters on parishes and counties and increasingly by professional forestry consultants like Robert Monteath.

Almost always, though not invariably, the trade remained in the hands of local contractors, when it was not being directly run by the landowners themselves. There were exceptions – in 1760 a group of Glasgow merchants acquired a twenty-four-year contract for the Menteith woods, along with

---

[80] *Lochiel Inventory*, pp. 81–2.
[81] NAS: SC 54/12/10, reg. 17 Feb. 1725.
[82] Cullen, *Anglo-Irish Trade*, p. 78; J. M. Lindsay, 'The use of woodland in Argyllshire and Perthshire between 1650 and 1850', unpublished University of Edinburgh Ph.D. thesis, 1974, p. 402.
[83] M. L. Anderson, *A History of Scottish Forestry* (Edinburgh, 1967), 2, p. 85.

those in Strathgartney and later some in Buchanan.[84] Further, some eighteenth-century local contractors were men of some substance, like James and John Fisher, merchants of Inveraray, who respectively in 1716 bought woods extensively from the Earl of Breadalbane in Lochaweside, and in 1768 bought extensively from Patrick Campbell of Ardchattan, in order to strip bark and make charcoal.[85] Another example was apparently a small landowner in his own right, James Drummond of Croftnappoch, who in 1754 offered the Forfeited Estates Commissioners £1,700 sterling to be paid over ten years for woods in the parish of Comrie, Perthshire: he had had a contract for the same woods twenty years before and handled woods on other estates. The factor commented of him:

> Everybody in this neighbourhood that have woods choose to deal with him, as he is a discreet man and no complaints of him from tenants about carriages or any abuse committed upon the wood.[86]

When prices rose at the end of the century, haggs tended to be sold annually or on short contracts to take advantage of inflation,[87] and some landowners probably always organised their own cutting and peeling in order to deal directly with the end purchaser of the bark.[88]

The price boom was fuelled by a combination of war after 1793 and rapid industrialisation. War cut off or reduced imports of the foreign tanbark that was becoming of increasing significance to the tanning industry, and demand was fuelled by enhanced needs for military shoes and harness, and more generally by a rising population. The new factories also demanded leather belting for engines and machinery on an unprecedented scale, and an ever-increasing supply of black cattle driven from the Highlands provided the hides. The bottleneck, reflected in rising price, was in the woods.

It was nevertheless much easier to supply tanbark than charcoal, because tanbark was easy to transport. Whereas charcoal was reduced to useless dust if carried by joggling horses more than about ten miles, tanbark was sturdy stuff: 'a good horse will take fully a ton on a cart, so that the carriage from an estate thirty or forty miles inland is but trifling'.[89] In fact, there was boom in the tanbark coppice market in a belt across Scotland from Argyll,

---

[84] Lindsay, 'Use of woodland', p. 512.
[85] NAS: SC 54/12/10, reg. 4 June 1722; SC 54/12/27, reg. 7 Feb. 1770.
[86] Anderson, *History of Scottish Forestry*, 1, pp. 465–6.
[87] Lindsay, 'Use of woodland', p. 512.
[88] Robert Monteath wrote his *Forester's Guide and Profitable Planter* (London, 1826 and subsequent editions) as though this was always an option.
[89] Monteath, *Forester's Guide* (London, edn 1836), p. 152.

Dumbarton and Stirling to Perthshire, with a good deal of activity also in the south-west.

Professor Anderson noted the sales of coppice that appeared in the *Edinburgh Advertiser* from 1790 to 1814, over 200 in all.

Table 9.3 Sales of coppice advertised in the *Edinburgh Advertiser*, 1790–1814

	1790–1804	1805–14
Argyll	23	20
Perth	36	39
Dunbarton, Stirling, Clackmannan	14	9
Renfrew, Ayrshire and Lanark	9	1
Dumfries and Galloway	12	2
Fife and Kinross	10	—
Lothian	11	8
Eastern Borders	3	1
Aberdeen and Banff	1	1
Inverness	1	1
Sutherland	—	1
Totals	120	83

Source: M. L. Anderson, *A History of Scottish Forestry* (London, 1967), 2, pp. 87–93. Anderson states there were 118 sales in the first period, but he seems to detail 120; either 84 or 81 in the second period, and he details 83.

The prominence of Argyll and Perthshire is very notable, but data from an Edinburgh newspaper would undoubtedly grossly understate the sales from Dunbarton, Stirling and the Clyde which would be more likely to appear in the Glasgow press, and probably also understate sales from the Solway which were no doubt focused to the south or to Ireland (though contemporaries complained of the neglect of coppice in Galloway).[90] The sales in the first period mopped up a surprising number of coppices in Fife, but possibly they were not sustainably managed, as they make no appearance in the second period. In neither period did the eastern Borders or the eastern Highlands make much of a showing, being just too far from the market. Perhaps, though, the Highlands had markets nearer at hand, at Inverness or Aberdeen, and were less likely to advertise a sale in the Edinburgh press.

Even earlier, and in places that had a lively market for the charcoal, bark had always provided the main return in coppice produce. Thus in 1751, the value of the Duke of Argyll's woods at Loch Fyne and Loch Tarbert was estimated at £2,492 sterling for a twelve-year cutting, 60 per cent in the

[90] Anderson, *History of Scottish Forestry*, 2, p. 136.

bark, 36 per cent in the charcoal and 4 per cent in the oak timber.[91] The bark remained much the most valuable component, but on well-grown coppice there came to be a valuable market for 'spokewood', small pieces of hard oak timber suitable for turning as wheel spokes, for which there was also an ever-increasing demand, as transport improvements and the industrial revolution multiplied the number of carts. In places where charcoal could never be marketed because of transport difficulty, this could make all the difference to the profitability of coppice. Monteath illustrated the value of an acre of coppice that would yield £167 sterling after twenty-five years: 67 per cent was the value of the bark, 30 per cent the value of the spokewood and 3 per cent the value of the brushwood. In this example, transport costs reduced the value by 20 per cent, so the spokewood could be said to more than cover the expense of getting the bark to the market.[92] This could be crucial in an inland county like Perth or Stirling, but it also made a difference to the way in which the Duke of Argyll would have contemplated the closure of the furnace at Craleckan in 1813.

As there was also always a market for large oak timber, for shipping, construction and specialised uses such as mill axles, most coppice woods were run at this time on a coppice-with-standards basis. Monteath's ideal coppice would apparently have contained 855 stools per acre, eight feet apart cut every twenty-four years, plus sixty standards, of which forty were left for two rotations (forty-eight years) and twenty for three rotations (seventy-two years).[93] To leave standards probably did make economic sense, though it partly involved guessing at markets beyond a human lifetime. Monteath swithered about this, and many of his calculations are ambiguous or contradictory. It certainly made psychological and patriotic sense, as landowners could feel that they were leaving oak for the dynastic and naval needs of the future.

Much of what exercised the improving literature of the time was the profit that could be made from woodland under coppice compared to any

---

[91] NLS: MS 17665, fo. 63. See also Walker, *Economical History*, 2, p. 291, who considered in 1808 that bark accounted for two-thirds and charcoal for one-third of the value of a coppice.

[92] Monteath, *Forester's Guide*, p. 365.

[93] *Ibid.*, pp. 152, 197. The uncertainty relates to the number of stools. Possibly Monteath meant that if standards are to be grown, the number of stools could only be half the 855 recommended. His calculation of net profit with standards was £240 per acre after twenty-four years; without, £132. But he was fairly cavalier with figures. Elsewhere (*Miscellaneous Reports on Woods and Plantations*, p. 15) he spoke of keeping forty maiden trees on each acre for two cuttings of twenty years each, but they would be worth only £2 a tree when cut and add £40 to the £50 per acre that the bark would be worth, making a yield of £90 per acre, or £4. 10s. a year.

alternative use. David Ure, considering Dunbartonshire in 1794, believed that there were in his county 11,800 acres of wood, of which 6,200 were natural woods:

> The value of natural woods in this place, is very considerable. The annual profit of woods, in which the oak prevails, is estimated at £1 sterling per acre, when cut in twenty years. This is a much higher rent than could be got for the ground if put to any other purpose; for in general it is unfit for tillage; and if laid under pasture, it would not give much.[94]

Similar points were made by other commentators such as John Naismith, considering Clydesdale in the same year, who put the value of coppice after thirty years at £20 to £30 an acre, and Andrew Whyte and Duncan Macfarlan in 1811, considering Dunbartonshire again, who put the value of an acre of oak coppice at twenty-four years at £30.[95]

All this paled before the claims of Robert Monteath, writing in the later 1820s on the merits of 'natural woods':

> By having yearly cuttings, very ordinary woodlands will pay an annual rent from £5 to £10 per acre, from natural oak, for any length of time, without the expense of planting, but keeping good the fences.[96]

We may suspect an element of creative accounting in his calculations, as he was a forester looking for employment as a consultant to the landed classes, but his lowest estimates were well above the rate of inflation between 1794 and 1825. No crop, not even wheat, would pay so handsomely, declared Monteath, becoming carried away by his enthusiasm, and he claimed to know proprietors who made £4,000 to £7,000 a year from their oakwoods.[97]

Whatever the hyperbole, the basic proof of the profitability of coppice management at this period is seen in the enclosure of woods and the exclusion of stock from individual haggs for periods of five to seven years or more, which amounts to a shift in land use from pasture (with a minor

---

[94] D. Ure, *General View of the Agriculture of the County of Dumbarton* (London, 1794), p. 83.
[95] Cited in Anderson, *History of Scottish Forestry*, 2, pp. 270–2.
[96] Monteath, *Miscellaneous Reports*, p. 15. Elsewhere he opined that 'woods judiciously planted and reared, will pay an annual rent of from £6 to £10 per acre': R. Monteath, *A New and Easy System of Draining and Reclaiming the Bogs of Ireland* (Edinburgh, 1829), p. 3. See also the calculations in n. 93 above.
[97] Monteath, *Forester's Guide*, pp. 151, 154; Monteath, *Miscellaneous Reports*, pp. 131, 145.

return from wood produce) towards forestry (with some allowances still made for stock). This occurred throughout the areas most affected by rising prices, so that tenants often complained of shortages of grass and winter shelter, and of lacking access to necessary wood supplies themselves (see above, p. 118). Even those not personally affected sometimes commented on the adverse consequences. Thus in Argyll, John Smith in 1798 thought farmers were discouraged by having to bring timber from Norway to construct their own houses while surrounded by woods that were reserved for charcoal and bark, and in 1845 the minister of Muckairn complained that the interests of silviculture were put before the interests of agriculture, in a district which needed as much low ground as possible for wintering stock.[98]

The ideal coppice in the eye of most experts, however, was not one where all animals were excluded all the time, though some writers advocated that.[99] Normally animals were kept out from a freshly cut hagg for five to seven years of a rotation that varied from twenty to thirty years, so that a third to a sixth of the woodland was unavailable for pasture at any one time, though on an individual farm this might deprive the tenant of wood pasture and winter shelter for a considerable period.[100] There was, however, much variation. The minister of Ardnamurchan explained in 1838 that in his parish in some cases the young wood was enclosed for seven to ten years (and the tenants compensated), in others for fifteen to twenty years, and the woods then only pastured lightly, and in yet others the animals were excluded altogether:

> The last method is most subservient to appearance, that first mentioned to profit; for, when profit is the object, the sacrifice of so much low land, where wintering is so much wanted, is thought bad management.[101]

So here was a serious attempt at integrating forestry with farming.

In general, the stricter exclusion of animals was considered very favourable to the survival and health of the wood, as contemporaries were unanimous in their condemnation of the damage that unregulated grazing had done in the past, and continued to do wherever enclosure was not practised.

---

[98] Smith, *Agriculture of Argyll*, p. 131; NSA (Edinburgh, 1845), 7, pp. 150, 520.
[99] Smith, *Agriculture of Argyll*, pp. 130–1.
[100] James Robertson, *General View of the Agriculture in the Southern Districts of the County of Perth* (London, 1794), pp. 97–8; Smith, *Agriculture of Argyll*, pp. 130–1; Monteath, *Forester's Guide*, p. 175, advocated exclusion for ten years.
[101] Quoted in Anderson, *History of Scottish Forestry*, 2, p. 100.

Within the coppice, experts advocated a strict regime of eradicating everything but oak, and that bare patches should be thickened up by setting acorns, often of English provenance. At the height of the boom, completely new oakwoods were often planted. In Dunbartonshire, for example, David Ure cited among the 'many gentlemen' who were planting oakwoods Herbert Buchanan of Arden, who had set 15,000 oak seedlings at the Strone of Achindennan, and in the north-east James Donaldson cited the Earl of Moray's planting of 614,000 oaks at Darnaway, and his neighbour's planting of 200 acres with 80,000 oaks, though in this area it was not clear they were ever intended for coppice rather than as timber trees.[102]

At the conclusion of the Napoleonic Wars, imports became easier again. Monteath's statement in 1829 that he had it on good authority that 10,000 tons of foreign oak bark a year was being imported into Ireland, 'as much into Scotland and I should suppose twice that quantity into England', while not exactly having the ring of statistical accuracy, at least conveys the sense of the market coming under pressure.[103] Prices fell, rapidly in the 1830s, until in 1847 it was reported that they had dropped from £16 a ton in 1822 to £5. 10s. a ton, 'consequently reducing the value of oak coppice plantations in the same ratio', that is, by about two-thirds.[104] As we shall see in the next chapter, from this point onwards oak coppice became progressively neglected and the balance of rural land use in the coppice districts swung back in the direction of grazing and woodland neglect. But in its day, the tanbark boom, like the ironmasters' efforts before it, had done much to ensure the survival and continuity, but also to change the character, of the oakwoods of western and central Scotland.

---

[102] Ure, Agriculture of Dumbarton, p. 84; James Donaldson, *General View of the Agriculture of the County of Nairn* (London, 1794), p. 26.
[103] Monteath, *Bogs of Ireland*, p. 3.
[104] Anderson, *History of Scottish Forestry*, 2, p. 272.

CHAPTER 10

Woodland management in an industrial economy, 1830–1920 and beyond

In the course of the nineteenth century, the woodland cover of Scotland, as officially recorded in various inquiries and censuses, did not vary greatly, though it varied somewhat.

Table 10.1 'Official' statistics of the percentage of land under wood in Scotland, 1812–1908

Date	Acres	% of land cover
1812	907,695	4.76
1872	734,488	3.85
1880	811,703	4.26
1884	812,000	4.26
1888	829,000	4.35
1891	830,000	4.35
1894	946,493	4.96
1901	878,765	4.60
1904	937,372	4.91
1905	868,409	4.55
1908	875,000	4.59

Source: M. L. Anderson, *A History of Scottish Forestry* (London, 1967), 2, p. 394

These statistics were described as 'quite unreliable' by M. L. Anderson,[1] and though the quality improves from the 1880s they cannot all be true, or at least cannot all be using the same definition of woodland. For example, the figures for 1894 and 1904 respectively imply a planting rate of 38,000 acres and 20,000 acres a year over the previous three years, whereas contemporary estimates of planting rates at this period are about 3,000 and 4,400 respectively.[2] The official statistics state that throughout the nineteenth

---

[1] M. L. Anderson, *A History of Scottish Forestry* (London, 1967), 2, p. 396.
[2] *Ibid.*, p. 395.

century only between 4 and 5 per cent of Scotland was wooded. We argued earlier, however, that official statistics had always underestimated reality, that the real figure for 1812 should have been around 9 per cent, and for 1900 possibly over 6 per cent, the difference being mainly due to definitions of tree cover that omitted wood pasture and upland scrub (see pp. 65–8 above). We further suggested that semi-natural woodland cover accounted for perhaps 7 per cent of Scotland in 1812, but only about 3 per cent or a little more in 1900, the main causes of loss being the spread of sheep farming, with its concomitant heavy grazing and muirburn, and of forestry itself.

No-one in the nineteenth century asked how much woodland was semi-natural and how much was plantation, but in 1908 an investigation into the species of trees grown in Scotland purported to show that Scots pine was by far the commonest conifer, and birch and oak by far the commonest broadleaf tree, though much of the pine and some of the oak would have been planted. Earlier, in 1884, it had been supposed that 75 per cent of the woodland was conifer, of which 60 per cent was Scots pine, 30 per cent larch and 10 per cent Norway spruce and silver fir: presumably that refers to woodland with some kind of commercial value, and discounts what would have been dismissed as birch or oak 'scrub'. Recent calculations based on the first Ordnance Survey maps of Scotland from the later nineteenth century, suggest that about 53 per cent of the woods portrayed therein which survive to this day were plantations at that time and the rest were semi-natural. Probably by 1914 barely half the woodland area in Scotland would have been ancient semi-natural wood.[3]

These changes were a reflection of economic history. By 1830, Scotland was becoming an industrial country, with nearly a third of her population already living in towns of 5,000 inhabitants or more, and many even in the countryside dependent on the textile and mining industries for their income. Eighty years later, this process had gone much further, and nearly 60 per cent of people lived in substantial towns. The population of Scotland, 2.4 million in 1831, had doubled by 1911: a hundred years before it had been a little over a million. In the west of Scotland before the First World War, the massive shipbuilding and engineering industries on the Clyde, backed up by steel and coal production on a huge scale, supplemented but did not supplant the textile industries of Glasgow, Paisley, Dundee, Aberdeen and the Borders. The ideology of free trade, dominant over agricultural and other forms of protectionism by the 1840s, ensured that Scotland was part of a global market, exporting industrial goods and importing raw

---

[3] Anderson, *History of Scottish Forestry*, 2, pp. 394–5; S. N. Pryor and S. Smith, *The Area and Composition of Plantations on Ancient Woodland Sites* (Woodland Trust), 2002, pp. 2, 10.

materials, including wood, unhindered. The coming of steamships and the construction of a railway network made it cheap to send goods to the consumer and to receive raw materials inland at low cost. New technologies, not least in the chemical and transport industries, constantly found new uses for old materials.

It was a new world, where demand for wood and wood products greatly increased, but also one where it often proved easier in the long run to satisfy that demand by imports from North America and northern Europe, where wood was generally cheaper and of better quality. The fortunes of the woods also depended on alternative land uses in the countryside. The highly efficient intensive farming of the Lowlands largely rode out the storm of globalisation to the end of the period, maintaining its reputation and profits. In the Highlands, the shift from upland peasant farming to a capitalist sheep monoculture, associated with depopulation and clearance, altered the face of the land and destroyed old ways of management. When sheep lost their profitability in the last quarter of the nineteenth century, the grouse moor and the deer forest became the dominant form of land use, with a coastal fringe of crofting. By all these developments, the Scottish woods were profoundly affected.

Of the many changes, one of the most significant was the decline of oak coppice. At the start of the nineteenth century it had been regarded as probably the most valuable of all types of native woodland, but by the early twentieth century it was considered as almost without value. The return for coppice produce was already dropping by 1830 (see p. 249 above), but the real crash came after 1840, as the course of tanbark prices demonstrates.

Table 10.2 Price of oak bark per ton, 1830–1903

Date	Price (£.s.d.)
1830	14.0.0.
ca. 1841	18.0.0.
1855	6.7.0.
1857	6.0.0.–6.3.0.
1863	7.0.0.–6.4.0.
1868	6.6.0.
1869	5.13.0.
1871	6.0.0.
1875	7.0.0.–6.10.0.
1903	4.0.0.

Source: M. L. Anderson, *History of Scottish Forestry*, 2, p. 391; *TSAS*, 2 (1860), 4 (1868), 6 (1871).

Behind the dizzy fall was, initially, the ever-increasing volumes of cheap foreign bark, including highly effective mimosa, then the arrival of chemical substitutes for oak tannin.[4] The other main coppice products were charcoal and spokewood. The steadily tightening grip of coal as the smelting fuel of the iron trade eventually led in 1876 to the closure of Bonawe furnace after years of uncertain operation, though the parent company at Ulverston in Lancashire went on manufacturing charcoal iron marketed as 'Lorn iron' to fill a specialised niche for many years later. Small-scale charcoal manufacture elsewhere, making special fuels and raw material for gunpowder, seems to have been given up about the same time: it ceased at Dalkeith in 1878.[5] A market for spokewood survived for longer, but American competition was being felt in all branches of the smallwood trade by 1893 and the value had been reduced 'to the lowest margin': nevertheless, cleft spokes for wheels could still command a quality premium over American prepared spokes.[6] By 1906, however, demand for homegrown oak had declined for carriage and wagon wheels and survived mainly in colliery and contractors' cart spokes, and in the arms of telegraph poles: hope was held out, not very convincingly, that motor car wheels would revive the market.[7]

Under these circumstances, the practice of coppice management steeply declined. Although, even before 1830, there had been obvious competition for land use between coppice and sheep grazing, the heartland of the practice in Argyll, Dunbartonshire, Stirlingshire and Perthshire still maintained the woods in good order for at least the next two decades, as can clearly be seen in the volumes of the *New Statistical Account*. The minister of Inveraray, for example, spoke of 9,000–12,000 acres in his parish being under wood, much of it never planted, and 'very important' to the local economy. The woods were 'extensive and valuable, covering almost all the subalpine district', and in the five years 1832–1836 over 400,000 new trees had been planted on land already enclosed and already 'supplied with many seedlings and stools'. Of these, about 40 per cent were oaks put in to fill gaps and to replace worn-out stools; the main produce of the woods was tanbark.[8] These were the old semi-natural woods, somewhat modified, but still in the full flow of sustainable management.

Forty years later, in 1881, in the same parish, Thomas Wilkie, 'forester at Ardkinglas, Inveraray', reflected on a very different situation. There had

---

[4] W. M'Corquodale, 'On the conversion of coppice into a more remunerative crop', *TSAS*, 4 (1868), pp. 47–50.
[5] R. Baxter, 'On the best kinds of wood for charcoal and processes of charring', *TSAS*, 8 (1878), pp. 246–9.
[6] A. T. Williamson, 'The utilisation of smallwood for turnery and other purposes', and 'The manufacture of home-grown timber', in *TRSAS*, 13 (1893), pp. 145–61.
[7] A. Spiers, 'Some notes of the home timber-trade in the east of Scotland', *TRSAS*, 19 (1906), pp. 66–72.
[8] *NSA*, 7 (Argyllshire, 1843), pp. 13–14.

indeed, he wrote, been a time around 1840 when 'large returns were derivable from oak coppice'. Oak bark had been worth £18 a ton, and the return per acre had been 'eight times its present value'. Now times had changed: prices had collapsed, costs risen and further chemical substitution for tanbark was in prospect. It was now not worth enclosing the woods, but better to realise the fullest returns from grazing and sheltering stock. He himself had gone over to a system of allowing the stock in immediately instead of waiting for seven years after a coppice was cut, while retaining old standards and reserving some new ones from the cut-over stools that would otherwise be browsed off. He was left with 350 trees to the acre, but should the economic circumstances change it would be easy to enclose again and revert to coppice. Should they not, a fourth of the trees would be cut in seven years time, and half the remainder in fourteen and twenty-one years respectively. In effect, he had converted coppice to high forest, or more accurately to a managed wood pasture, though on a final rotation planned to end in 1902.[9]

Something of this sort was happening all over Scotland. As early as 1868, William M'Corquodale, the forester at Scone, Perthshire, recommended the conversion of oak coppice to larch and pine plantations, not only because the value of bark had sunk so low but also because the price of labour had risen so high: he recalled when 6d. to 8d. a day was good remuneration for able peelers, 'whereas they cannot be had now under 1s. 2d. to 1s. 6d. per day; and good axemen cannot be had under 3s. a day'.[10] In the same year that Wilkie wrote, 1881, Robert Ross, assistant forester at Darnaway, one of the most notable ancient oak forests in Scotland, wrote that 'oak, as a profitable crop, may now be said to be a thing of the past in Scotland', and that 'on this estate there is a very large extent of oak forest, much of which would pay far better under a crop of larch or Scotch fir'. Nevertheless, in the hope that 'the depressed state of the bark market may be of short duration', he was not prepared altogether to abandon oak at the present.[11]

The decline of coppicing was indeed protracted. At much the same time as these experts were advocating change, others were writing as though it still had a future.[12] Andrew Gilchrist in particular, in 1872, attributed the unprofitability of oak coppice to neglect and bad forestry practice, though in fact falling profits were the root of neglect rather than the other way

---

[9] T. Wilkie, 'On the system of oak coppice management recently adopted', *TSAS*, 9 (1881), pp. 270–2.
[10] M'Corquodale, 'On the conversion of coppice'.
[11] R. Ross, 'On the peeling and harvesting of native barks', *TSAS*, 9 (1881), pp. 58–62.
[12] A. Gilchrist, 'On the treatment and management of oak coppice in Scotland', *TRHAS*, 4th series, 6 (1874), pp. 118–32; J. Brown, *The Forester* (5th edn, Edinburgh, 1882), pp. 621–33.

round.[13] In 1896 the noted Prussian forester Adam Schwappach reported that between Callander and Balmaha he saw 'numerous woods of oak coppice', managed, he also thought, very badly, on a rotation of twenty-one years to provide tanbark and light timber.[14] In this case, as in others, the existence of a local pyroligneous acid works made all the difference to continuing to coppice. These works produced acid for use in the linen and cotton textile industries, notably in the printfields, and also produced vinegar. Seven were in operation as early as 1815, and the closest to the Montrose coppice woods was a long-established concern at Balmaha that did not close until the interwar years. It was described in the *New Statistical Account* as consuming 700 tons of smallwood a year, and in 1911 as taking 1,200–1,500 tons, peeled oak preferred.[15] There were such acid factories in Glasgow and near Perth, and several in Stirlingshire. James Tait in 1884 mentioned them at Stirling, Denny and Falkirk: 'the oldest firms employed in this work are those of Messrs William McLaren & Sons, and Mr James McAlley, Grahamston'. Returns to woodland proprietors fell over time: some were being paid £1.10s. a ton for acid works smallwood in the 1820s, but Balmaha in 1845 paid only 7 shillings a ton. Nevertheless it could swing the margin between profit and loss.[16]

Such local advantages were, however, comparatively rare. In the woodland census of 1905, only 23,000 acres of coppice were reported, less than 3 per cent of the total woodland area,[17] whereas a century before it may have amounted to as much as a fifth. The remainder had mostly been converted to pasture, or to conifer plantation, singled to one-stemmed trees and left, or just abandoned as outgrown coppice. Some had been converted first to high forest and then to game coverts. Wilkie in 1889 recommended planting, in such circumstances, an underwood of privet, current, barberry, laurel, hazel, rhododendron, cotoneaster, sweet gale, gaultheria and elder, with a few silver firs.[18] Coppice was last cut at Dunkeld in 1912, and the Forestry Commission census of 1947 revealed just eighty-nine acres of

---

[13] A. Gilchrist, 'On the pruning of oak coppice', *TRHAS*, 4th series, 4 (1872), pp. 197–204.
[14] A. Schwappach, 'Report of a visit to the forests of Scotland in August 1896', *TRSAS*, 15 (1898), pp. 11–21.
[15] J. M. Lindsay, 'The use of woodland in Argyllshire and Perthshire between 1650 and 1850', unpublished University of Edinburgh Ph.D. thesis, 1974, p. 431; *NSA*, 8 (Stirlingshire) (1843), p. 96; Lord Lovat and Captain Stirling of Keir, *Afforestation in Scotland: a Forest Survey of Glen Mor*, *TRSAS*, 25 (1911), p. 21.
[16] J. Tait, 'The agriculture of the County of Stirling', *TRHAS*, 4th series, 16 (1884), pp. 177–8 (we are grateful to Mairi Stewart for this reference); Lindsay, 'Use of woodland', pp. 431–3.
[17] Anderson, *History of Scottish Forestry*, 2, p. 394.
[18] T. Wilkie, 'Report upon the rearing of underwood for game coverts in high forest', *TRSAS*, 12 (1889), pp. 371–3.

managed coppice-with-standard remaining in Scotland, surely the last gasp of an ancient practice.[19]

Birchwoods, however useful they might be to local people for building material and other purposes, had traditionally had slight market value. Their bark had a lower tannin content than oak, their charcoal was of less worth, and they were hard to manage on a coppice rotation. In the nineteenth century, however, the invention of the circular saw made it easier to deal with bulk quantities of timber of small diameter such as birch staves and alder poles, and the demand for bobbins in the textile mills created a new interest in their birchwoods on the part of landowners. The Glasgow journalist Robert Somers described the mill established at Salen in 1840 'to promote the consumption of birch': it employed destitute boys from Glasgow who worked in return for food and clothing, manufactured about 1,400 tons of wood yearly at a rate of 75,000 bobbins a day, and paid the local landed proprietor 7s. 6d. a ton for the wood laid down at the mill door. It burned down in 1854 and was not rebuilt.[20] Other such mills existed in the Highlands both in the north and in the south, as at Pitlochry and near Aberfeldy. J. Hutton described a bobbin factory at Lochaber in 1878, six miles from the sea and 160 miles from the thread factory (presumably in Paisley) that was the final consumer. It had been operating for many years converting 'many hundreds of tons of birch into bobbins', and paying, formerly 5s. a ton, now 7s. 6d.[21] Apparently the last Highland mill to survive into the twentieth century was at Drumnadrochit near Inverness, but by then Scottish birch was regarded as of inferior quality and bobbin-makers in the Lowlands used Swedish birch made into pre-formed bobbin blocks that could be turned at the mill.[22]

In the industrial areas, the thread mills at Paisley plainly found a use for Scottish wood either turned locally or imported from the north of Scotland, but the cotton mills of the west of Scotland came to buy bobbins and shuttles from Lancashire.[23] The east of Scotland, however, used local bobbins and shuttles, whether for cotton, linen, or (especially) jute. In the 1860s and

---

[19] Anderson, *History of Scottish Forestry*, 2, p. 322; H. L. Edlin, 'Coppice-with-standards and the oak tanning bark trade in the Scottish Highlands', *Scottish Forestry*, 9 (1955), pp. 145–8.

[20] R. Somers, *Letters from the Highlands, or The Famine of 1847* (London, 1848), pp. 151–2; Sunart Oakwoods Research Group, *The Sunart Oakwoods: a Report on their History and Archaeology* (n.p., 2001), pp. 11, 60.

[21] Mairi Stewart, *pers. comm.*; J. Hutton, 'On the woods and plantations of the Mackintosh estate in Brae Lochaber', *TSAS*, 8 (1878), pp. 233–40.

[22] A. Highland, 'Gateside Mills: the Scottish bobbin and shuttle trade in the British and international setting, 1860–1960', unpublished University of St Andrews Ph.D. thesis, 1989, pp. 40–1.

[23] *Ibid.*, p. xiv.

Fig. 10.1 Turning bobbins at Gateside Mills, Fife, after the Second World War: close your eyes to the flying woodchips. From A. G. Highland, 'Gateside Mills', unpublished St Andrews Ph.D. thesis, St Andrews University Library, with kind permission of Mrs Highland.

1870s, for example, we know of bobbin and shuttle mills at Luncarty and Logie Almond in Perthshire, Monymusk in Aberdeenshire, at Gateside in Fife and in and around Dundee.[24] Gateside employed twelve men in 1871, twenty men and five boys by 1881, and ninety people, including management and clerical staff, by the time of the Second World War (Fig. 10.1). It was one of the larger concerns: Ordie Mill at Luncarty was the smallest, and employed a workforce of just six men in the twentieth century.[25]

These Lowland mills used birch, often imported, for the bobbins, and, initially, local supplies of beech for the shuttles, which were subject to much more stress as they flew across the looms and therefore demanded a close-grained wood. Scottish beech, though not native, was superior in quality and strength to that grown further south. From the 1880s, however,

---

[24] Highland, 'Gateside Mills', pp. 53–5; Anderson, *History of Scottish Forestry*, 2, pp. 332–3.
[25] On Gateside, Highland, 'Gateside Mills', pp. 54–5; on Ordie, oral evidence from G. R. Macfarlane conveyed by A. Cooke.

it was increasingly supplanted in shuttle-making by imported American persimmon (first used by John Ireland, shuttle-makers of Dundee). Hickory, cornel and boxwood from the United States were also used, the first named for the picking arms that propelled the shuttle across the loom. Scottish sycamore was favoured for the rollers on the spinning frames and for the flanges of bobbins used in the wet spinning trade, as in the Irish linen industry.[26]

The Scottish bobbin and shuttle industry around Dundee, Forfar, Fife and Perthshire organised itself in 1900 into a Bobbin and Shuttle Manufacturers Association, essentially to fix prices and terms of employment. By 1914 the market for its products in the native Dundee jute industry was becoming eclipsed by the market in Calcutta. Dundee by 1913 had 13,500 looms, but India 36,000; already in 1900, about 40 per cent of Gateside's sales went overseas.[27]

In slack times, mills like Gateside sold naves for wagon wheels (the central parts into which the spokes and axle-trees fitted), ladder rungs, clothes pegs and fencing stobs, from the hardwood offcuts.[28] But up and down the country there were mills never involved in the bobbin or shuttle trade which also regularly produced such turnery ware, along with barrel staves, buckets, brooms, bungs and so forth, often from birch, but also from ash, oak, alder, beech, sycamore and other hardwoods. Two most interesting survivals are at Finzean in the Forest of Birse, one still manufacturing wooden buckets. Willow and other pliant twigs were readily made up into wicker baskets as a home or workshop craft, much in demand in textile districts, and alder and birch charcoal was especially suitable for the manufacture of gunpowder before the advent of nitroglycerine at the end of the century made this redundant. There were a number of gunpowder mills in the Lothians, for instance at Penicuik, and also in the Borders, as at Jedburgh. No fewer than five such works were founded in southern Argyll between 1835 and 1845, using alder in particular, but also birch, hazel, rowan and willow: species hitherto of little commercial value. The presence or otherwise of a nearby gunpowder works, like that of a pyroligneous acid factory, could make all the difference to the prospects for profit from a wood (Figs 10.2 and 10.3).[29]

Sometimes, though, wood products of little intrinsic value might be shipped by railway hundreds of miles. From the 1890s birch twigs were

[26] Highland, 'Gateside Mills', pp. 40–2.
[27] *Ibid.*, p. 103.
[28] *Ibidem.*
[29] Anderson, *History of Scottish Forestry*, 2, p. 333; Lindsay, 'Use of woodland', pp. 429–31.

Fig. 10.2 The bucket mill at Finzean, Aberdeenshire, a nineteenth-century mill restored by Birse Community Trust and still operating. © The Trustees of the National Museums of Scotland.

used in steel manufacture in Lanarkshire and the north of England, made up into brooms to brush the red-hot steel free of scale as it was tapped from the furnaces. This earned modest sums for providers from Inverness to Argyll, even in remote areas like Sunart.[30]

In fact, the opportunities provided by industrialisation initially widened the market for many native hardwood trees, and Gilchrist as late as 1876 was urging proprietors to make more use of coppice of trees other than oak, from what he considered to be 'about a fourth of part of the woodlands of Scotland that yield little or no return per acre'. He was not suggesting that coppice be planted on any land capable of growing larch, Scots pine or Norway spruce, but that existing woods should be utilised better and odd corners considered for such coppice use. There was, he said, 'in almost every district of Scotland', a strong demand for birch to make gunpowder,

---

[30] Anderson, *History of Scottish Forestry*, p. 333; *Sunart Oakwoods*, p. 60.

Fig. 10.3 The pyroligneous acid works at Balmaha, Loch Lomondside, ca. 1900. Note the large stacks of wood waiting to be processed. Stirling and District Libraries.

clogs and bobbins, with the thinnings going to gardeners in the cities. Likewise for alder there was 'always a ready sale'. He recalled how in 1859:

> We cleared a crop of natural coppice from the banks of the Nethen and one of its tributaries in the county of Lanark; this crop consisted for the most part of alder and birch, and was sold for the manufacture of gunpowder at 18s. per ton peeled. They had grown about twenty-five years, and the estimated yield per acre was about £16.[31]

There was similarly a strong market for ash, 'one of the most profitable coppice trees in Scotland', selling its rods and wands for the manufacture of crates and the older wood for carts, barrows and handle wood. There was hardly a native tree in Scotland, as well as beech and sycamore, that did not have a potential sale. He cited prices of bobbin wood selling at Paisley: birch 22s. to 26s. per ton; alder at 20s. to 24s. per ton, sycamore at 18s. to

---

[31] A. Gilchrist, 'On natural coppice wood, of other species than oak', *TRHAS*, 4th series, 8 (1876), pp. 210–19.

20s. per ton. At Edinburgh, birch and alder sold for 15s. to 20s. per ton, ash and elm at 14s. to 18s. per ton, and so on.[32]

Unfortunately for the producers, almost all these hardwood products came to be faced in the last quarter of the nineteenth century with severe and increasing foreign competition. Gilchrist's optimistic picture contrasted with the review of the market by A. T. Williamson in 1893. American bobbins were being sold at half the price of domestically manufactured ones, so birch was hard pressed to find any market unless there was a bobbin mill on the doorstep and even then prices were very low. Most oak timber found its way to the railway wagon shops: only the straightest and best could compete with American imports. Oak wheel spokes still commanded a premium above American products, and had a use in gates, palings and ladders. Ash was still used for bobbins, and for wheels, hay rakes, spades and shovels, and much went for cart and wagon shafts, but ash smallwood prices had fallen by half and the larger pieces were in short supply, so that American imports were increasing there as well. So it went on: only sycamore seemed capable of maintaining its value.[33]

Thirteen years later, a similar survey was published by A. Spiers, and the picture was worse than ever. The invention of nitroglycerine had destroyed the market for birch in the gunpowder mills, and such use as was left for home-grown birch was diminished by the import of larger and better-quality birch timber from Russia and Sweden. Ash had been affected by the coming of motor transport and rubber tyres – fewer carriages were built, and those that were lasted much longer, so that not one horse carriage was manufactured where ten had been ten to fifteen years ago. The import of prefabricated American wheels had eroded the remaining market, so that two-thirds of the wheelwrights in the east of Scotland had gone out of business. Railway wagons were going over to American oak because it was larger, and oak spokewood was used mainly in colliers' and contractors' carts and on telegraph poles. The metal brake-blocks on railway and other wagons had replaced those of willow, poplar and lime. Beech was still used for tools and shuttles, but metal machinery, including metal cogwheels, and the use of stronger belting, had lessened demand. Even golf clubs were now being made of American persimmon and dogwood. Large sycamore, though, was again still saleable.[34]

Between them, these reviews of the market painted a picture of deepening woe, a discouragement to landowners to keep or to manage their natural

---

[32] *Ibidem.*
[33] Williamson, 'The utilisation of smallwood' and 'The manufacture of home-grown timber'.
[34] Spiers, 'Some notes of the home timber-trade'.

broadleaf woods except at minimal levels of expense. On the other hand, by emphasising the diversity of hardwood use until almost the end of the century, they also help to explain why some forms of coppice management and active broadleaf forestry survived for as long as they did.

What about the coniferous woods, in particular the semi-natural forests of Scots pine? Certainly at the start of the reign of Queen Victoria most of the big native pine forests were recovering from a burst of very heavy exploitation that had begun under wartime conditions late in the eighteenth century and continued until around 1830 (see above, Chapter 8). John Grigor in his report on the Speyside forests in 1839 reported on Abernethy, Duthil, Rothiemurchus and Glenmore, as well as on the plantations of Castle Grant – all had been heavily exploited, but Glenmore and Rothiemurchus appeared devastated by felling. On the other hand (a fact seldom emphasised when Grigor's accounts were quoted by later writers) he also noted much regeneration. Thus at Duthil he commented that though a young plant could seldom be found 'near the remains of the old trees', there were 'extensive masses of them . . . rising along the borders of the forest'. Similarly, at Glenmore, 'along the outside of this forest, particularly at the west end, and on the east of Rothiemurchus forest, the young wood, to the extent of several square miles, is fast advancing'. Within Rothiemurchus itself, 'young natural pines are growing extensively'.[35] Others noted the same thing in this forest (see below pp. 310–11).

On Deeside, likewise, while writers like Selby in 1842 wrung their hands over the devastation of ancient trees in Forest of Mar, others noted the care 'now taken to promote the growth of many thousand young timbers which will follow in succession those fit for the axe'.[36] These ancient woods were resilient to felling episodes especially if there was space at their edges to (as the eighteenth-century observation on Mar put it) 'shift their stances'. They were less resilient to heavy grazing within and around them, and when proprietors in Deeside and at the Black Wood of Rannoch enclosed the forests to keep deer within them instead of beyond them, it was much more of a threat to sustainability than the woodcutter.[37] Even Selby noted the deleterious effects of 'the great increase of herds and flocks', which, once

---

[35] J. Grigor, 'Report on the native pine forests of Scotland', *TRHAS*, 2nd series 12 (1839), pp. 122–31.

[36] P. J. Selby, *A History of British Forest Trees, Indigenous and Introduced* (London, 1842), p. 303; Professor Barlow, 'On the strength of the fir in the Forest of Mar', *TRHAS*, 13 (1841), p. 122.

[37] A. Watson, 'Eighteenth century deer numbers and pine regeneration near Braemar, Scotland', *Biological Conservation*, 25 (1983), pp. 289–305; Anderson, *History of Scottish Forestry*, 2, p. 328.

allowed in any numbers into the forest, prevented 'the renovation of its timber, by constantly browsing upon and destroying the seedling plants as they spring from the ground'.[38]

In no case, however, did any ancient pinewood actually disappear in these years, and although many were neglected or misused by being exposed to excessive grazing, others, including many of the most famous like Abernethy and Rothiemurchus, continued even as deer forests to be the object of a certain level of forestry management designed to perpetuate them and to produce, at least intermittently, a cash income (see below, pp. 312–17). This often involved planting new Scots pine adjacent to or perhaps among the naturally regenerating trees, not always of native origin: more seriously, perhaps, it involved introducing non-native species such as larch, Austrian pine or Norway spruce. On the other hand the merits of the native tree and of natural regeneration were often extolled, and the great forests retained much of their ecological integrity.

The owners of Scottish woodlands between 1830 and 1914 wrestled with a globalisation that called into being new markets growing beyond anything imagined in an earlier age, and then, after an interval, largely destroyed them. At first, it seemed to provide an opportunity for new profits. In the age when Britain was the workshop of the world, with expanding population and exploding cities, her exports reaching the corners of the globe, carried with clockwork regularity by the new steamships and the new railways, wood remained absolutely essential to economic growth and social provision. The four main Scottish cities, Glasgow, Edinburgh (with Leith), Dundee and Aberdeen totalled 565,000 people in 1841, but 1,529,000 in 1911. Except in the New Town of Edinburgh, it was rare to find a house built, or at least left substantially unaltered, before the start of Queen Victoria's reign, and the rebuilding of rural Scotland was at least as complete. Stone and slate they may have been on the outside, but the interior structures needed quantities of timber from somewhere. At the beginning of the nineteenth century fewer than 10,000 men were employed in Scottish coal-mining: at the beginning of the twentieth there were some 110,000 employed in 455 collieries, each one of which needed pit props, a purpose for which rough Scottish wood was considerably better fitted than for house construction. The total of cotton spindles in the United Kingdom grew from 10 million in 1833 to 55.7 million by 1913, of power looms from 85,000 to 786,000, and if the local Scottish cotton industry was less dynamic than that of Lancashire, the native woollen and jute trades were extremely

---

[38] Selby, *British Forest Trees*, p. 404.

vigorous. All needed bobbins and shuttles. The Clyde shipbuilding industry led the world on the eve of the First World War, building a third of British tonnage and almost a fifth of all the tonnage constructed on the globe. The hulls and decks were steel by then, but extraordinary quantities of wood were still needed for shipyard scaffolding. Besides, thousands of small boats were built for the fishing industry, still largely wood and sail; steel hulls were rare even among the new steam drifters until the decade before 1914. This provided a market for timber in innumerable small harbours around the coast, where boats were constructed at little yards on the beach. The Scottish herring catch rose from around 250,000 barrels in 1840 to around 2 million in 1910, each one constructed out of wooden staves, withy hoops and a wooden bung. The whole internal economy was knitted together by the coastal steamer and the railway, especially the railway. By 1900 it extended to some 2,000 miles of track in Scotland, the nucleus of which had been laid during and after the mania of the 1840s. The track needed wooden sleepers, which presented an opportunity to the owners of native pine forests when it came near their property, and the trains needed wooden waggons, not least to carry the enormous quantities of coal.

The initial impact of railway transport on timber prices was especially favourable in areas hitherto remote from the main markets. In 1870, it was reported that prices in Aberdeenshire had risen by about a quarter since the coming of rail and that in parts of the north of Scotland they had doubled over twenty years. Overall, hardwoods in such places particularly gained in value. The fall in demand for oak in shipbuilding had been compensated for by extended use in house fittings, the manufacture of furniture, barrels, spokes, and loppings for pyroligneous acid. There was a lively demand for ash in making wheels, barrels and coaches, for beech, poplar, willow and larch for pit wood, alder for gunpowder, larch also for sleepers and boats. Conversely, in the vicinity of large towns prices had dropped, as the advantage of a close location was undermined: 'at the present day, the price of wood of home-growth is pretty much equalised over the whole country'.[39]

D. F. MacKenzie in 1884 reported that there were nearly 600 wood merchants in Scotland, of which 370 were involved exclusively on the home trade, mostly country firms selling on to timber brokers and middlemen in the towns. He thought that they gave employment to 10,000 hands, excluding foresters, and that 80 per cent of the power they used was steam. This was a remarkably vigorous trade, and of relatively recent origin. MacKenzie

---

[39] R. Hutchison, 'On the economic uses and comparative value of different descriptions of timber grown in Scotland', *TSAS*, 5 (1870), pp. 109–13.

Fig. 10.4 Muirhead Bone's etching of the Clyde shipyards, showing the huge quantity of wooden scaffolding required. Frontispiece to J. H. Muir, *Glasgow in 1901*. St Andrews University Library.

attributed it to a variety of causes: the extension of the railways (he reported that the Highland railway south of Forres carried about 30,000 tons of timber in 1882 alone), the invention of the portable steam engine, and also the great storm of 1860, which had reduced prices in the short term by between 30 and 70 per cent and so encouraged people of small capital like clerks and sawmillers, even carters, to enter the trade. So in place of a few firms there arose 'an entirely new race of wood merchants', some of whom had since made their fortunes.[40]

On the other hand, globalisation brought about foreign competition on a completely new scale and eventually introduced it into market niches hitherto protected by specialisation and distance. At the opening of the nineteenth century imports into Britain were running at the level of 200,000 loads a year – a load was roughly equivalent to a ton. They were to increase perhaps thirty-fold in the next century. During and after the Napoleonic Wars heavy duties were imposed on Scandinavian and Baltic timber, but much lighter ones on Canadian and other empire producers: to some extent this provided an opportunity, as we have seen, for heavy exploitation of the native forests of Scots pine on Speyside and Deeside, as it was slightly easier to compete with wood that 'had to bear the long, slow Atlantic haul' than with what came over from Norway and Russia. Nevertheless, the quantities imported doubled even by 1831 and doubled again in the next ten years.

Tariff reform, beloved of the Victorians, reduced the duties on 'foreign' wood from 65 shillings a load in 1819, to 55 shillings in 1821, to 15 shillings in 1851 and to 2 shillings in 1860, and abolished the duty entirely in 1866. The result was a swing back to European sources of softwood, and, by the end of the century, a considerable impact from American sources of hardwood. As J. H. Clapham commented of Britain in the 1880s, 'economic life could not have gone on for six months without imported timber, and at certain seasons of the year, not for three months'.[41] Nor did the growth of trade cease in the new century. Imports valued at £20 million in 1890 were worth £26 million by 1909.[42] On the eve of the First World War it was said that 90 to 95 per cent of the wood consumed was imported. Only 3 per cent of the softwood and 18 per cent of the hardwood were produced from home supplies.[43]

---

[40] D. F. MacKenzie, 'Scottish timber and its uses', *TSAS*, 10 (1884), pp. 189–98.
[41] J. H. Clapham, *An Economic History of Modern Britain* (Cambridge, 1939), esp. 1, pp. 237–8, 331; 2, pp. 5, 220.
[42] Anderson, *History of Scottish Forestry*, 2, p. 394.
[43] Anon., 'Prices of home timber', *TRSAS*, 27 (1913), p. 223; W. T. Hall, 'Utilisation and exploitation technique in Scottish forestry', *Scottish Forestry*, 8 (1954), p. 176.

Table 10.3 Imports of timber into the UK in the nineteenth century

Date	Loads (1,000)
1800	200
1831	546
1844	1,318
1853	2,404
1873	3,416
1886	5,100
1890	6,321

Source: J. H. Clapham, *An Economic History of Modern Britain* (Cambridge, 1939), 1, pp. 237–8; 2, pp. 5, 220; M. L. Anderson, *A History of Scottish Forestry* (London, 1967), 2, p. 394.
Note: Previous to 1842, the figures given relate only to sawn timber of a certain size, and underestimate total imports by at least one-fifth. A load was approximately equivalent to one ton (Clapham, *Economic History*, 1, pp. 238, 478; 2, p. 220).

Under these circumstances the price of Scottish-grown wood was naturally determined by the level of imports, and seldom seemed to Scottish producers enough to encourage confidence in the future, or even in many cases to provide an immediate return. Those who had once taken advantage of a new opportunity, as when the railway track was first laid down or buyers of bobbins and spokewood came knocking at the door, now found that the moment had passed and Americans or Scandinavians were supplying the particular market. On the other hand, even 5 or 10 per cent of such a huge market left a space for the well-placed producer with something to sell at the railhead, if price and quality could be addressed: between a tenth and a fifth of the mining timbers were home grown.[44] The dilemma was how to plan for the future when trees took a generation or more to mature and Scotland was such a small player in a big game.

On many estates, reality was faced early and forestry either given up or reduced to a nominal level. Tyninghame, once so valuable because accessible to the Edinburgh market, lost its comparative advantage and the trees were underplanted with game cover: it was typical of hundreds of estates similarly placed.[45] The woods of Lochtayside were too far from the railway to reach much of an external sale and the Earls of Breadalbane, once careful foresters for local use, preferred to keep them as sporting coverts or to allow farmers unrestricted access for animals.[46] In grouse and deer country further north, many estates found that it paid better to rent the land for

---

[44] Hall, 'Utilisation', p. 180; Anderson, *History of Scottish Forestry*, 2, pp. 457–8.
[45] Wilkie, 'Report upon the rearing of underwood'.
[46] M. Stewart, 'The utilisation and management of the semi-natural woodlands of Lochtayside, 1650–1850', unpublished University of St Andrews M.Phil. thesis, 1997.

shooting than to grow trees, or, more simply, they came into the hands of new owners whose only interest was sport however much it cost them. The Forest of Mar became one of the earliest deer forests, followed by Rothiemurchus and Glenmore in 1859, Kinveachy (Dulnain) in 1864 and Abernethy in 1866. At Glentanar in 1895, twelve gamekeepers were employed but no foresters.[47] In some Deeside woods and at the Black Wood of Rannoch, the deer were enclosed within the wood, rendering the chances of regeneration negligible.[48]

Foreigners and native commentators alike saw the dominance of the sporting interest as a main reason for poor forest management in Scotland. The French forester Louis Boppé considered it subjected to 'the fancies of sportsmen from all parts of the world' and the Prussian Adam Schwappach declared that the greatest factor opposed to good forestry was sport.[49] 'As long as deer forests pay their owners fabulous rents, there will be no incentive to any great expansion of wood forests in the Highland Counties', said David Nairne in 1891, lamenting that even on the Lovat estate where 'an excellent system of forestry' had continued for half a century, planting had fallen off and continuity been lost.[50] Yet, he added, this had been 'for economic reasons', and indeed the landowners' decisions were neither wilfully neglectful nor irrational, but a considered response to the dilemma of an uncertain future. If the situation was to change, Abernethy, Rothiemurchus, Glentanar and the Lovat estate could turn back again to commercial forestry, since in each case the mature trees were still there. The Lovat estate famously did so. In other cases, though, woods, perhaps especially birchwoods, were eradicated to provide more sheep grazing, and even some of the Duke of Atholl's famous plantings of larch at Dunkeld were converted to pasture.[51]

The logic of the dilemma of an uncertain future positively encouraged landowners to raid their woods when the going was good and a short-term market opportunity opened, but discouraged them from spending much further money or effort on maintenance until the crystal ball was more easily read, especially when deer or sheep were an alternative. It was not a situation conducive to sustainability in any sense of the word.

---

[47] N. A. MacKenzie, 'The native woodlands: history, decline and present status', in C. Gimingham (ed.), *The Ecology, Land Use and Conservation of the Cairngorms* (Chichester, 2002), p. 111; J. Atterson and I. Ross, 'Man and woodlands', *ibid.*, p. 123.

[48] Watson, 'Eighteenth century deer numbers'; Anderson, *History of Scottish Forestry*, 2, p. 328.

[49] Anderson, *History of Scottish Forestry*, 2, p. 388.

[50] D. Nairne, 'Notes on Highland woods, ancient and modern', *Transactions of the Gaelic Society of Inverness*, 17 (1891), pp. 205, 220.

[51] Anderson, *History of Scottish Forestry*, 2, p. 327.

How then, do we explain, in such dismal circumstances, the paradox of the emergence for the first time, and then the steady strengthening, of an organised movement to encourage forestry in Scotland, and then to lobby for state intervention in its support? In 1854, the Scottish Arboricultural Society was founded, and was to prove, in various guises, a body of extraordinary influence. The roots of the Society ran deep into an old Scottish mania for planting. As early as the sixteenth century, lairds had decorated the environs of their towerhouses with trees, and in the seventeenth century the Scottish admirers of John Evelyn were enthusiastic planters with a clear economic and aesthetic purpose. For the Improving movement of the eighteenth century, forestry was a true passion, consuming great landowners and small. 'In my opinion, Planting ought to be carried on for Beauty, Effect and Profit', wrote the fourth Duke of Atholl, who, with his two predecessors planted over 21 million trees on some 15,000 acres of ground.[52]

To embellish, to impress and to make money continued to be the motives of planters throughout the first half of the nineteenth century, and beyond. Landowners increasingly vied with one another to try new species. Larch was famously the choice of the Dukes of Atholl, but when the great plant explorers returned with seed from North America, attention turned to the Pacific Northwest. David Douglas, once an apprentice gardener at Scone Palace outside Perth, was alone responsible for introducing between 1824 and 1834 well over a hundred plants and trees to British gardens and woods, most notably for forestry the Sitka spruce and Douglas fir, while his compatriots William Drummond, William Murray and John Jeffrey sent back the lodgepole pine, the western red cedar, Lawson's cypress and western hemlock. Grown first in ornamental arboreta such as Scone, Dawyk, Murthly and Ardkinglas, where some of the trees from the original seeds can still be seen, the economic importance of these discoveries was only slowly appreciated.[53]

Part of the explanation, then, for the rise of the Scottish Arboricultural Society was the widespread enthusiasm for collecting, planting and growing trees, something not completely rational in economic terms – profit had, after all, only been the third motive of the Duke of Atholl. Another part, increasingly as time went by, was an awareness of developments abroad, both in Germany and France where foresters had a much more systematic and

---

[52] S. House and C. Dingwall, '"A Nation of Planters": introducing the new trees, 1650–1900', in T. C. Smout (ed.), *People and Woods in Scotland: a History* (Edinburgh, 2003), pp. 128–57; J. Fowler, *Landscapes and Lives: the Scottish Forest through the Ages* (Edinburgh, 2002).

[53] House and Dingwall, '"a Nation of Planters"'; A. Mitchell and S. House, *David Douglas, Explorer and Botanist* (London, 1999).

economically driven attitude to a much larger resource, and in the Empire, where Scottish forest administrators like Hugh Cleghorn had a big influence on the development of Indian forestry. Cleghorn was to be an enthusiastic and influential member of the Society in the 1880s, and the training of foresters to work in India and Africa became a significant part of the new forestry departments' work in the universities in the twentieth century.

A third motive for the success of the Society, however, clearly was economic. As we have seen, the extension of the railways created something of a bonanza in the Scottish timber industry, and early recruitment was not difficult: by 1879 the Society had 800 members.[54] Thereafter, as foreign imports tightened their grip and invaded new corners that had once been profitable, there were enough members to want to fight their corner by lobbying government and reminding the public how little wood there was in Britain compared to its continental rivals. There were also enough estates sufficiently well placed for access to the market that they were prepared to continue planting: the rate of new woodland establishment between the 1860s and 1913 was still around 4,000 acres a year.[55] The leaders of the society in the early twentieth century themselves embodied the traditional and the modern. Lord Lovat was of old family with a long forestry tradition and an estate near the railhead. Sir John Stirling Maxwell was a newcomer to Highland landownership, with a deer forest at Corrour in the wildest, highest open country beyond Rannoch, where he experimented with planting Sitka spruce on deep peat. These plantations were, he said, 'intended to improve the landscape and provide shelter for deer', but they had enormous meaning for the future.[56]

This, however, is to anticipate. The Scottish Arboricultural Society began small, but was publishing what was to become an influential journal within two or three years. It won the patronage of Queen Victoria in 1871 and a new charter with the right to call itself the Royal Scottish Arboricultural Society (later Royal Scottish Forestry Society) in 1887. In 1884 it hosted a remarkable International Forestry Exhibition in Edinburgh visited by half a million people, and stimulated the first Parliamentary Select Committee on forestry in the UK, reporting in 1887. By 1889, one of its aims, to secure the teaching of forestry in Scottish universities was achieved by the establishment of a lectureship at Edinburgh.[57]

---

[54] Anderson, *History of Scottish Forestry*, 2, p. 367.
[55] House and Dingwall, '"a Nation of Planters"', p. 155.
[56] [J. S. Maxwell], *Loch Ossian Plantations: an Essay in Afforesting High Moorland* (privately printed, 1913 and 1929); A. Scott, *A Pleasure in Scottish Trees* (Edinburgh, 2002), pp. 85–6.
[57] Anderson, *History of Scottish Forestry*, 2, pp. 367–8, 415.

This post, which later became a Chair of Forestry, was initially held by Dr William Somerville, who shortly afterwards moved to Newcastle. In 1906, Somerville secured appointment as one of nineteen members of the Royal Commission on Coastal Erosion and Afforestation. When it reported three years later it came up with startling proposals that were widely credited to Somerville, described as 'perhaps the foremost authority on forestry in the kingdom'. They included setting up a state-funded body with powers of compulsory purchase of private land, and a target to afforest no less than 9 million acres of Great Britain at the rate of 150,000 a year.[58] This frightened everyone, including the Royal Scottish Arboricultural Society, which urged instead a considered survey of the country to see what could and could not be afforested, and led the way themselves in 1911 by sponsoring a survey of the Great Glen.[59]

In the first fourteen years of the twentieth century, the Society was indeed extraordinarily vigorous in publication and lobbying. It succeeded in persuading government to undertake at least some formal state responsibility for encouraging forestry as part of the remit of the new Development Commission of 1909 and of the Board of Agriculture for Scotland in 1914, though in practice the new bodies did little. The grandees of the Society were exasperated by the limited character of their success. Lord Lovat said in 1914 that they had seen, since the foundation of the Society, sixty years of private endeavour and sixty years of organised inactivity on the part of the state. Sir John Stirling Maxwell, the President, also eloquently vented his sarcasm:

> We have heard of the open door of the Development Commission – it is a very peculiar piece of furniture. It undoubtedly stands open and inside is a beautiful silence and a beautiful abstention from action of any kind so far as we are concerned. When we approach that door – it is invariably slammed in our faces.[60]

Such was the frustration of the Royal Scottish Arboricultural Society, but for them it proved to be but the darkness before the dawn. In that very same year, as the First World War swept Europe, Britain was exposed as extremely vulnerable to an interruption of timber supplies by enemy action, more nearly brought to her knees, according to Lloyd George, by

---

[58] *Ibid.*, pp. 371–2; W. Somerville, 'Forestry in some of its economic aspects', *Journal of the Royal Statistical Society*, 72 (1909), pp. 40–63.
[59] Lovat and Stirling, *Afforestation in Scotland*.
[60] Anderson, *History of Scottish Forestry*, 2, pp. 369–78, quotation on p. 378.

shortage of wood than by shortage of food. According to returns by the landowners themselves, some 149,000 acres were felled during the war in Scotland, out of an estimated 875,000 acres, or 17 per cent of the whole. Particularly hard hit were pinewoods on the estates of upper Speyside and upper Deeside, but the war also utilised former oak coppice around Dunkeld, for example, which had not been cut for years. In 1916 the Acland Committee was appointed to consider proposals for state-aided afforestation, and the proposals of the Royal Scottish Arboricultural Society bulked large in their findings. In 1919 the Forestry Act established the Forestry Commission with a purposeful agenda to bring about enough afforestation to save Britain from peril in another war. Lord Lovat emerged as the first Chairman, serving from 1919 to 1928, and Sir John Stirling Maxwell followed as the third Chairman, from 1929 to 1932. The Scottish imprint could not have been firmer.[61]

It is easy to think now that there was something inevitable about the arrival of the Forestry Commission, but the despair of Lovat and Stirling Maxwell in 1914 reminds us that this was not so. The main planks of the argument that the Royal Scottish Arboricultural Society and others used to justify the intervention of the state in the forestry sector, either by direct purchase of land or by some form of subsidy or tax advantage, was that Britain was among the least forested countries in Europe, and that before long there would occur an international timber famine that would leave a country so dependent on exports very vulnerable. Neither argument cut much ice outside the little world of forestry enthusiasts. A commentator, R. H. Rew, said at the meeting of the Royal Statistical Society called in 1909 to hear a paper by William Somerville in the aftermath of the report of the Coastal Erosion Commission:

> Dr Somerville probably thought they ought to be ashamed that the United Kingdom had only four per cent of its area under woodland, as compared with other countries, which made a very much better show. But that class of comparison always left him calm. It might be applied to many other products of which the proportion to the total area was very much less here than in other countries. For instance, France had twelve per cent of its area under wheat, whereas we had only two per cent; but he did not suppose Prof. Somerville would necessarily draw any deduction from that fact.[62]

[61] *Ibid.*, pp. 493–9; D. Foot, 'The twentieth century: forestry takes off', in Smout (ed.), *People and Woods in Scotland: a History*, pp. 158–94; G. Ryle, *Forest Service: the First Forty-five Years of the Forestry Commission of Great Britain* (Newton Abbot, 1969), pp. 19–48; Fowler, *Landscapes and Lives.*

[62] Comments appended to Somerville, 'Forestry in some of its economic aspects'.

As for a future famine in wood supplies, that was considered a possible but also a somewhat distant prospect. It was certainly not enough to jerk politicians from their usual short-termist frame of mind, and probably nothing would have done so had the First World War not demonstrated so suddenly and so starkly the strategic vulnerability of relying totally on foreign supplies.

The main specific aims of the Forestry Commission in its first period were to build up a strategic timber reserve, then from the 1960s to produce timber cheaply as a form of import substitution, and (throughout) to support the economy, and the employment opportunities, of rural areas. It is questionable if it was successful in any of these aims. The Second World War came before the modest plantations of the 1920s and 1930s had matured, and afterwards the Zuckerman Committee of 1957 pointed out that, as the next war would be over in a couple of nuclear flashes, there was no point in maintaining a strategic reserve. The Treasury then set the Commission the task of production forestry at lowest possible cost, which led the forest industry, public and private, to increase the proportion of Scotland under wood (according to official statistics) from 6 per cent (which was all that had been achieved by 1960) to 17 per cent by the end of the century. Almost all the increase until the final decade was achieved by planting non-native conifers, overwhelmingly Sitka spruce, which incidentally reduced the proportion of the land under ancient semi-natural wood to about 1 per cent of the surface area. Still 80 per cent of British timber needs were met by imports, and the perpetual need for production at lowest possible cost meant that the aim of the early Commissioners to stem the flow of population from the land by providing enhanced local employment had largely been frustrated. In the 1990s, the remit of the Forestry Commission changed towards the encouragement of a more holistic, multi-purpose forestry, combining production, recreation and nature conservation, a completely new agenda, which, greatly to its credit, it pursued with enthusiasm.

The approach of those in command of policy for most of the twentieth century, therefore, was for the most part, focused on non-native species and dismissive and destructive of the native woods. This was partly because the Forestry Act of 1919, as M. L. Anderson shrewdly observed, 'was not a true forestry policy but a policy of afforestation'.[63] Even when the Forestry Commissioners did see a wider purpose in its work, as in the creation of Forest Parks from 1936 and the softening of forest design under the guidance of Sylvia Crowe in 1963, they did not permit recreational or aesthetic priorities seriously to interfere with the task of afforestation with non-native

---

[63] Anderson, *History of Scottish Forestry*, 2, p. 499.

trees, since they were what grew best and achieved the targets for new forest cover, faster.

For most of the twentieth century, then, we were accustomed to a forestry culture highly inimical to the conservation of native woods. This was not merely a product of the modern official mind, since to a considerable degree it was inherited from the period before 1914. Most of what was to follow was foreshadowed in articles in the *Transactions of the Royal Scottish Arboricultural Society* in the pre-war years, and sprang from the particular market circumstances of the previous decades.

First, experience before 1914 showed that hardwoods now seldom paid, and the kind of sustainable management which, despite some shortcomings from a purely ecological perspective, had kept the native oakwoods going for centuries, no longer produced a sufficiently saleable product. Much the same was true of birch and alder. Though their preservation had been less systematic, tenants had always cared to keep enough for domestic use: but rural builders and tool makers now used imported wood. Nevertheless, the ground which they occupied indicated to the forester that other timber crops could grow there, and, however little value they might have, they were still worth something on the market to put towards the cost of replanting. They might be worth cutting but they were not worth keeping.

This was all very clearly expressed by Lord Lovat and Captain Stirling of Keir, in their influential forest survey of the Great Glen in 1911, which in many ways was to form a blueprint for future action after 1919. There were in the survey area, they said, 14,000 acres of birch and 3,400 acres of oak and scrub ('the remains of coppices formerly worked in regular rotation'), compared to 14,800 acres of conifer wood. The broadleaf trees still had some small value, the birch for brooms, bobbin wood and sawn blocks, the oak as firewood and perhaps as cross-pieces on telegraph poles, and a pyroligneous acid works might be tempted north to utilise the remainder. But, as a priority, they should be replaced by conifers for which the market prospect was much greater.[64]

Over the next fifty years, such remarks were commonplace, though usually bereft of the notion that oak and birch had any value whatsoever. E. P. Stebbing of Edinburgh University in 1916 spoke of the need to plant conifers, 'just what we could grow, and grow successfully, on our derelict and waste lands', especially on areas felled during the war and on 'the areas at present occupied by worthless scrub, (of which there are extensive tracts in Scotland)'.[65] A. Murray in 1917 recommended that the old oak scrub, once

---

[64] Lovat and Stirling, *Afforestation*, especially pp. 69–80.
[65] E. P. Stebbing, *British Forestry: Its Present Position and Outlook after the War* (London, 1916).

coppiced, should be cleared for firewood or distillation, and the main question for the Forestry Commission thereafter seemed to be whether it would be better to kill hardwood trees by ring-barking or uprooting.[66]

These attitudes were much in evidence in the 1940s, when the Forestry Commission was confronted with proposals to establish National Parks (in the event only in England and Wales), and nature reserves under the care of the proposed Nature Conservancy. At a meeting in January 1944, the Commissioners met to discuss the (as yet unpublished) Dower Report, and were deeply hostile. The Chairman, Roy Robinson said that Mr Dower had misunderstood that the public preferred open moors, for in reality they preferred to walk among trees, and that some of the proposals would make it impossible for the Commission to carry out their task. John Bannerman, a leading Scottish Commissioner, said that the effect of the recommendations would be to sterilise large sections of the countryside. But they saw that they had to tread carefully and to make their voices heard where it counted.[67]

By this time, however, it was becoming evident that some of the old native woods had friends, and some of these friends might even have influence. This also had been foreshadowed before the First World War, in the work of the early ecologists associated with Dundee, Robert and William Smith and Marcel Hardy and also in the work of the British Vegetation Committee which in 1910 published a classification of English woods into six ecological categories between the 'primitive' and the 'plantation', using the phrase 'semi-natural' for the first time to describe old managed native woodlands.[68] William Smith, secretary and co-founder of the committee, introduced this work and emphasised its applicability north of the Border, in a paper to the Royal Scottish Arboricultural Society the following year. The *Transactions* for 1911 also carried a notable article by G. P. Gordon examining survivals of 'urwald' (what we would now call ancient woodland) of birch, pine and oak in various parts of Scotland, mentioning among other broadleaf woods, Pressmennan, Dalkeith Park, Lochwood in Dumfriesshire and Loch Long and Loch Goil in Argyll.[69] By the 1940s, leading British ecologists like Arthur Tansley (who had been with the British Vegetation Committee from the start), Frank Fraser Darling and

---

[66] Anderson, *History of Scottish Forestry*, 2, p. 432.
[67] NAS: Forestry Commission (Scotland), FC 9/1.
[68] J. Sheail, *Seventy-five Years in Ecology: the British Ecological Society* (Oxford, 1987), pp. 6–10, 22–35; Anderson, *History of Scottish Forestry*, 2, p. 315; C. E. Moss, W. M. Rankin and A. G. Tansley, 'The woodlands of England', *New Phytologist*, 9 (1910), pp. 113–49.
[69] W. G. Smith, 'The vegetation of woodlands', *TRSAS*, 24 (1911), pp. 6–23, 131–8; G. P. Gordon, 'Primitive woodland and plantation types in Scotland', *TRSAS*, 24 (1911), pp. 153–77.

James Ritchie had a fairly clear idea of some of the prime sites they would like to see incorporated into nature reserves under the care of a new Nature Conservancy, and the Forestry Commission was on its guard.[70]

At first they were inclined to be disingenuous. Mr Gosling represented their views to the National Parks Committee in June, 1946, stating that at the Commission's latest National Forest Park at Glentrool they were trying to maintain traditional scrub woodland.[71] More revealing of their true underlying attitude was a letter to his superiors from the Scottish Director, Sir Henry Beresford-Peirse in March, 1949, warning that in nature reserves woodland would be neglected, and that 'there is a possibility of large stretches of poor scrub being left much in its present form when the area could grow timber of far greater value and quantity'.[72]

In the event, the Commission need hardly have worried. Only about a dozen National Nature Reserves were created with significant resources of oak, birch, alder, hazel or ash wood, most of them small, though all remarkable and precious, and until 1981 there was no effective protection from afforestation or anything else for sites merely declared Sites of Special Scientific Interest. The conversion to conifer plantation went on apace: perhaps 40 per cent of birchwoods in Highland Region were lost in the four decades after 1945.[73] There were protests, and not all were directed at the Forestry Commission. Frank Fraser Darling in 1968 deplored the practice of the Department of Agriculture giving a 50 per cent grant to cut down a birchwood and turn it into an indifferent pasture soaked in fertilisers:

> Any Highland natural woodland should be valued in terms of tourism and recreation, for let it be quite clearly understood, the recreational value of Highland land is now likely to be its greatest commercial value.[74]

There was a voice ahead of its time.

In respect to the native woods of Scots pine, the situation was slightly,

---

[70] J. Sheail, *Nature in Trust: the History of Nature Conservation in Britain* (Glasgow, 1976), pp. 150–1, 164–7.

[71] NAS: Forestry Commission (Scotland), FC 9/2. Meeting with Mr Gosling, representative of the Forestry Commission, 26 June 1946.

[72] NAS: FC 9/3. Letter from Director of Forestry Scotland, 4 March 1949.

[73] The point is hard to establish, but in the 1947–9 census the Forestry Commission recorded 50,896 ha of broadleaf wood and scrub and devastated woods, while the Friends of the Earth Survey of 1987 recorded 24,281 ha of birchwood. Birch is by far the commonest broadleaf in the region. I am indebted to Neil MacKenzie for this information.

[74] F. Fraser Darling, 'The ecology of land use in the Highlands and Islands', in D. S. Thomson and I. Grimble (eds), *The Future of the Highlands* (London, 1968), p. 52.

though not very markedly, different. It was certainly not seen as the ideal tree for the new programme of conifer expansion. Larch had been a general favourite in the nineteenth century, and the virtues of Norway spruce had also been early recognised. But Sitka spruce was quickly identified as the tree *par excellence* for the twentieth century, both by Stirling Maxwell's experiments at Corrour and in an article by J. D. Crozier in 1910, significantly entitled 'Sitka spruce as a tree for hill planting and general afforestation'. It was, said the author, the best of the introductions for producing timber on high, moisture-holding land.[75] It needed the invention in the 1950s, of the Cuthbertson plough that could tackle deep peat, and the era of mechanical drainage that followed, fully to prove the truth of his words.[76]

Nevertheless, Scots pine had always had its admirers as a tree well suited for the thin glacial soils where it naturally grew best. Prominent among them was Lord Lovat, whose estate had been among the few in Scotland systematically to utilise the natural regeneration of pine in the late nineteenth century, and which had attracted the approving attention of a number of British and foreign foresters, including the great Sir Dietrich Brandis, Inspector General of forests in India.[77] Besides, there were aesthetic considerations. Romantics as well as foresters had long admired the sense of timelessness among the open glades with the largest trees, and sometimes, as in David Nairne's account of 1891, they were combined into one persona. Visiting Glen Mallie forest on the Lochiel estate, his sense of wonder was awoken:

> We came across trees of striking grandeur. The most notable, principally on account of its magnificent ramifications, is named 'Miss Cameron's tree', or more poetically, 'The Queen of the Forest'. It appropriately stands amidst the most rugged beauty of the primæval forest, guarded by the massive and umbrageous proportions of its juniors.

He measured it at eighteen feet circumference a yard from the swell of the roots, twenty-four feet at a point where 'it bifurcates into seven enormous limbs', four of which totalled forty-seven and a half feet in circumference, and declared that its height and spread could not be calculated.[78] It had been a famous tree for at least a hundred years before that, but no trace of it now remains.

---

[75] J. D. Crozier, 'Sitka spruce as a tree for hill planting and general afforestation', *TRSAS*, 23 (1910), pp. 7–16.
[76] Foot, 'The twentieth century', pp. 177–8, 183.
[77] Nairne, 'Notes on Highland woods', pp. 202–3.
[78] *Ibid.*, p. 216.

Such feelings of respect were only added to by the researches of the early ecologists. Gordon in 1911 counted among his 'urwald' the ancient and spectacular forests of Glen Lui, Rothiemurchus, Ballochbuie, Glenmore, Abernethy and Loch Arkaig.[79] When two forestry academics from the University of Aberdeen, H. M. Steven and A. Carlisle, in 1959 wrote their remarkable monograph, *The Native Pinewoods of Scotland*, they built on both foundations. It was a minute account of each ancient pinewood in Scotland, of the character and morphology of the trees and of the associated flora and fauna. But the most quoted passage in the book comes from the preface, and recalls the feelings of Nairne nearly seventy years earlier:

> Even to walk through the larger of them gives a better idea of what a primeval forest was like than can be got from any other woodland scene in Britain. The trees range in age up to 300 years in some instances, and there are thus not very many generations between their earliest predecessors about 9,000 years ago and those growing today; to stand in them is to feel the past.[80]

It is interesting that Professor Steven did not feel at all the same about birchwoods. In 1954 he commented that there had long been large areas of natural birch in the Highlands, most of it 'poorly stocked or scrub': the Forestry Commission and some private owners had converted some of these to conifer forest, 'but this has only touched the fringe of the problem'.[81]

All things considered, therefore, it is understandable that the Forestry Commission apparently felt a little more gently towards Scots pines than towards native broadleaves. There was an episode in 1935 when R. A. Galloway, who for forty years had served as Secretary of the Royal Scottish Arboricultural Society (now become the Royal Scottish Forestry Society), contacted his opposite number in the newly formed National Trust for Scotland to suggest that the two organisations collaborated to obtain a remnant of old pine forest to ensure its preservation. The attempt floundered because no landowner could be persuaded to sell to the NTS, but in 1936 the Forestry Commission told Galloway that they were:

> Entirely sympathetic to the Society's proposals that some remnants of the Natural Forests of Scotland should be retained in their original

---

[79] Gordon, 'Primitive woodland'.
[80] H. M. Steven and A. Carlisle, *The Native Pinewoods of Scotland* (Edinburgh, 1959), p. v.
[81] H. M. Steven, 'Changes in silvicultural practice in Scotland, 1854–1954', *Scottish Forestry*, 8 (1954), p. 53.

state and they have already set aside an area of forty-two acres in Glenloy, near Fort William.

They further intimated that they would consider other areas for special protection in Ghiusachan, Glenmore, Glengarry and Achnashellach, but they intended to retain ownership and to ensure that 'the stock of trees upon them will receive all proper attention henceforward', while preserving them from the grazing pressures of game and domestic stock.[82]

There is no indication of what in practice this promise was worth. It certainly did not inaugurate an era of publicly acknowledged forest reserves, though at least in Glen Loy the original forty-two acres remained in a wonderfully pristine state. Mr Gosling in 1946 told the Douglas Ramsay Committee that the Forestry Commission liked to preserve fragments of old woodland, and that had they been able to buy Glen Affric (as they later did) they would have intended 'to manage a good section of these woods to maintain their present character rather than to run them entirely as a commercial proposition'.[83] Two years later Beresford-Peirse was asked by James Ritchie how the Commission would treat the Black Wood of Rannoch:

> I told him that we would fence it, drain it and try for natural regeneration of the pine, but I pointed out that the conditions were not very favourable over much of the area, and that in the end we might have to abandon any idea of regenerating the wood naturally and that we might have to plant parts of it with other species more favourable to the site.[84]

This is exactly what happened, and the present Black Wood exists surrounded by swathes of Sitka spruce, though (unlike in some other woods) the main stands of old pines were not underplanted.

Beresford-Peirse later turned up on the Scottish advisory committee of the Nature Conservancy in Scotland, and pressed the case at every opportunity for giving the Forestry Commission primacy over the fledgling conservation body. There was a well-recorded spat over Coille na Glas Leitir, the ancient pinewood in Scotland's first National Nature Reserve established in 1951 at Beinn Eighe in Wester Ross. First, the Commission wanted to buy the wood under the noses of the Conservancy, and had to be

---

[82] R. A. Lambert, 'Preserving a remnant of the old natural forest: the joint venture of the NTS and RSFS in the mid-1930s', *Scottish Forestry*, 51 (1997), pp. 31–3.
[83] NAS: FC 9/2. Meeting with Mr Gosling, 26 June 1946.
[84] NAS: FC 9/3. Reports on a conversation, 23 January 1948.

reminded that the latter had legal precedence in this case. Next, they wanted to manage it: Beresford-Peirse told John Berry, his opposite number, that the Forestry Commission would be happy to meet the Conservancy's needs if they want only indigenous species to be used, but 'possibly larch could be planted'.[85] When this kind offer was declined, the Commission demanded that it be allowed to afforest the area of the NNR beyond the wood, and, to placate them, the Conservancy did ultimately allow the afforestation in 1960 of some 400 acres with 10,000 Scots pine of non-local origin, 27,000 lodgepole pine and 10,000 Sitka spruce. The dispute was important because, despite this concession, it finally established that ecologists as well as foresters had the right to manage woodland on behalf of the state, something the Forestry Commission had been anxious to avoid.[86]

The Commission's management of their own ancient pinewoods in the post-war period did not impress Steven and Carlisle. They noted that, apart from some limited experiments with natural regeneration and seed-sowing, they 'have tended to manage the native pinewoods in their ownership in the same way as other woodlands', and in particular that they had used non-native strains of Scots pine in forests such as Glenmore, and planted other species of conifer in 'some of the native woodlands'. Their plea for gentler treatment was only modest:

> Where they are extensive, and there are large areas with only scattered native pine, it is only reasonable that at least parts of such land should be used for normal productive forestry, but the most distinctive parts of the native pinewoods should be maintained in their natural condition.[87]

Even private woodland owners were not exempt from pressure by the Commission: the laird of Rothiemurchus had to go to government ministers to defend his own woods from 'improved' management.[88]

A similar unease to that of Steven and Carlisle was evident in M. L. Anderson's critique in 1967 of national forest policy. Good policy, he said, should have a protective as well as a productive aim, and should encompass 'the replacement of some forest cover of a suitable type to benefit agriculture and the general commonwealth'. He had soil amelioration as well as

---

[85] J. L. Johnston and D. Balharry, *Beinn Eighe: the Mountain above the Wood* (Edinburgh, 2001), pp. 26–48, 67–70.
[86] *Ibidem*.
[87] Steven and Carlisle, *Native Pinewoods*, p. 300.
[88] W. H. Murray, *Highland Landscape: a Survey* (Edinburgh, 1962), p. 16.

amenity in mind, and was thinking in particular of the benefits of returning to broadleaf and mixed planting.[89]

No new spirit, alas, came upon the leadership of the Forestry Commission until the 1980s, when, largely as a result of initiatives pursued by George Peterken, forestry expert in the Nature Conservancy Council, and Oliver Rackham, Cambridge botanist and forest historian, the Commission became converted first, to the notion of ancient woodland (the validity of which was strongly denied by the Director General as recently as 1978) and secondly, to the proposition that it deserved to be properly identified and protected. A conference at Loughborough on 'Broadleaves in Britain' in 1982 was the key moment, the point when 'the Forestry Commission recognised that they were out of step with the rest of forestry thinking, and their attitudes changed considerably thereafter'.[90]

Government fiscal rules were also changed later in the decade, away from providing the big tax breaks for private forestry companies that had led to the afforestation of the Flow Country in Caithness and Sutherland and an enormous public row about the consequent damage to nature conservation interests in a unique area. The ethos of the Forestry Commission and of British forest policy was transformed with extraordinary speed. A new emphasis emerged on holistic forestry, balancing production, amenity and conservation, inventories were made of ancient woodland in England and Wales and in Scotland, grants were available for community forests and farm woodlands, and the planting of native broadleaves and Scots pine was particularly encouraged. Most remarkably of all, the Commission threw itself with enthusiasm into the task of restoring woods in their ownership badly treated in the past rush for production at all costs. At Glenmore, Glengarry, Glen Affric and elsewhere the ancient pines emerged, battered but more or less intact once their shroud of Sitka spruce had been removed: on Lochlomondside, Sunart and elsewhere the woods were restored to broadleaf. The speed and enthusiasm of this change of heart bears out what we have learned from more than one confidential piece of oral history, that many among the staff of the Conservancy, who could not bear the direction that things had taken in previous decades, were only too glad to throw the mighty machinery into reverse. By the 1990s, the attitudes bred by the problems of forestry before 1914 were finally put to rest.

---

[89] Anderson, *History of Scottish Forestry*, 2, pp. 562–3.

[90] J. Tsouvalis-Gerber, 'Making the invisible visible: ancient woodlands, British forest policy and the social construction of reality', in C. Watkins (ed.), *European Woods and Forests: Studies in Cultural History* (Wallingford, 1998), pp. 220–42.

CHAPTER 11

Rothiemurchus, 1650–1900

At this point in our account, the focus changes to consider much more closely how management decisions worked out on the ground, in four case studies of a chapter each: one is in the heartlands of the ancient pinewoods of Strathspey, one in the far north, one in Argyll and one on Skye. In each case, market forces from outside unleashed big changes. None of the consequent management responses was a good model for sustainability, but the impact varied according to location and climate as much as to market opportunity. Of the woods that we consider, the most famous and still outwardly the most pristine is Rothiemurchus, where in the words of two modern romantics, 'it is still possible . . . to sense the past reaching back through the shadowy trunks to the untouched wildwood of long ago'.[1] Its history, however, is of a wood continuously exploited.

Rothiemurchus, together with the equally famous forests of Abernethy and Glenmore (Map 11.1), probably in the Middle Ages formed one contiguous block of woodland that touched over the Pass of Ryvoan, as they still almost do on Roy's Military Survey of the 1750s, which shows Abernethy stretching further south than its current limits. Rothiemurchus, may have at one time stretched a finger up Glen Einich, past Carn a' Phris-ghiubhais (the Hill of the Pine Bush – or clump), now bare moor, to within 10 kms of Glen Ghiusachan, the vanished pinewood at the head of Glen Dee. All that had apparently gone by 1750, possibly, like other woods in the area that are identifiable only by a name, the victim of climate change at high altitudes. For what it is worth, Rothiemurchus forest itself appears more densely wooded within its bounds on Roy's map than on late twentieth-century maps, though there were always open spaces for farms and cultivation. On Gordon lands within the adjacent forest of Glenmore, there were no fewer than twenty summer shielings listed in a document of 1679, and it is reasonable to imagine all the Speyside forests, like the forest of Mar, as at least grazed seasonally by cattle, though we know nothing of stocking densities. Estate maps of 1762 and 1789 at the Doune, also suggest that much

---

[1] H. Miles and B. Jackman, *The Great Wood of Caledon* (Lanark, 1991), p. 18.

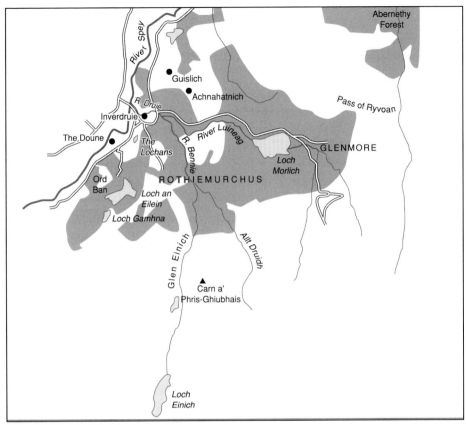

Map 11.1 Rothiemurchus Forest and Glenmore.

of the ground was not as densely wooded as Roy depicts. Though not in reality pristine, Rothiemurchus shares most of the ecological characteristics of its neighbours, and does indeed support a remarkable flora and fauna that reaches back to the wildwood.[2]

---

[2] In general on the Cairngorms, C. Gimingham (ed.), *The Ecology, Land Use and Conservation of the Cairngorms* (Chichester, 2002); on the pinewoods, H. M. Steven and A. Carlisle, *The Native Pinewoods of Scotland* (Edinburgh, 1959); on Rothiemurchus, T. C. Smout and R. A. Lambert (eds), *Rothiemurchus: Nature and People on a Highland Estate, 1500–2000* (Dalkeith, 1999), where an earlier version of this chapter appears on pp. 60–78, and T. C. Smout, 'The history of the Rothiemurchus woods in the eighteenth century', *Northern Scotland*, 15 (1995), pp. 19–32. See also H. L. Edlin, 'Place names as a guide to former forest cover in the Grampians', *Scottish Forestry*, 13 (1959), pp. 63–76, but Simon Taylor *pers. comm.*, points out that Edlin mistranslated Phris-ghiubhais as 'prized pine trees', and that pine names at high altitude could refer to bog pine remnants as well as to living pine. For the Glenmore shielings, Alasdair Ross, University of Stirling, 'Transhumance in the Cairngorms of Moray', forthcoming.

Of the three Speyside woods, Abernethy has attracted most historical attention, and Glenmore has been noticed especially as the source for eighteenth-century shipbuilding timber. Rothiemurchus had been the least studied before our research into the papers at the Doune, its past being known mainly from the brilliant autobiography of Elizabeth Grant, daughter of the spendthrift laird of the early nineteenth century, with its snapshot of the opportunist exploitation of the age.[3] Rothiemurchus, however, unlike the other woods of Speyside and indeed those of most parts of Scotland, has remained in the hands of one family for over four centuries, since Patrick Grant of Muckrack in 1580 finally succeeded in expelling the Shaws. It thus provides an opportunity to examine quality of management by one family over many generations.[4]

The earliest references to Rothiemurchus nevertheless date to long before the Grants. In 1226 it was granted by the Crown to the Bishop of Moray to be held 'in forest', with the customary penalties of £10 sterling for felling timber or hunting without episcopal leave. It is clear that the clerics were interested in sport, for they reserved for themselves all eyries of hawks (probably goshawks), but they were also interested in timber. In 1383, the bishop employed a head forester to help to regulate the conflicting claims of hunting, grazing and felling. A document of 1537 sold the life-rent of Rothiemurchus to George Gordon, constable of Ruthven, on the condition that he yearly provided the bishop, as his predecessor had done, with 160 pine trunks big enough to serve as joists, to be delivered to a yard on the Spey for floating downriver. Already the wood had more than local use.[5]

These, however, are just chance mentions. It is only from about the middle of the seventeenth century that the records of the Doune become plentiful enough to tell anything like a continuous story. In the century and a half from about 1650, exploitation of the woods was characterised by the spasmodic attempts by outsiders, or by the laird's family, to reach a southern market. These efforts failed to bring sustained profit, and probably had little impact on the woods. It was also characterised by the continuous trafficking in wood by local people, either buying from the laird or, more

---

[3] E. Grant, *Abernethy Forest: its People and its Past* (Nethy Bridge, 1994); R. Summers, *The History and Ecology of Abernethy Forest, Strathspey* (RSPB Internal Report, 1999); J. Skelton, *Speybuilt: the Story of a Forgotten Industry* (Garmouth, 2nd edn, 1995); Elizabeth Grant, *Memoirs of a Highland Lady* (edn London, 1978), esp. chapters 9, 10 and 16.

[4] T. C. Smout, 'The Grant lairds of Rothiemurchus', in Smout and Lambert (eds), *Rothiemurchus*, pp. 14–21.

[5] G. W. S. Barrow, 'The background to medieval Rothiemurchus', in Smout and Lambert (eds), *Rothiemurchus*, pp. 1–6; N. A. MacKenzie, 'The native woodlands: history, decline and present status', in Gimingham (ed.), *Cairngorms*, p. 109.

probably and commonly, operating with permission from the laird and selling on, downriver or overland. This exploitation succeeded. It was broadly sustainable but small scale, though so persistent that it is likely to have had a substantial effect on the structure and character of the wood.

The nineteenth century, by contrast, was dominated by spasms of heavy exploitation by the landowners or their creditors, including two episodes approaching clear-fell, but also accompanied by the first efforts of the owners to protect the forest from domestic stock. At this time local people had no stake in the forest except as employees – fellers, haulers, sawyers or floaters – and the marketing of the wood was in the hands of the estate. The effect on the wood at this time was very severe, and appeared to contemporaries as catastrophic, but in the long term Rothiemurchus proved to have remarkable resilience. Since 1914, apart from episodes of felling in the two world wars, the emphasis has been on conservation. This was true both before and after the establishment of the Cairngorms National Nature Reserve, of which the woodlands form an important component. The emphasis in this chapter will be on the period 1650–1900.

The Rothiemurchus pines were thus already familiar to the outside world and utilised for the market by the middle of the seventeenth century. An account of the parish produced for Sir Robert Sibbald around 1682 speaks of a sawmill and a 'great firr wood of two miles in length, but very broad in respect it runneth up into many burnes', and one for the adjacent parish of Duthil speaks of the population neglecting working the land, 'being addicted to the Wood which leaves them poor'.[6] The situation in Strathspey was probably much as in Mar, in Glentanar, Glencarron, and other places where local people gained part of their living from cutting deals and spars in the woods and floating them downriver or taking them by horse to the nearest market.

The first outsider who thought he might turn a penny from the woods was an Englishman, Benjamin Parsons, described initially as a merchant of Aberdeen and doubtless an incomer with the Cromwellian Union, who in 1658 secured interests both in Rothiemurchus and in Abernethy. In Rothiemurchus he obtained a lease of 'the whole fir woods' for thirteen years, with permission to build sawmills on the local streams. By 1664 Parsons had taken over the whole of Abernethy, and was also described as master of the sawmill at Garmouth. By 1671 he had acquired additional interests at Kincardine and even at Glencalvie in Ross-shire, and was building a ship at Garmouth, no doubt from Strathspey pine. Nevertheless, he was soon in

---

[6] Arthur Mitchell (ed.), *Geographical Collections relating to Scotland made by Walter Macfarlane* (Edinburgh, Scottish History Society, 1906), 3, pp. 240–2.

deep financial trouble, pursued for outstanding debts by the laird of Grant in 1669, and by the laird of Rothiemurchus and Dr Samuel Collins of London in 1671. Next year he is described as a merchant of Leith and nothing further is heard of him in the north. It remains a mystery how much timber he was able to extract from Rothiemurchus in his thirteen-year tack, but there is no evidence that it was a vast amount.[7] Rothiemurchus wood was purchased for repairs to the Tolbooth at Nairn in 1673, so the withdrawal of Parsons from the scene did not bring trade to an end, though the account of the parish of Duthil around 1682 puts the floating of timber on the Spey as something in the past.[8]

Immediately after the Act of Union, two other external partnerships appeared, in 1709 of one Captain Brodie, 'indweller in Edinburgh' and two partners surnamed Grant, and the other in 1710 of John Gage, merchant in Banff and Donald Tayliour, merchant in London – further proof of the attraction of English capital to the deluded idea that there was money in Scots pine. The laird ultimately preferred the Banff–London partners to the Edinburgh-based ones. Gage was still busy five years later, when he gave a discharge to the laird for the work of no fewer than 300 men to clear rocks in the Bennie, a remarkable testimony to the number of men the laird of Rothiemurchus could apparently call together from his estates to perform direct labour.[9]

These engineering operations on the burns were no doubt related to the same partnership's contract with the Duke of Gordon to cut pine timber in Glenmore, for which permission ('a servitude') was needed for floating through Rothiemurchus ground. When the Banff–London partnership gave up, apparently frightened off by the 1715 rebellion, the Duke of Gordon was able to replace them for a decade in Glenmore by a new English partnership from Durham and Stockton, for whom he built a new sawmill on Rothiemurchus ground 'at the upper end of the waggon road'.[10] Correspondence of the same period, however, seems to imply that the Rothiemurchus estate had then taken the exploitation of their woods back into their own management.[11]

---

[7] Doune MSS: 48; J. Munro, 'The golden groves of Abernethy: the cutting and extraction of timber before the Union', in G. Cruickshank (ed.), *A Sense of Place: Studies in Scottish Local History* (Edinburgh, 1988), pp. 160–1.
[8] We are indebted to Dr Pat Torrie for this information; Mitchell (ed.), *Geographical Collections*, 3, pp. 240–2.
[9] Doune MSS: 82, 83 and correspondence of 1715. It is, however, possible that 300 man-days is implied rather than the labour of 300 individual people.
[10] SRO: GD 44/29/6/24 and 2/10.
[11] See Doune MSS, correspondence, especially a letter of William Stewart, Garmouth, 1720.

The next speculative adventure at Rothiemurchus was more ambitious than ever, and still more ill-fated. In 1738, James Grant, eldest son of the ageing Patrick Grant, the sixth laird, encountered an Englishman named John Lummis, who is named in other documents as a 'wood surveyor' of Sheriff Hailes in Shropshire or of Preston in Lancashire. Lummis is implicated elsewhere in bargains that went badly wrong because the wood could not be extracted or because it did not live up to expectations (see above, pp. 208–9). He enticed a company of Whitehaven men to buy pine from Donald Cameron of Lochiel in 1739, a venture which ended in the courts, and a partnership from Preston and Liverpool to buy pine from Roderick Chisholm of Strathglass in 1742, which also ended in futility and litigation. Lummis in addition had personal interests in woodland at Callart on Loch Leven that he wanted to sell for charcoal manufacture.[12]

In this case he persuaded James Grant to write him a letter after his visit to Rothiemurchus detailing the timber on the estates, 'as you have seen and experimented'. Grant claimed in the letter that the estate contained 20,000 acres covered with firwoods, that each acre contained nearly 50,000 trees, that therefore there were 100 million trees available (no-one seems to have noticed that the sum should have amounted to a billion trees).

Even allowing for this, there appear to be wild elements of exaggeration in these claims. Later in the same document the wood was said by Grant to be fifteen or sixteen miles in circumference, a dimension roughly corresponding to claims in 1808 that it covered about sixteen square miles, and 1780 that it covered 'not less than twenty square miles': but Alexander Wight in 1784 reported it as five miles long and three broad.[13] Probably all these estimates were on the high side, as lawyers on those occasions also wished to emphasise the scale and density of the Rothiemurchus woods, and claimed that 'over the whole of which, except in partial openings, the trees grow as close to each other as their respective sizes enable them to stand': but estate maps, admittedly not complete, suggest in 1762 and 1789 that much of the ground was moor, pasture or only lightly wooded. There

---

[12] For Lummis, see NLS: Haldane papers, letters from Donald Cameron dated 2 July 1739 and 21 April 1740 (we are indebted to Professor Allan Macinnes for these references); SRO: RD 2/151, 21 April 1742; J. Munro (ed.), *The Inventory of Chisholm Writs, 1456–1810* (Scottish Record Society, Edinburgh, 1992), p. 167.

[13] Doune MSS: *Memorial and queries for John Peter Grant, Esquire of Rothiemurchus for the Opinion of Counsel* [hereafter *Memorial for J. P. Grant*] (London, 1808), pp. 1–2; *Memorial for Sir James Grant Bart . . . and others . . . against Alexander, Duke of Gordon*, Edinburgh, 22 April 1780 [hereafter *Memorial for Sir James Grant*], p. 41; see also, *Memorial for Alexander, Duke of Gordon . . . against Sir James Grant*, Edinburgh, 1780 [hereafter *Memorial for Duke of Gordon*]; Alexander Wight, *Present State of Husbandry in Scotland* (Edinburgh, 1778–84), 4, p. 210.

is the further point that the woods were by no means exclusively pine: James Robertson the botanist described the Speyside woods in 1768 as 'fir [i.e. pine] and birch', and the Rothiemurchus woods specifically in 1771 as having 'abundance of excellent fir and birch, together with some hazle and poplar'.[14] There was probably much greater diversity than has developed in the woods since. So 6,000 to 8,000 acres of mixed wood might seem to be a more faithful outside estimate of the extent of the Rothiemurchus forests around 1738.

That, however, is incidental. James Grant went on to say that the wood would be found to contain at least 5 million trees that would yield timber six inches square measured six feet above the ground, and that 300,000 of the largest would yield timber twelve inches square at that height: 'the least dimensions of the trees are sixty foot in height'. If the trees were felled over a period of thirty years, they would be worth £1,000 a year to the estate, but as he was burdened by a wife and children he could hardly come to England to 'travell and roam around for proper traders', so he was willing to let them go for a thirty-year contract at a mere £500 a year. Would Mr Lummis please try to find some English merchant who would take them off his hands?[15]

These sums were ambitious, considering that it was later said (admittedly only by a lawyer) that at this period 'any man would be permitted to go into his woods and manufacture at pleasure, for a season, for payment of the small sum of 40 lib. Scots'.[16] The offer to allow felling for £500 sterling a year should be seen in the light of the optimistic operations of the York Buildings Company, the English speculators who had contracted with Sir James Grant of Grant in 1728 to buy 60,000 trees from Abernethy and to extract them over fifteen years for a payment of £7,000, with equal lack of success.

The letter was obviously designed by Grant and Lummis in collaboration, and it worked. A Hull merchant partnership of Andrew Perrott and David Field rose to the bait, though they were cautious enough to send their own 'servants' to view the wood on their behalf: these were later to claim that the Grant foresters took them 'twice or thrice through the same places of purpose to make them believe there was a great abundance of large trees in

---

[14] Doune MSS: Plans of the Mains of the Down of Rothemurchas, surveyed by William Henderson, 1762; Plan of the Estate of Rothiemurchus, surveyed by Archibald Tait, 1789; D. M. Henderson and J. H. Dickson (eds), *A Naturalist in the Highlands: James Robertson, His Life and Travels in Scotland, 1767–1771* (Edinburgh, 1994), pp. 102, 167.
[15] Doune MSS: 600, Copie of the reasons of reduction, Perrot and Field against Grant of Rothiemurchus, 1742.
[16] *Memorial for Sir James Grant.*

the wood', a statement which the Grants naturally and indignantly denied. Nevertheless, at the time Perrott and Field were satisfied that 'the wood is as good as the best Norway wood', and agreed to buy and cut at a fixed rate 40,000 trees anywhere in the wood, half of them to be ten to twelve inches square at ten feet above the ground, and half to measure five to seven inches, the buyers agreeing to pay $^{1}/_{2}$d. an inch extra for any of the second sort larger than specified. The bargain was later varied to allow the merchants to cut 40,000 trees of any size they chose.[17]

Within two years, however, Perrott and Field gave up, after paying Grant £400 for pines and some other sums: they had not been able to find 500 trees of the larger dimension in the whole wood, and claimed that there were not 2,000 in the wood even six inches square at ten feet above the ground. Altogether they had felled 9,000 trees in two years. Alleging that they had been 'egregiously cheated and imposed on' they stopped work and demanded redress for 'such a notorious imposition brought about by the most unheard of cunning and deceit'. When they complained to Patrick Grant that there were too few marketable trees in his woods he (in their words) 'pulls off the mask' and said that it was no concern of his that the trees were small, since they had agreed to pay for 40,000 irrespective of their size. They had already paid £400 sterling to Patrick Grant for two years' work on the pines, as well as paying another £167 for alder trees, lending James Grant £104, and spending £1,600 on the capital and running expenses of the work: even the farm and accommodation promised them as a base for operations had proved inadequate, and they had had to pay extra to obtain proper housing and facilities. Now, they said, the laird of Rothiemurchus was demanding £4,000 for non-fulfilment of contract.

Patrick Grant for his part, rejecting the portrait of himself and his son as villains, said that the Hull merchants had cut 9,500 trees, the 'best of his woods', without the least complaint about their size and quality, and claimed there were still many large trees in the wood. Their problem was, he claimed, that they were trying to escape from a contract not as profitable as they had hoped for reasons that had nothing to do with the wood resources at Rothiemurchus. But, not wishing to be vindictive, he was now willing to release them from the contract for a payment of £500 on 1 May 1742, and £500 on 1 November 1745.[18]

No doubt the truth lay between the two extremes so dramatically portrayed by their lawyers' pleadings, but even before legal proceedings commenced the merchants had privately written to the laird to say that the

---

[17] Doune MSS: 600, Letter from Andrew Perrott and David Field, 6 December 1738.
[18] Doune MSS: 600, various.

wood was too young and too small to cut with profit and advising him allow it to stand some years longer to allow it to increase in value.[19] That the laird was not on the strongest ground is also indicated by his immediate offer in court to settle for a payment of £1,000 instead of the £4,000 he had originally claimed.

This was the last time that the Grants of Rothiemurchus invited outsiders to exploit their woods, but their experience at this point was not very different from the owners of Abernethy, Grant of Grant, or the owners of Glenmore, the Duke of Gordon, or, for that matter, from owners of pinewoods in the north and west like the Chisholms in Strathglass or the Camerons of Lochiel. As it was summarised in 1781 by legal presentation in a dispute with the Duke of Gordon about rights to float wood through the Rothiemurchus estate down to the Spey:

> A number of English adventurers since the Union have been allured from the prospect of manufacturing wood that paid no [import] duty, and relying probably on the unfavourable opinion that had been conceived in England of the want of industry in the inhabitants of the Highlands of Scotland and a total ignorance of the use of any sort of machinery or management of water carriage, have induced attempts from time to time to manufacture Highland woods to be sent to England to be consumed there; though, so far as the memorialist can learn, few if any of these adventurers have had cause to boast of the profits they have just acquired.[20]

The rather superior tone of the Rothiemurchus lawyers at this juncture conceals the fact that the Grants themselves had again recently dabbled in the English market and burned their fingers, but this time more personally. William, brother to Patrick the eighth laird, had followed a prosperous medical career in London, and in his sojourn in the capital became convinced of the commercial possibilities of selling hollowed pine logs to the London water companies, in competition with the English elm that they usually used. In 1770 he persuaded his reluctant brother to allow the operation of a boring mill on the estate and the despatch of an experimental cargo of logs, bored and unbored, via Garmouth.

William was full of energy and optimism about the venture, his brother always less so. After a year the doctor made arrangements to go it alone, and for over two years cut, bored and shipped timber to London on his own

---

[19] Doune MSS: Letter from Andrew Perrott and David Field, 14 February 1741.
[20] Doune MSS: 343.

account: he provided about 4,000 pipes in all, 'good, bad and indifferent'. In the end the price of elm fell so low that Rothiemurchus wood became uncompetitive, and in 1774 the venture was abandoned.[21]

The small size of the trees, however, had, just as with Perrott and Field, been a continuous handicap to the boring mill. William Grant had difficulty in supplying the optimal size, trees of ten inches in diameter, from which pipes could be made of three or four inches diameter. Taken in conjunction with the complaints of the Hull merchants, it seems that few of the Rothiemurchus trees were of the dimensions that Steven and Carlisle tell us is normal in twentieth-century Scottish pinewoods, usually between five and six feet circumference at breast height.[22]

If they were so small, what was the reason? It seems unlikely that the English adventurers from 1658 were able to despoil them so thoroughly, as their operations were intermittent and unsuccessful. Much more likely as an explanation is the activities of local people, cutting frequently wood of modest size that would be convenient to simple tools and carriage by horseback, and which would mainly be used for building small houses. The Rothiemurchus estate accounts are not complete enough to give a full picture of this activity, but we know, for example, that in 1766 there were 106 buyers of wood 'at the back of Lochinnellan', all local men apart from one from Cromarty. Between 1769 and 1771 two-thirds of the wood sales was in 'sparwood', totalling 13,823 pieces: most of the rest was in deal plank, 'milnsawn deals' predominating over 'handsawn deals'. Only the 'milnsawn deals' involved individuals at Garmouth, the only sales outwith the local area, though we know there was a lively trade downriver and overland by horseback, in pieces of timber destined for the outside market. On average, wood sales of this sort realised £370 per annum, more than twice the rental of the farms of Rothiemurchus and nearly twice as much per year as the total of £400 that Perrott and Field actually paid for their two years' operations in 1739 and 1740.[23] One can understand the reluctance of the laird to support his brother's boring mill if that speculation was to interfere with his steady income from country trade.

After the collapse of the boring mill, exploitation of the woods then reverted to a more traditional pattern of others acting as cutters and merchants, exemplified by the conditions of sale of 'two firr wood bargains' in 1778 – one consisting of 3,000 trees sold at 3s. each, allowing the laird £25

---

[21] Doune MSS: correspondence of 1770–4. For more detail, see Smout, 'History of the Rothiemurchus woods', pp. 25–6.
[22] Steven and Carlisle, *Native Pinewoods*, p. 77.
[23] Doune MSS: 143, wood sales 1766, 1769–71.

to build a proper sawmill, the purchasers to have six years to cut and saw the trees, but to pay for them in three; the other involving 2,000 trees at 2s., taking four years to manufacture but three to pay, the purchasers to build the whole sawmill at their own expense 'on any part of the Bannie they please above the present saw mill'. Another contract of similar date between Patrick Grant and two Scottish operators, probably local men, John Carmichael and James MacGregor, allowed them to buy £150 of timber in the woods. It specified that they were first to take 'the decayed wood or the wood which is not growing commonly called "garracks"', which was to be marked and valued by Grant's foresters; thereafter they could choose good growing trees, 'and the best trees they shall choose at 3s. sterling per tree until the said sum ... be made up'. They were allowed to rent a sawmill, but were obliged to build 'a clever good house of stone and lime', with wooden roof joists and a timber floor, windows, a chimney that did not smoke, and a roof of heather thatch. They were also to construct a timber bridge over the burn.[24]

Then, in 1779, the estate became embroiled in a long dispute with the Duke of Gordon over the right to float timber down the Spey: they were partners in the case with other proprietors on Speyside, but especially Sir James Grant of Grant, who owned Abernethy. The laird of Rothiemurchus claimed to be selling £1,500 worth of wood a year, to go downriver: the claim was ridiculed by his opponent's lawyers as a 'pompous sum', perhaps a slip of translating Scots money as sterling (i.e. only worth one twelfth of the claim!).[25] Indeed, surviving records make it hard to see that the wood was ever worth more than a few hundred a year, probably less than £500. It is true that Alexander Wight was told in the early 1780s that 'the wood is let to undertakers who pay about £1,200 sterling yearly', but he was probably only repeating what Patrick Grant told him, to reinforce his claim in the courts.[26]

The outcome of this dispute probably damaged the timber trade of the Rothiemurchus estate by making it illegal to float between 15 May and 26 August, which made it difficult to catch the most profitable market. Certainly according to estate accounts the trade was only worth £256 by 1786–7.[27] By then the rental of the farms had climbed to £420, and in 1787

[24] Doune MSS: 178, Note of condition of sale of two firr-wood bargains, 1778; 143, Contract between Patrick Grant and John Carmichael and James MacGregor.
[25] Doune MSS: *Memorial for Sir James Grant* and *Memorial for Duke of Gordon*, both Edinburgh, 1780. See Smout, 'History of the Rothiemurchus woods', pp. 27–8, for more details.
[26] Wight, *Present State of Husbandry*, 4, p. 210.
[27] Doune MSS: 190.

Patrick Grant drew up an unusual deed of entail intended to halt the exploitation of woods for a generation. It would not be lawful for his heirs 'to fell, cut, sell or dispone of any of the fir woods' for twenty years, 'except the misgrown woods called garrocks, and decayed wood, or wood that shall fall or be thrown over or damaged by wind, storm or other accident'. At the conclusion of the period, the woods could be sold; but the money raised from the sale, up to a total of £40,000, but in any case not less than £10,000 or the sale could not take place, had to be placed in a fund to buy land as contiguous to the present estate as possible.[28] The entire land of the estate was also entailed at the same time.

Patrick's motives must have been mixed. No doubt he mistrusted the wisdom and frugality of the young heir, his nephew, whose ambitions and financial incompetence were indeed to prove disastrous. No doubt he was exhausted by litigation with his neighbour, with whom he had had another expensive dispute about rights to float wood down the Bennie in 1780. Probably he considered that the woods needed resting, and would be more valuable if they were allowed to grow larger timber: Perrott and Field had told the estate as much as long ago as 1740. He surely noted the changing ratio of the value of his woods and his farms, and considered it a good strategy to use the former in due course to buy more of the latter. Finally, he was in debt himself by the time of his death.

The entail did not immediately end the sale of wood. There is an account of deals, spars and oars sold to fourteen people at Rothiemurchus in 1795, worth £293. It is hard to believe it was fully legal, and that all the wood came from garrocks or windthrow as allowed by entail. But there were no further sales recorded in the later extant rentals of 1802, and of 1806–7.[29]

There are few detailed indications of eighteenth-century woodland management. We know that the woods were unfenced, that the Grants appointed foresters who were probably tenants given a general responsibility of care for the wood, that from time to time the Baron Court met in order to punish general offences within the forest, such as muirburn at the wrong season, or allowing the cutting, burning or barking of young trees, or cutting 'green wood' for local use without permission, or taking 'candle fir' from the pines. The court also, as early as 1705 punished poaching.[30] Interestingly, two prosecutions of 1705 were for 'muirburn' or 'burning of the woods' at the farmtoun of Altnahatnich, 'the moor' of which in the plan of 1789 was

---

[28] Doune MSS: *Memorial for J. P. Grant* (London, 1808), pp. 1–2.
[29] Doune MSS: 224, 310.
[30] Doune MSS: correspondence: Baron Baillie Court Record, 1705–6; 149, Baron Court held in Rothiemurchus, 1772–3.

described as 'mostly covered with young birch and firr wood', and in 1842 when the area was next mapped it was portrayed almost completely as mature wood.[31] In this case it looks as though the rules of the estate had helped to ensure successful natural regeneration over more than a century.

It also seems probable that the normal method of forestry exploitation was to fell the trees before they became inconveniently large, and to do so selectively rather than to clear-fell. Later managers decried this as leading to damage and chaos, and to an inordinate amount of dead wood left lying all over the forest, choking regeneration, but it made sense to peasant woodmen looking for small building timber that could be carried by horseback. It did not rule out the reservation of some big trees that might have been grown on for particular purposes. Certainly the house of the old Doune contained pine baulks sixteen feet by twelve inches by three inches and the new Doune, built towards 1813, baulks seventeen feet long and a foot square. They must, in their day, have come from large fine trees.[32] But these, it appears, were the exception, at least by the eighteenth century.

In the first part of the nineteenth century, the exploitation of the Rothiemurchus woods was far more intensive than at any previous period, partly because it was driven by a greater economic need and partly because there was more opportunity. John Peter Grant, the ninth laird and father of Elizabeth Grant, was consumed by a fatal ambition to succeed on the national political stage, but handicapped by a constitutional inability to live within his means – even though these had been, initially, substantial by inheritance and by marriage (Fig. 11.1).[33]

By 1807, he was feeling the pinch, just as his political career got into its stride. It was the year in which he narrowly failed to take the Parliamentary seat of Morayshire despite having spent, according to his daughter, up to £20,000 on the canvas: by standing against the second son of his Chief, he had alienated the Grants of Grant into the bargain.[34] It was also the year when he over-invested in improvements to the Hertfordshire estate of Twyford which he had inherited from a relative, and sold it after a disastrous speculation in cattle. Thirdly, it was the year when the twenty-year moratorium on the felling of the Rothiemurchus woods ended: he could now fell them if he could get £10,000 for the first sale, but under the entail all the money for the foreseeable future would have to be put to the purchase of nearby farms. Such conditions did not suit John Peter at all. In 1808 he

---

[31] Doune MSS: Plan of the estate of Rothiemurchus surveyed by Alexander Tait, 1789; Plan of the low grounds of Rothiemurchus, surveyed by H. Morrison, 1842.
[32] David Warren, *pers. comm.* Mr Warren worked on the recent restoration of the Doune.
[33] Smout, 'The Grant lairds', p. 15.
[34] Grant, *Memoirs*, pp. 50–1.

Fig. 11.1 John Peter Grant, laird of Rothiemurchus, who felled most of the forest in the early nineteenth century to fund his election expenses and other extravagances.
National Galleries of Scotland, with permission. Present whereabouts unknown.

obtained a private Act of Parliament that enabled him to vary the entail on the woods. In essence, he was allowed to buy from the estate for £10,000 most of the woods, to apply some of that money to liquidating debts left by the previous laird and with the remainder (£5,042. 6s. 4d.) either to invest in land or heritable securities that would become part of the entailed property, or to provide security for doing so. This complicated arrangement released capital immediately, particularly as John Peter chose not to pay out the sum, and the security he gave for doing so (the estate of Kinloss which he had bought from part of the proceeds of the sale of Twyford) ultimately turned out to be inadequate. In effect, from 1809, he was free either to sell the timber or to use it as security for more borrowing, or both. One small section, round Loch an Eilein, Loch Gamhna and the Lochans, was reserved and remained legally protected from sale because the family considered it an essential part of the 'pleasure grounds of the manour of the Doune', perhaps one of the earliest examples of old Scots pines in their natural state being preserved for their amenity value, though they had already attracted the admiring attention of visitors drawn to the 'grandeur and sublimity of such venerable extensive woods'.[35]

The economic climate was also exceptionally suitable for a more intensive exploitation of the pinewoods. The Napoleonic Wars had cut Britain off from easy communication from Norway and the Baltic, whence came most of the softwood used in the towns and industrial areas, and after 1815, when peace returned, relatively heavy import duties were placed on foreign wood in an attempt to stimulate the Canadian timber trade by imperial preference. This had the incidental effect, at least for a time, of also maintaining the wartime profit margins for homegrown Scottish wood.

Elizabeth Grant left a famous account of the activity at Rothiemurchus in this period, describing the rhythm of work in the seasons, the great flushes of water that carried the logs along the burns to the Spey, and the daring art of the floaters who took the rafts downriver to the seaports at its mouth.[36] She also explained how when the estate took charge of the operations of felling, floating, sawing and rafting, they began to replace the old single-saw sawmills on the burns. Similarly, investment was increased in sluices and embankments. As early as the 1770s, Patrick Grant had put sluices on the lochs to enable timber to be flushed down the streams, to the approbation of his brother William – 'I like vastly the idea of making sluices on all the lakes: that has something noble as well as useful in it'. Before 1813 this system had been extended as far up as Loch Einich, and Elizabeth Grant describes a fatal accident in this 'very high and stormy glen'

---

[35] Doune MSS: *An Act for enabling the heirs of entail in possession for the time being of the lands and estate of Rothiemurchus . . . to contract for the sale of the fir woods, 1808*; see also Doune MSS: 211. The purchase sum originally £11,500 but was reduced on the exclusion of the woods round the lochs. For visitors, see R. A. Lambert, 'In search of wilderness, nature and sport: the visitor to Rothiemurchus, 1780–1820', in Smout and Lambert (eds), *Rothiemurchus*, pp. 32–59, quote on p. 34.

[36] Grant, *Memoirs*, pp. 151–7.

when the young keeper of the sluice gates went to open them in a winter storm.[37]

As the years passed, the affairs of John Peter became more and more tortuous, and the need to keep up the flow of wood more compelling. In 1811 the woods had been conveyed as security to a trust of creditors, consisting of a cousin, James Grant of Burnhall, WS, and banker, and two or three others, for forty years or until the debts had been liquidated. They had the right to fell trees above nineteen inches in circumference at five feet from the ground. The income of the wood was to service the debts, and John Peter remained in charge. In 1817 the trust apparently consented to the consignment of £6,500 worth of wood to John Baxter, surgeon in Edinburgh, 'after making allowance and deductions of all fair and necessary expenses' for its manufacture and floating to Garmouth, apparently as security for a further loan. He probably received it, as he does not appear among future creditors. That the woods could have yielded such a sum would have seemed incredible to his father and uncle back in the 1770s, but it was apparently possible in the post-war boom created by the change in duties.[38]

In 1820, an alteration in the trust arrangement placed William Patrick, heir to John Peter and then aged twenty-two, in charge of the practical arrangements for managing the wood. He gave up the practice of law, and devoted himself to managing Rothiemurchus on the spot. Elizabeth remarked with a customary touch of acerbity that 'my father was to proceed as usual: London and the House in spring, and such improvements as amused him when at home'.

According to his sister, William Patrick successfully revolutionised the exploitation of the woods. He brought to an end 'the general felling of timber at whatever spot the men so employed found it most convenient to them to put an axe to a marked tree', and he introduced instead a system of rotational clear-fell:

> William made a plan of the forest, divided it into sections, and as far as was practicable allotted one portion to be cleared immediately, enclosed by a stout fencing, and then left to nature, not to be touched again for fifty or sixty years. The ground was so rich in seed that no other course was necessary. By the following spring a carpet of inch-high

---

[37] Doune MSS: correspondence, 1770; *Statement of J. P. Grant of Rothiemurchus, 1824*; Grant, *Memoirs*, p. 157.
[38] SRO: CS 96/2617, p. 20; Doune MSS: 470. It is possible that the 1817 transaction was not approved, as the contract at the Doune is an unsigned draft.

plants would be struggling to rise above the heather, in a season or two more a thicket of young firs would be found there, thinning themselves as they grew, the larger destroying all the weaker. Had this plan been pursued from the beginning there would never have been an end to the wood of Rothiemurchus.[39]

He next centralised the provision of horses to work in the wood from the Doune itself, employing them out of season in carting deals to Forres and bringing back meal, or in carrying coal from Inverness: on the estate 'the little bodies and idle boys with ponies were got rid of'. In a similar way the operations of the lesser mills were concentrated on one large building at Inverdruie, equipped with 'a coarse upright saw for slabbing', several 'packs of saws which cut the whole log up at once into deals'. A smaller compartment of the large mill was fitted up with circular saws for the purpose of preparing the thinnings of the birchwoods for herring-barrel staves – 'it was a mere toy beside its gigantic neighbour, but a very pretty and a very profitable one, above £1,000 a year being cleared by this manufacture of what had hitherto been valueless except as fuel'. Indeed, sales of birch between 1769 and 1771 had averaged only £10 each year.[40]

William Patrick also gave an independent account of how he had managed the forest in a report to the creditors in 1824, and, though not being so precise about the division into sections, confirmed much of what Elizabeth recollected. He explained that before he took over, the workforce lived and laboured all over the forest and took whatever they found most convenient for cutting, virtually unsupervised – 'the forest was picked according to their fancy'. Consequently much was left to rot in the woods, obstructing regeneration, and was prepared so badly that great expense was incurred at Garmouth fitting it for the market. Nor could it be dragged to the water until the forester had come and accounted for it, and this meant long delays while the timber deteriorated. Much was cut into deals, when floating logs would have been more profitable, and workers were paid in advance, with meal, for work that they sometimes failed to undertake.

He explained that when he took over in 1820 his first action was to clear the forest of all the scattered, cut wood that was preventing regeneration all over the forest – 'there was hardly a spot where the axe had not been at work', and all the profits of the following year were consumed in this tidying operation. He then centralised cutting in one place at a time, under the eye of a foreman, preparing each tree according to its qualities. Similarly

---

[39] Grant, *Memoirs*, p. 254.
[40] *Ibid.*, pp. 254–5; Doune MSS: 143.

he abolished the old mills dispersed through the forest, sending much down the Spey in log (a quirk of the import duties made it more profitable to do it this way) but still manufacturing at one big mill at Inverdruie, 'the whole trimmings of the timber and all the small rubbish in the forest'. Workers were paid weekly by the foreman, in money or meal, on piece rates for work actually performed. Less reliance was placed on the horses of the small tenants, and more on the more powerful animals kept at the Doune, partly because in late springs like that of 1823 the tenants needed their animals to cultivate the land at exactly the time when it was necessary to get the wood to the Spey to beat the floating moratorium of 15 May.[41]

The occasion for the report of 1824 was mounting discontent among the creditors. James Grant of Burnhall, clearly pressed by his own financial problems, accused John Peter Grant of earning £15,000 to £20,000 from the woods and applying it to service new debts rather than to paying old ones. John Peter managed to convince the trust that this was not the case. He admitted to having sold £9,000 worth of wood at Garmouth in the year from 1 September 1822, but explained that 40 per cent of that sum was swallowed up by the expenses of manufacturing the timber and floating it to Garmouth. The remainder had been paid to the trust, as testified by the accounts. James Grant apologised, but the administration of the trust passed to a new man, Patrick Borthwick, whose Leith merchant partnership had contracted to sell the timber at Garmouth on terms apparently less expensive to the trustees. Nevertheless it is clear from this point that they were more suspicious of John Peter, and more doubtful that they would ever see their money back, than ever before: a reconstituted trust, headed by Borthwick, secured powers to sequester his personal moveable effects including the plate, library and furniture at the Doune, and all unentailed assets on the estate, should it become necessary to limit their own losses. Meanwhile William Patrick remained in charge of the wood. The total debts due to the trustees in 1824 were said to amount to about £65,000, not so much incurred by losses in the woods as by the extravagant political debts and lavish lifestyle of the laird.[42]

All this raises the question of what the forest was actually worth, and what was sold from it at this period. Papers relating to the affairs of James Grant in 1829 speak of the wood having been valued at 'between £60,000 and £90,000 a few years ago' and state that 'these woods have produced about £10,000 per annum' in the past, but 'owing to the depression of the

---

[41] Doune MSS: *Statement of J. P. Grant, 1824*.
[42] Doune MSS: *Statement of J. P. Grant, 1824*; 'Trust disposition and assignation, J. P. Grant to Patrick Borthwick, 9 June, 1824'.

times for some years back, the sales of that wood have been very limited and little of the debt has been paid'.[43] Sir Thomas Dick Lauder in 1834 maintained that it had once yielded large profits over many years, sometimes over £20,000 a year, but he was less likely to be well informed than James Grant and his lawyers.[44] There are no extant estate accounts to guide us, only occasional statements like the apparently reliable figure of £9,000 gross yield for the wood in 1822–3 – not a year likely to have been particularly outstanding.

The figure of £10,000 mentioned by James Grant's lawyers was presumably intended to be net – that is, the profit available to the trust after the cost of manufacturing in the woods and floating to Garmouth had been taken into account. John Peter Grant put the cost at 40 per cent, roughly equally divided between manufacturing and floating, and his figure was not challenged. Later estimates put the figure at 25 per cent or 30 per cent, but in 1831 they appear to have exceeded 50 per cent.[45] At all events they appear to have been large. But whether the sums mentioned are net or gross, they are extremely substantial compared to eighteenth-century yields, even allowing for changes in the value of money, and also very much larger than anything obtained from the wood later.

It is also worth noticing that if £10,000 net really was obtained in one year from a wood only estimated at six to nine times that value, William Patrick's plan to cut on sustainable rotation would have been quickly overtaken by events – as indeed proved to be the case. But the wood was all the creditors had by way of a regular income from the estate: the value of the farm rental by 1824 was apparently not in excess of £250 a year, or about one-twentieth of the probable minimum net yield from the wood. This was an extraordinary contrast with the eighteenth century: in 1769–71 the farms had yielded on average about half the net yield of the wood, and in 1786–7 the rental of the farms considerably exceeded the net yield of the wood. In essence, if the creditors were ever to see their loans repaid, they had to get as much out of the wood as they could as quickly as they could.

In 1827, John Peter's financial house of cards finally collapsed, when he lost his seat in Parliament: 'without this shield, his person was not safe'.[46] The trust could not seize and sell the lands of Rothiemurchus because they were entailed, but they sequestered the moveable assets of the estate. The

---

[43] SRO: CS 96/2617, p. 26.
[44] See his edition of W. Gilpin, *Remarks on Forest Scenery and Other Woodland Views* (Edinburgh, 1834), 2, p. 175.
[45] Doune MSS: 315, 392.
[46] Grant, *Memoirs*, p. 282.

diary of the mill at Inverdruie recorded supplying twenty-one deals 'for packing boxes for the Doune family'[47] and the family made a rapid exit to India, fending off the tradesmen as they went. A few months before this dramatic *dénouement*, the workers in the sawmill had complained through their foreman about non-payment of wages. They were in turn upbraided by William Patrick for their disloyalty in making such a complaint, in an address which he had printed in Gaelic and English:

> I will never hear from any individual a particular complaint, without inquiring into it . . . but I tell you distinctly that I will not listen to any reports affecting you, that I hear from other people nor to any general complaint that you may think fit as a body to make.

Many of the workers, he went on, had worked on the estate for generations, or been for years in domestic service:

> Men who have imbibed an affectionate regard to our family almost with their mother's milk . . . When such men go wrong . . . when any of them shew a temporary estrangement of their regard to their hereditary proprietors, I will confess that it gives me pain; but it is a pain of the nature of which a father feels at the misconduct of his children; it leads me rather to feel anxious to restore you to the right feeling you have departed from, than angrily to leave you to the mischievous effects of your own folly, and to cast you off forever.[48]

It was a virtuoso exercise, from one not yet thirty years old, in the art of blaming the victim.

The departure of the family by no means brought the employment of the hapless workforce to an immediate end, as the trustees were more anxious than ever to get some of their money back, a situation complicated by the bankruptcy of James Grant of Burnhall two years later. Felling continued throughout the 1830s, though in 1831 the trustees reported on difficulties due to low prices brought about by competition from Canada and Norway, compounded for 'Rothiemurchus and other natural timber of Scotch growth' by the 'increasing use, in the north country, of planted fir timber which is now sold at a price which would yield to the creditors almost no return upon the Rothiemurchus timber, owing to the expense necessarily incurred in bringing it from the forest to the mouth of the

---

[47] Doune MSS: uncatalogued diary of the sawmill.
[48] Doune MSS: 369, *W. P. Grant's address to the forest workmen and sawmillers, 1827.*

Spey'.⁴⁹ That year the rental of Rothiemurchus (it would have included the rent of the Doune to the Duchess of Bedford, which had been leased as a Highland hideaway where she could entertain her great friend Edwin Landseer) was £1,107, while the net yield on wood sales was £1,872 (£3,840 gross). Over the following three years, sales remained at a similar level, averaging £1,600 net (£2,100 gross), the costs of manufacture and floating temporarily dropping as a proportion of the price, but the actual quantities of wood marketed also apparently falling as the forest became exhausted. In 1836, the creditors obtained permission to sell the remaining trees round the lochs, which had been originally reserved in the arrangements of 1809: they were sold for £2,500, but after deductions only realised £420. Finally, a document of 1839 revealed that over the previous five years the woods had yielded an average of £1,526 per annum (£2,875 gross), little enough compared to the past, but now more than double the yield of the rents (including those of the Doune) in that year.⁵⁰

By then, however, the forest had been largely but not completely clear-felled. In 1834 Sir Thomas Dick Lauder described Rothiemurchus and Glenmore as 'now equally denuded of all their forest timber', but regenerating fast and in a way that was certain to produce valuable wood in due course:

> The young saplings come up as thick as they do in a nurseryman's seed bed, and in the same *relative* degree of thickness do they continue to grow, till they are old enough to be cut down. The competition that takes place between the adjacent individual plants, creates a rivalry that increases their upward growth whilst the exclusion of the air prevents the formation of lateral branches or destroys them soon after they are formed. Thus Nature produces by far the most valuable timber, for it is tall, straight, of uniform diameter throughout its whole length, and free from knots ... the large and spreading trees are on the outskirts of the masses.⁵¹

In 1839, John Grigor, the well-known nurseryman and seedsman of Forres, reported in similar terms that the old forest is 'much exhausted',

---

⁴⁹ Doune MSS: 319.
⁵⁰ Doune MSS: 315, 319, 356, 365, 481. The 'reserved' woods were bought by a trustee on behalf of Sir John Peter Grant himself, and the sum of £420 was not distributed to the trustees but added to the entailed sum of £5,042. 6s. 4d. Presumably the trustees found that, legally, they could not touch this residue: on the other hand some of the £2,189 'deductions' may have gone into their pockets. We have not been able to unravel this transaction properly.
⁵¹ T. Dick Lauder in Gilpin, *Remarks on Forest Scenery*, 1, p. 175.

though 'many beautiful clumps of pines still remain'. Like Dick Lauder, he was struck by their slender and tall quality – 'the trees almost stand of a uniform size, measuring at 6 feet high 4½ feet in circumference, and nearly of the same girth to the height of about 35 feet', going on to achieve an average height of about seventy feet. The stumps of felled trees of similar dimension were, from their rings, 120 to 125 years old, which would take their origin back to the decade after the Act of Union. He also noted that the marshy hollows within the forest, 'now too wet for pine', were crowded with the roots of smaller trees, that the close-growing trees on the knolls crowded out other vegetation, and that young natural pine was growing well.[52] These accounts may be compared with internal reports on the forest of 1831, which described the old reserved wood as containing 11,248 trees with an average of twelve to thirteen cubic feet of timber per tree, valued other old timber available for felling at £6,267 gross, and also referred to the 'young wood of Rothiemurchus ... growing before 1808' as containing 23,500 trees over thirty years old (but none in this category over fifty to sixty years old). The wood was divided into lots, evidently reflecting William's reorganisation, but there is no sign that they were to be felled in rotation. Presumably by 1839 when Grigor wrote, little was left except in the category of the former reserved wood, along with regrowth from recent felling and some isolated older clumps.[53]

Finally, when Elizabeth Grant returned on a nostalgic visit to Rothiemurchus in 1846, she found the old sawmill all but deserted, took a walk 'through the young fir forest' and spoke of 'a young forest to replace the fallen': when two years later the income of the estate was likely to be impounded by new creditors, she wrote bitterly of the sequestration of 'the few trees fit for the axe in that once fine forest', estimating their value at almost £9,000.[54]

This valuation may well have been roughly based on another detailed report on the forest, drawn up in April 1848 by Neil MacLean of Inverness. He calculated the number of the 'old reserved trees' – the survivors reserved from the sale of 1809 – at around 10,000, yielding on average about fifteen cubic feet of timber per tree. He estimated the current price of pine timber at Garmouth at 1s. 2d. per foot, from which 30 per cent had to be deducted for cutting, dragging and floating down the Spey, and recommended that a

---

[52] J. Grigor, 'Report on the native pine forests of Scotland', *TRHAS*, 2nd series, 6, pp. 125–6. Steven and Carlisle, *Native Pinewoods*, p. 122 seem to have misread Grigor: we can see no sign that he found regeneration thirty years old and thirty feet high.
[53] Doune MSS: 315, 392.
[54] Elizabeth Grant, *The Highland Lady in Ireland: Journals, 1840–50* (ed. P. Pelly and A. Tod, Edinburgh, 1991), pp. 241–2, 402.

third of the trees be taken out now, as they were approaching maturity. These, he thought, would realise £2,000, so the value of the whole stand was presumably £6,000. The value of the rest of forest he could not accurately state, but pointed out that there could be a potential market for thinnings from the regenerating forest, though given the current price of wood this would probably not repay the effort. He commented on the 'general spread of natural fir seedlings in this locality' as 'most remarkable', and said it seemed only necessary to enclose a piece of ground in the forest 'to give rise at once to a thick and healthy crop of young trees'. On these grounds he predicted that in fifty or sixty years time the forest would be of far greater value than ever it had been in the past. Already, interestingly, what we now call 'granny pines' were seen as worth preserving for aesthetic reasons: 'such of the old branchy trees as are more picturesque than useful for timber will not I suppose be touched under any circumstances'.[55]

By this time, William Patrick had once more been managing Rothiemurchus for six years, albeit by correspondence from India, whence he had gone with his father and resumed a successful career as a lawyer. New trustees headed by his sister Jane's father-in-law William Gibson Craig had paid off the debts from the old trustees and restored the estate to family custodianship. William Patrick returned suddenly from India in 1848 disgraced by implication in the massive collapse of the Union Bank of Calcutta and bankrupt again.[56] 'William is not a man I would like to have any dealings with – speculating, arrogant and I am afraid selfish', said Elizabeth, yet the trustees allowed him to remain in charge of the estate until it was again sequestrated around 1855, and he was certainly back in the saddle once more from 1861 until his death in 1874.[57] It is safe to say that all the major investment decisions on the estate for three decades after 1842 were his, and he proved a vigorous and proactive land manager, embanking the Spey, building a bridge across the river and constructing many miles of road apart from his forest works.[58]

His approach to the forest was summed up in his own words as 'to attend to systemateck method of turning the prime timber into money, and of keeping up a continual succession of such prime timber' – no different, in fact, from his aims in 1820. However, in 1842:

[55] Doune MSS: 392.
[56] Smout, 'The Grant lairds', p. 17.
[57] See Grant, *Highland Lady in Ireland*, pp. 372–6, 417, and Smout, 'The Grant lairds', pp. 17–18, for details.
[58] Doune MSS: 211.

When I acquired dominion over Rothiemurchus so much of the prime timber had been felled and so little regard paid to the reproduction of it that I determined to put a stop for a time to the manufacture and to give all my care to growing material.[59]

He set about this by renewed planting and enclosing. Some planting had taken place before, as there are references to planting and also to larch as early as 1824. When Lord Cockburn visited Aviemore in April, 1839 and fantasised about the house and woods he would like to create on 'one of the grandest inland places in Scotland', he said:

We must begin by clearing the country of at least nineteen parts out of every twenty of that abominable larch with which it pleased the late Rothiemurchus, as it still pleases many Highland lairds, to stiffen and blacken the land.[60]

William may well have been responsible for those plantings, and in 1842 he had a plan drawn up for planting Cambusmore, which was probably at least partly carried out while his brother was in temporary residence at the Doune. In the winter of 1846–7 he had, by correspondence from Calcutta, planted almost half a million trees, of which 433,000 were larch and 39,000 Weymouth pine, and the remainder a blend of hardwoods – 2,000 oaks, about 1,000 each of Norway maple, 'tan willows' and beech and a thousand assorted elms, 'planes or limes', rowans, Italian black poplars and Italian white poplars, as well as 6,000 hawthorns for hedges.[61]

Simultaneously, he started 'removing obstructions to the natural growth of trees for which the place is wonderfully adapted', by securing the forest from 'the depradations of domestick beasts pasturing among young trees'. Again, this was not totally novel, as before 1824 'a large flock of sheep' kept on the farm had been 'dismissed, for the preservation of the woods'. Now, however, he entirely abandoned all hill or unenclosed pasture on the estate and enclosed all the lower or farm lands with three stone dykes, two along the estate marches running up from the Spey to points above the cultivated land, where they intersected with a 'Forest Dyke' that kept the farms and the natural wood apart. All but 900 yards of this had been completed when work was temporarily ended because of the new bankruptcy.[62] William

---

[59] Doune MSS: W. P. Grant's notebook.
[60] H. Cockburn, *Circuit Journeys* (Edinburgh, 1888), pp. 39–40.
[61] Doune MSS: 230; Plans of farms, 1819 (with additions).
[62] Doune MSS: W. P. Grant's notebook; *Statement of J. P. Grant, 1824*.

Patrick was exceedingly proud of this achievement, though Elizabeth slightingly referred to it as 'that absurd wall of such enormous extent over the mountains and round the woods of Rothiemurchus'.[63] He protected his new plantations by temporary fencing to keep out red deer and roe deer 'which have free scope in the forest', but at this point they were probably at low density. The early decision to keep the hill ground free of sheep probably accounts for the fact that Rothiemurchus, at Creag Fhiaclach, is apparently unique in Great Britain as a place where a natural tree line can still be seen.

There is little information about the forest in the 1850s, but average wood sales over the years 1851–3 reached only £214, compared to an enhanced estate rental of £1,134. From the early 1860s, William was busy again with planting, not discouraged by a letter from his agent in 1856 which mentioned that, although natural regeneration was doing well, the previous attempts at planting had 'in a great measure failed, particularly the hard woods'.[64] In 1861, he planted (or replanted) about 3,000 hardwoods around the Doune. In 1862 he noted: 'this season I planted the Moor of Callert in I believe the third time. My brother John planted it with larch in one of the early 1840s. When I saw it the appearance was wretched and would have been better without a plant.' He decided, since Scots pine was regenerating on the western part of the moor, to plant more pine among the surviving larches and to fence with tarred pine posts and wire.[65]

The following year he planted the eastern shore of Loch an Eilein, near to a thriving larch plantation, with a mixture of larch, Scots pine and Corsican pine. Scots pine was already regenerating there, but the ground was 'mostly full of strong heather... there are frequent patches without any plants at all – I intend to plant the whole thing up to about four feet apart'. He had great faith in Corsican pine ('Laricco') as an infill for sparse Scots pine, believing it to be second only to larch and quicker in its growth than Scots pine. He bought 100,000 two-year-old seedlings of Corsican pine in 1863, but there is only a trace of it on the estate now.[66]

At that period he was planning timber sales to the railway companies 'as our principal market', arguing that iron and stone sleepers had both been tried and found wanting, and that one-sixth of creosoted timber sleepers needed to be replaced yearly, thus creating a constant market. He believed that a well-grown pine would provide four sleepers and a wooden block.

---

[63] Grant, *Highland Lady in Ireland*, p. 417.
[64] Doune MSS: 319.
[65] Doune MSS: 479.
[66] *Ibidem*.

Over 1864–5 the estate did indeed sell 16,727 sleepers to the railway, newly arrived in Strathspey and pushing ahead to Inverness.

Timber sales in the 1860s were stimulated by the railway in other ways, because it largely freed the business from expensive and inconvenient floating down the Spey to Garmouth. Anything loaded on a train waggon could go north and south to its market within hours. In 1863 some 30,000 trees were sold to a Forres merchant as sparwood for £2,625 – they were perhaps thinnings. The sale of the sleepers realised £2,369, and in the same year £500 worth of wood went to Garmouth. In the period 1862–4, £4,982 worth of wood, some of it larch, and including 60,000 pit props, was sold from Loch an Eilein, Achnahatnich and Guislich, to a wood merchant at Crieff.[67]

All these bargains added up to about £2,600 a year, perhaps modest compared to the exploitation earlier in the century, but still enough to strain the productive capacity of the forest. By 1869 the heirs made it a condition of agreeing to a variation in part of the entail that 'Mr [William] Grant shall engage to cut no more wood. All the wood that is of an age to be fit for cutting I understand has been cut and none remains that could be cut in Mr Grant's lifetime without grave waste.'[68] The woods seem to have come close to being clear-felled once more, though plainly clumps and individual trees remained.

How good a steward of the wood was William Patrick? There is no sign that he ever tried to revive after 1842 the scheme which he devised as a young man in 1820 for dividing the wood into portions to be cut in rotation. Such a scheme is inherently more difficult in pinewoods than in oakwoods because of the unpredictable densities of regeneration.[69] Nor is there (we may think fortunately) any indication that his schemes for planting non-native softwoods met with much success, though some of the larch may have done so and been felled subsequently. Large individual larches can still be seen in the forest. His main contribution to the survival of the native forest was to build the enclosure dyke and abolish cattle and sheep grazing in the wood. He continued to be strict in keeping stock out, as a tenant farmer found to his cost in 1866, when a flock of sheep broke into a plantation of larch and Corsican pine at Cambusmore.[70] At least, as his brother and successor put it in 1883, 'because sheep were abolished for the sake of the wood before deer were introduced', the estate was given a clean

---

[67] Doune MSS: 481.
[68] Doune MSS: 328.
[69] For instance, see J. M. Lindsay, 'The use of woodland in Argyllshire and Perthshire between 1650 and 1850', unpublished University of Edinburgh Ph.D. thesis, 1974, pp. 241, 284–5.
[70] Doune MSS: 392, 'Precognition in the matter of destruction of young trees'.

bill of health before the Napier Commission: no-one appeared before the Commissioners at their meeting at Kingussie, and their secretary wrote to the laird that 'I suppose we may conclude there are no dissatisfied tenants on this estate.'[71]

That is not to say that there had not been many economic hardships for those involved. In 1824 John Peter had reported to the trustees that total employment from the woods included a professional forester, a foreman at the sawmill, a superintendent acting as grieve, a meal store-keeper, a clerk who kept the books, between thirty-six and forty woodmen (of whom twenty were at work at any one time when the season allowed) and ten mill hands, apart from small tenants employed casually to draw the wood.[72] Elizabeth Grant on her return in 1846 found at the mill only three men and a boy, in place of 'a mob of workmen', hauliers and floaters as well as sawyers: 'the forest being gone no mill work is ever done beyond what is required for fencing'. She was taken aback by the social condition of the people, no better, she thought, in the last twenty years and 'far behind our Irish, not within half a century so much up to the times'.[73] This, however, was at the start, not at the conclusion, of William Patrick's main period of management. Some at least of the employment returned, at least intermittently. In October and November 1862 he was employing thirty workmen, eighteen regularly, on extracting 'lime and sleepers'.[74]

Little more active management seems to have taken place in the forest before the end of the century. On William's death in 1874 they were surveyed again in great detail, in a report running to forty-three pages. The net worth of the woods that could then be cut was only £2,341, and many of them were described as neglected. The report recommended little more than thinning and care; trees north of the old boring mill were described as 'a fine specimen of the old wood and might be kept for particular estate purposes', and those at the west end of Loch an Eilein comprised 'a number of full grown trees of the old forest still left which might be cleared away with advantage as they will never be of more value than at present and many of them are doing damage to the young wood'.[75] Presumably some old trees were still kept for their aesthetic worth, but it was hard to find even token wood of good quality for use. When in 1877 the Doune was rebuilt it was panelled in Canadian pine delivered by railway to Aviemore.[76]

[71] Doune MSS: 534.
[72] Doune MSS: *Statement of J. P. Grant, 1824*.
[73] Grant, *Highland Lady in Ireland*, pp. 242–7.
[74] Doune MSS: 481.
[75] Doune MSS: 489.
[76] David Warren, *pers. comm.*

A separate report of 1874 concerned William's dyke separating the forest from the farmland. It was described as 6,415 yards long, stone-built with earth backing, four and a half feet high with a row of six-inch turf sods on top, and surmounted by a slight wooden paling supported by posts driven into the dyke: except for its great length (over three and a half miles) it was a typical woodland enclosure of its time. The surveyor described it as now decayed and no longer a proper fence for keeping the deer in or the sheep out. He recommended repair, and replacement of the turf and wooden topping by wire. It is this reconstructed dyke that the visitor sees in places along the edge of the forest today.[77]

In the final quarter of the nineteenth century the primary use of the forest was as part of a sporting estate: in 1892–3, for example, the rental from the shootings was £1,585, from the farms £1,300, but there was no mention of any income from the wood. Indeed, at this stage the wood could never have produced, in a sustained way, any income to equal that from sporting rents. In 1896 the game bag for the year amounted to 92 stags (average weight of seventy was fourteen stone), 54 hinds, 14 roe, 108 brace of red grouse and 8 brace of black grouse.[78] In 1914, a guide to Scottish sporting estates put the yield of Rothiemurchus at 80 stags and 200–300 brace of grouse, besides other highland game.[79] Presumably there had always been a few deer in the woods, but they were certainly reinforced by introductions from elsewhere some time before 1883.

Nevertheless, by the end of the century, following an extensive fire at the south-east of Loch an Eilein, felling was begun again. In 1899–1900, the woods yielded £501 for small quantities sold to a buyer in Pitlochry, compared to income from rents and farms amounting to £3,564: the yield of the wood increased to £2,197 in 1900–1 (to buyers in Pitlochry and Aberdeen) and stood at £1,557 in 1901–2, again to Pitlochry. Sending wood south had become much easier since the advent of the railway in the 1860s, but it was only intermittently worthwhile to do so. The last time floating was used was in 1903, when trees were sent down the River Luineag using the sluice at Loch Morlich.[80]

We have not attempted to trace the history of the wood in the twentieth century. It was marked by heavy felling episodes in both world wars, carried out on government instructions by Canadian workers, though

[77] Doune MSS: 489.
[78] Doune MSS: 483.
[79] Anon., *Poacher's Guide* (n.p., n.d. but 1914). There is a copy in St Andrews University Library.
[80] Doune MSS: 335. Steven and Carlisle, *Native Pinewoods*, p. 122.

some of the lower pinewood at Abernethy and Rothiemurchus escaped 'almost unscathed' because they were used as training areas for the military. Otherwise there were low levels of extraction though a disastrous fire broke out in the forest south of the Glenmore road in 1959, and only now is the devastated area showing good regeneration.[81] There was a stubborn disinclination by the lairds to make the land occupied by the old forest available for the Forestry Commission or for commercial planting. W. H. Murray in 1962 mentioned a 'recent' proposal by the Forestry Commission to 'fell the old Caledonian pines of Rothiemurchus', which had been modified 'only at the intervention of the Prime Minister, acting on appeal by the Member of Parliament for Inverness-shire and by Colonel J. P. Grant'.[82] Exactly what this incident was we have been unable to discover, but the Forestry Commission was involved in compulsory purchase proposals in other places in the 1950s.[83] In 1954, the declaration of a National Nature Reserve in the Cairngorms, and subsequent nature reserve agreements signed between the estate and the Nature Conservancy and its successor bodies, strengthened the protection of the ancient wood, which is now valued as a prime site of surviving native pine forest and guarded by European as well as by British legislation. It is also, of course, an integral part of the new Cairngorms National Park and enjoyed by many tens of thousands of visitors every year.

No-one, however, can study the history of Rothiemurchus and remain under the beguiling illusion that it is the untouched wilderness that it can seem to the visitor on a quiet summer day. Rather, it is the outstanding example of a pine forest's ability to recover from heavy economic use, providing the ecological processes that enable it to do so remain intact. The Rothiemurchus woods are there because they were permitted to regenerate, but not because they were spared the woodman's axe.

---

[81] Steven and Carlisle, *Native Pinewoods*, p. 122; B. M. S. Dunlop, 'The woods of Strathspey in the nineteenth and twentieth centuries', in T. C. Smout (ed.), *Scottish Woodland History* (Edinburgh, 1997), pp. 182–3.
[82] W. H. Murray, *Highland Landscape: a Survey* (Edinburgh, 1962), p. 16.
[83] G. Ryle, *Forest Service: the First Forty-five Years of the Forestry Commission of Great Britain* (Newton Abbot, 1969), pp. 124–5.

CHAPTER 12

The Navy, Holyrood and Strathcarron in the seventeenth century

If small pockets of pine on the islands in Loch Assynt in Sutherland are to be discounted as ancient woods, as Steven and Carlisle have argued, then the county of Ross possesses the two most northerly surviving ancient Scottish pinewoods, at Rhidorroch, on a catchment flowing westwards into Loch Broom, and near Strathcarron, flowing eastwards into the Dornoch Firth (Map 12.1). We are concerned here with the latter: for Rhidorroch, see pp. 205–6 above.[1]

The exploitation of Strathcarron is age-old. Barbara Crawford has cogently argued that the Vikings were drawn south into the area by their need to secure timber for boat-building, and Norse elements in the place-names of Strathcarron, especially Amat, Alladale and Diebidale, are powerful evidence of their interest in natural resources sometimes many miles from the sea but central to their control of it.[2] Their importance was no less evident in the later Middle Ages.

In the foundation charter of the Abbey of Fearn by Farquhar, Earl of Ross, which was confirmed by his successor, John, in the fifteenth century, the monks of Fearn were granted the 'use of timber and trees throughout our whole earldom of Ross and particularly the use of timber and trees in the parish of Kilmure'. This grant was again confirmed in 1529 in a papal bull which listed the possessions of the abbey.[3] In about 1590, the cartographer Timothy Pont noted the 'gryt firr woods' of 'Amad na heglisse' in 'Stra Charroun' and, some time in the middle of the seventeenth century, a slightly less laconic description was written by Robert Gordon of Straloch,

---

[1] H. M. Steven and A. Carlisle, *The Native Pinewoods of Scotland* (Edinburgh, 1959), pp. 206–7.
[2] B. E. Crawford, *Earl and Mormaer: Norse-Pictish Relationships in Northern Scotland* (Groam House Museum Annual Academic Lecture, Rosemarkie, 1995), pp. 7–17.
[3] J. Munro and R. W. Munro (eds), *The Acts of the Lords of the Isles, 1336–1493* (Scottish History Society, Edinburgh, 1986), no. 90: '*et usum lignorum et arborum per totum nostrum comitatum Rossie et presertim usum lignorum et arborum in parochia de Kilmure*'; NAS: Bull of Pope Clement VII, J. and F. Anderson Collection, GD 297/189.

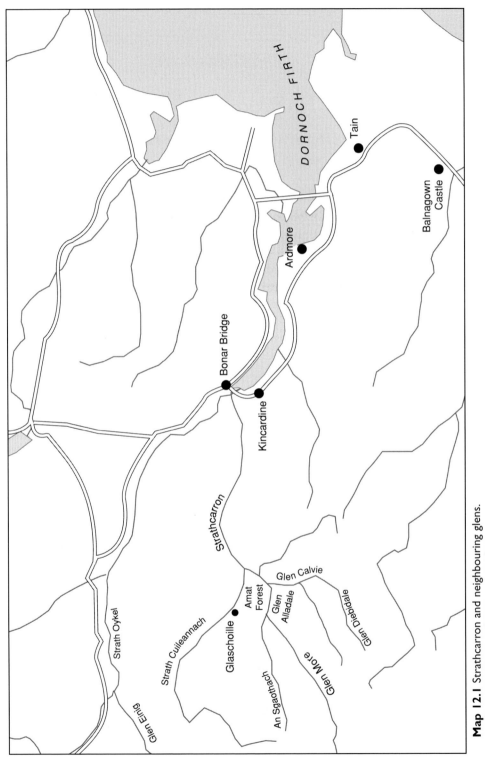

Map 12.1 Strathcarron and neighbouring glens.

who updated many of Pont's maps in preparation for their publication. He wrote that the River Carron flowed from among the mountains 'through hilly and wooded ground' into Strathcarron which 'is for the most part wooded and, being clothed with particularly tall firs, supplies neighbouring and distant places with timber'.[4] By the early seventeenth century, then, the woods of Strathcarron formed the basis of what was already a widespread trade in timber.

After 1603, the English began to take an interest in the potential of Scottish timber (see above, pp. 192–6). The first indication of this in Strathcarron was in 1612, when Alexander Ross of Invercarron and his heir entered an agreement with Mr Lambert Antony (surely an Englishman), Robert Munro of Taininch and John Abrahame, burgess of Inverness, to allow the lessees to fell and extract all the trees above one foot in diameter from the 'great fir and aiken wood of the north syd of Glencalvie', with liberty to build one or more sawmills. They failed, however, to pay their tack duty, and in 1615 the Rosses were granted legal permission to seize the mills that Lambert Antony, 'master of work', had erected. There were, however, English 'wood-smiths' (whatever that expression might mean) at work in Strathcarron in 1625, when John Wayner and a certain Laurence witnessed to sasine of the sawmills.[5]

Then there was the chance of interesting the Royal Navy. Around the late 1620s, Sir John Grant of Freuchie, who had acquired extensive interest in woods beyond his native Strathspey, invited Captain John Mason to visit Scotland to view woods in Strathspey, Lochaber and Glencalvie. After travelling to see them all, 'induring som hardshipe and undergoinge large disbursments', he expressed interest only in Strathspey, where, in 1630, he took a lease of Abernethy and other woods. Although with no apparent connection to the Crown, he then reported to the Commissioners of the Navy that the timber of Strathspey was suitable for naval purposes, yet there is no evidence that any naval timber was extracted as a result of this. A few years later, Mason sold part of his interest in the lease to the Earl of Tullibardine who had been a witness to the original contract and was the uncle of the laird of Grant's wife.[6] The tenuous connections which had been established with the navy thus appear to have vanished.

---

[4] A. Mitchell (ed.), Geographical Collections Relating to Scotland made by Walter Macfarlane (Scottish History Society, Edinburgh, 1906), 2, pp. 447, 546–7.
[5] NAS: RD 1/206 fos 361–2, also fo. 359; CS 7/295, fos 312–13; R. J. Adam (ed.), *The Calendar of Fearn* (Scottish History Society, Edinburgh, 1991), p. 222.
[6] W. Fraser, *The Chiefs of Grant* (Edinburgh, 1883), 1, p. 277, 3, p. 353; Steven and Carlisle, *Native Pinewoods*, p. 116; J. Munro, 'The golden groves of Abernethy', in G. Cruickshank (ed.), *A Sense of Place* (Edinburgh, 1988), pp. 152–62, especially pp. 157–8.

Not until the 1650s do the English navy appear to have seriously renewed their interest in the possibility of finding a source of pine for masts in Scotland. The outbreak of the First Dutch War in the summer of 1652 made the North Sea, across which England's naval pine came from the Baltic, a perilous supply line and sometimes virtually impassable. Moreover, it was a war which involved a good deal of naval combat and the resultant loss of and damage to ships meant that a ready supply of timber, especially for repairing masts and yardarms, was vital. In 1650, Oliver Cromwell's forces invaded Scotland, defeating a Scots army at Dunbar in September. Although many Scottish strongpoints were still unsecured by the English by the end of 1651, they effectively occupied southern Scotland.

The idea of seeking naval timber in the north was revived even before the outbreak of hostilities with the Dutch. Plans must have first been mooted in the last months of 1651 for, in January 1652, Andrew Sandelands, an agent of the English government, sent a report to London on the orders of Major General Deane, one of the commanders-in-chief of Cromwell's army in Scotland. The focus, initially, was again Speyside. Lord Tullibardine, as the holder of a third share of Captain Mason's lease (see above p. 196), was an obvious potential source of supply since the Abernethy timber was 'the best and most usefull in all Scotland for tarre', manufactured from pine resin and a commodity upon which the navy relied for waterproofing. Sandelands also suggested that the woods of Glen Moriston and Deeside would be good for masts, and timber and tar respectively. There was a problem, however. Tullibardine refused to help 'because noe privat man in these times is able to carry on the worke, not having power to command neyther the soldiery nor the country to affourd...assistance...The truth is, no Scotts man will undertake it because it is reputed a dis-service to the king and country' (Charles II had been crowned as King of Scots at Scone in January 1651). Sandelands therefore suggested that English or even Norwegian labour should be brought in. Even if this extra effort were made, with its associated expense, he believed that 'if the state will sett men a worke themselves they may make for every hundreth pound a thousand'. Any soldiers needed to protect the operations could thus be paid for easily and it was beyond doubt that troops would be needed because the Highlands remained firmly outwith the control of Cromwell's forces. The extraction of naval timber was thus envisaged as operating in tandem with the military campaign to reduce the north to obedience.[7]

---

[7] *CSP Dom 1651–2*, pp. 103–5; PRO: SP 18/23 nos 6 and 6i; C. H. Firth (ed.), *Scotland and the Commonwealth* (Scottish History Society, Edinburgh, 1895), p. 73; F. D. Dow, *Cromwellian Scotland, 1651–1660* (Edinburgh, 1979), chs 1 and 3.

There were further discussions of the matter in the English council of state in July, at the time of the outbreak of war with the Dutch. In the autumn, the king of Denmark closed off the route into the Baltic and seized cargoes of masts bound for England. By November, the Royal Navy had sustained a defeat at the hands of the Dutch off Dungeness, repairs were urgently required, the stores were almost empty and it would be months before relief supplies which had been sent for from the American colonies would arrive.[8] Things on the Scottish front therefore began to move again in December 1652 when the committee for trade and foreign affairs took the matter up and, in January 1653, it was discovered that, in the prisons of the Commonwealth of England lay Colonel David Ross, laird of Balnagown. He had been captured after the crushing defeat at Worcester of a Scottish army, invading England for Charles II, in September 1651.[9] They thus had in their custody a Scottish landowner with extensive pinewoods, so they diverted their attentions from Speyside to Ross, and specifically to Strathcarron. The committee for foreign affairs considered a report concerning Colonel Ross and his woods. They decided to consult directly with him and to proceed with plans to procure Scottish pine for masts by sending someone north to begin the work. By 15 January, a Dutch prize ship had been found 'for to goe upon the discovery of masts to Scotland'. Although attempts were still being made to come to an agreement with Tullibardine, and would be until they fell through in the summer of 1653, it seems that the woods of Ross were now to be the first target.[10]

Colonel Robert Lilburne, by now commander-in-chief in Scotland, corroborated Colonel Ross's accounts of his woods which he had supplied to the council of state, and urged that a beginning should be made on the Balnagown estate. Lilburne planned to send thirty 'pioneers' there and asked for a frigate to be fitted out for that purpose and for 'biscuit, cheese and axes' and other necessary materials to be sent to Tarbat Ness. A few days later, it was reported that a ship was to be sent to Scotland to fetch 'great masts'. It was to have naval convoy, four months' victuals, some arms and a crew of about ten, along with four men 'well skilled in the choosing and cutting of masts'.[11] Further delays beset the project and it was not until May that the ship was ready. On 5 May, an instruction was sent to Inverness, the nearest Cromwellian fortress, to provide money for labour, land carriage

---

[8] *CSP Dom 1651–2*, p. 351; R. G. Albion, *Forests and Seapower: the Timber Problem of the Royal Navy, 1652–1862* (Cambridge, MA, 1926), pp. 201, 206–8, 214–16, 234.
[9] Adam (ed.), *Calendar of Fearn*, p. 218.
[10] *CSP Dom 1652–3*, pp. 6, 61, 178, 453; PRO: SP 18/33 no. 85, SP 18/37 no. 165.
[11] *CSP Dom 1652–3*, pp. 76, 78, 82, 97, 125, 129, 163, 167, 178; PRO: SP 18/45 no. 38, SP 18/46 no. 3, SP 18/47 no. 174, SP 18/33 nos 49 and 85.

and the masts themselves, and ordering a guard to be sent out to the woods 'if the country is not quiet'. On 21 May, the ship was reported to be 'now bound thither'.[12] Again, the trail runs into the sand with no information in the English state papers as to what the fate of the operation was. The only evidence that any work was actually carried out comes from the estate accounts of Balnagown which, in the discharge section for 1652–3, recorded that £40 was accounted for 'to cut the mastis', £3 'for wreiting the contract' and £24 for 'biging 3 houses in the wood'. The records of the Baron Court of Balnagown also record a contract and the cutting, transporting and loading of the masts onto a ship.[13] Although it is possible that the woods of the Balnagown estate were producing masts for someone else, it seems more likely that these entries refer to timber operations for the English navy, especially since the landlord was then in an English prison and in no position to enter into commercial agreements with anyone else.

The difficulties presented by a hostile countryside were enough to render the project a failure. A rising in the Highlands in 1653–4 must have created significant difficulties for those trying to extract masts from the woods for the English navy. The strain which was put upon the state's resources by the financial demands of the Dutch war meant that the combined tasks of putting down a Highland rising and extracting timber were beyond the means of the regime.[14] Ironically, the Dutch war, which had made the English undertake serious attempts to seek naval timber in Scotland, put the enterprise beyond their capabilities. Moreover, the main fighting in the first Dutch war was over by the end of 1653, so there was soon no further need to seek Scottish timber for the English navy. The only wood which the occupying army usefully exploited appears to have been that used by the garrisons of the fortresses, notably at Inverlochy from which the neighbouring woods of Cameron of Lochiel were plundered.[15] In 1655, Thomas Tucker, an employee of the Cromwellian regime sent to Scotland to report on the state of trade, gave the impression that little had come of the government's attempts to extract timber from Ross. He reported that the timber trade out of Tain did not amount to much, comprising 'may bee a small barke once in a yeere from Leith, to fetch deales, which are brought down thither from the hills'. Typical of such small-scale trade was a contract of 1658 between John Ross of Little Tarrell and Patrick Smith of Braco for delivery to the shore of the firth near Ross's sawmill at Bonar, of two

---

[12] *CSP Dom 1652–3*, pp. 308, 343; PRO: SP 18/39 no. 19.
[13] NAS: Balnagown Castle MSS, GD 129/1/10/34; 'Balnagown Barony Court Register', GD 129/2/353, pp. 64–6.
[14] Dow, *Cromwellian Scotland*, ch. 4.
[15] NAS: GD 1/658/1 bundle 7; Albion, *Forests and Seapower*, p. 207.

hundred fir deals, thirty squared 'red fir trees' (twenty-two feet long and seven inches square) and thirty bolls of birch bark.[16]

The woods of Ross were not to be forgotten by the English, however, and a man whose family had been involved in the project in the 1650s, Phineas Pett, was to have some success in bringing Scottish timber south for the navy.[17] The outbreak of the Second Dutch War at the beginning of 1665 meant that alternative sources of timber were again required. In July 1665, estate papers record that masts were already being felled in the laird of Balnagown's woods. They may have been for the commercial market, rather than for the Royal Navy, for it was not until the autumn that Phineas Pett arrived in northern Scotland. On 8 September, David Ross of Balnagown, the son of the laird who had been captured at Worcester and had died in prison in England, received a letter from a friend in Inverness. Balnagown's correspondent reported that:

> Fryday last cam heir on[e] Captaine Pett on[e] of his majesties shippe buylders who came heir too Scotland of purpose to sight all the woods in Scotland and especially for masts. And he is mayd verie confident heir that your honour is the only person that his recourse wold be to any avail.

Pett had thus been directed to Balnagown who was urged to 'mak him wellcome' because he was 'verie deserveing in civilitie and much mor because of his master and his imployment [which is] so necessar for both king and countrie'.[18]

Phineas Pett must have been impressed by what he saw for, at the end of November, he entered into a contract with Ross of Balnagown, by which Balnagown agreed to sell to Pett 1,000 pines from Glencalvie, Glen Einig, 'Glaschekyllis and Mewliche, and certane other woods'. Glen Einig is in Strath Oykell, but the others were all apparently in Strathcarron. The contract was:

> For mastes non of quhich to be less then eighteen inches diameter . . . twell foot hieghe from the butt or ground end quhich ar to be delyvered and transported and brought to Ardmore and from thence to the shipp or shippes sydes that sall come for the same.

---

[16] 'Report by Thomas Tucker upon the Settlement of the Revenues of Excise and Customs in Scotland', in P. Hume Brown, *Early Travellers in Scotland* (Edinburgh, 1891), pp. 162–81, at p. 175; NAS: GD 190/3/151/42.
[17] See *CSP Dom 1652–3*, p. 537 which is addressed to 'Commander Pett'.
[18] NAS: GD 129/1/38, GD 129/1/41.

Each mast was to cost 30s. sterling and Balnagown undertook to transport them to the sea within twelve months. It was also stated that, for £100 sterling, Balnagown had sold to Pett an additional 140 trees for masts which were to be 'chosyn and marked out of the above named woods by the said Phineas Pett' who would 'be directed in and about the forsaid woods quhair the biggest and tallest tries are standing and groweing'.[19] These particular trees were to be marked by Pett within three months of the date of the contract. It seems that this arrangement was made in haste so that a prime example of the masts could be sent south immediately.

By March in the following year, Pett was able to write to the Navy Commissioners in connection with sending ships to the 'river of Tane' for the masts which were 'now in readiness'. He also reported that, as well as the trees which were ready to be loaded (presumably the 140 mentioned above), he had made contracts with Balnagown for a further 1,000, with MacKenzie [of Seaforth] for 300 and with the laird of Glenmoriston too, all of which had been confirmed by the Navy Board in January. He urged that these should be 'looked into and such dispatch given to each circumstance ... that a breach of their covenants may be prevented'. Although Pett was still trying to expedite the agreement with Seaforth in July, there is little evidence that anything came of these other contracts.[20] The contract with Ross of Balnagown for 1,000 masts was, however, definitely undertaken. By the end of April 1666, a servant of Phineas Pett had seen to it that 1,015 trees had been marked for felling and a further 145 were still to be marked. Also, 260 had already been felled, all but nineteen of which were waiting for shipment at Ardmore. There must have been some delay in procuring the ships for freighting the timber and it was not until July that Pett returned to Scotland with two flyboats for the masts and a large convoy for protection.[21] The customs books of Inverness show that the masts were successfully shipped out on 2 August. Although there is no record of money having been levied on the timber itself, custom was payable on 3,000 skins and forty-eight barrels of salmon on 'the kinges shipes' which 'caried the masts for London'.[22]

---

[19] NAS: GD 129/1/54; see also a set of notes from some lost Balnagown manuscripts made by Miss Rosa Ross Williamson, now held by Miss Rosemary MacKenzie, c/o Tain Museum.
[20] *CSP Dom 1665–6*, pp. 326, 515–16; PRO: SP 29/152 no. 61, SP 29/162 no. 32i. No trace of either of these contracts or of the letters relating to them has been found. Catalogues of the documents for Seaforth (NAS: GD 46) and Glenmoriston (National Register of Archives (Scotland) Survey No. 101) give no indication of any such papers. Only a lengthy trawl through poorly catalogued collections, impossible for the purposes of this study, could resolve this question.
[21] NAS: GD 129/1/47; *CSP Dom 1665–6*, pp. 515–16, 517–18; PRO: SP 29/162 nos 32 and 49.
[22] NAS: Inverness Customs Books, E 72/11/1.

Phineas Pett seems to have regarded his initial efforts as such a success that he even attempted to branch out into private enterprise, remaining in Scotland after the masts were shipped out. Sir William Penn, one of the Navy Commissioners, and Samuel Pepys, the diarist and clerk of the king's ships, contemplated using Pett as an agent for procuring timber in Scotland for the rebuilding of much of London in the aftermath of the Great Fire of 1666. Pepys wrote in his diary that he and Penn had 'walked in the garden by moonlight and he proposes his and my looking out into Scotland about timber and to use Pett there; for timber will be a good commodity this time of building the City. And I like the notion and doubt not that we may do good in it.'[23] Perhaps with a view to establishing himself in the long term as an agent in Scotland for supplying timber to England, and in spite of the fact that nothing came of the London enterprise, Pett settled himself at Inverness. In November 1666 he bought a substantial property in the Cromwellian fortress, or 'citidaill of Inverness . . . built be the lait usurpers' (Fig. 12.1) and, in the following year, he had a stout stone wall built around the property with two arched entries and a strong wooden door.[24] He was clearly planning to stay for some time and perhaps use his property, which lay near the harbour of Inverness at the mouth of the River Ness, as a depot for timber.

In February 1667, what seems to have been a further 247 Scottish masts along with some deals and other timber arrived in the Thames and were put into storage. A few pieces of timber had even been used to replace two yardarms and a capstan on one of the ships which had brought it south.[25] It was noted on their arrival that some of the masts were shorter than expected and it seems that, in spite of the fact that over 1,000 had been marked, Balnagown was unable to supply enough trees of sufficient size. In an attempt to fulfil his contract, he even resorted to contracting with neighbouring landowners for their trees. In January 1667, he made a bargain with Walter Ross of Invercarron to buy 150 trees from his part of the wood of Glencalvie for £84 sterling.[26] It is, of course, possible that this was done because all of the marked trees which had been bought from Balnagown's woods had been used but it seems more likely that the cause was the deficiency in the first 500 or so to be sent out. When the 247 masts mentioned above were put into the stores, it was noted that some of the 136

---

[23] R. Latham and W. Matthews (eds), *The Diary of Samuel Pepys* (London, 1970–83), 7, pp. 298, 300–1.
[24] NAS Register of Deeds, RD 2/20, 10–11, RD 2/21, 371–7.
[25] *CSP Dom 1666–7*, pp. 510, 513, 524; PRO: SP 29/191, nos 87, 103, 152.
[26] NAS: J. and F. Anderson Collection, GD 297/204/7.

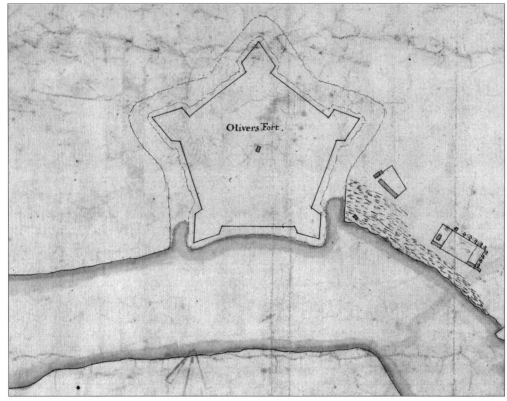

Fig. 12.1 The Cromwellian citadel at Inverness which Phineas Pett bought in 1666 to act as a secure wood-store for his purchases from Strathcarron. Reproduced by permission of the Trustees of the National Libraries of Scotland and the Ministry of Defence.

masts received off the ship *Milkmaid* 'were short in length' and the 111 which were unloaded from the *Franiken*, 'wanted . . . three hundred ninety one foote and a halfe in length', or an average of about three feet six inches per mast.[27] Clearly, there were not actually enough trees of the required length to be had in Balnagown's woods, or for that matter in those of his neighbours, in spite of what Phineas Pett's servant had thought.

Either due to a deficiency in the record or more probably due to the end of the Second Dutch War in 1667, no more shipping of masts appears to have taken place that year. There was, however, an outstanding contract with Balnagown to be fulfilled, albeit that his trees were not as large as had been claimed. In April 1668, a new agreement was made between Balnagown and Pett. It began by reciting the original contract of November

[27] *CSP Dom 1666–7*, pp. 510, 513; PRO: SP 29/191, nos 87, 103.

1665 but it went on to state that, not including the 140 masts which were covered by a separate agreement, only 445 masts of the agreed 1,000 had conformed to the terms of that contract and it conceded that a further 70 masts would be felled, making in all 515. The new articles of agreement also contain a puzzling piece of information. Although Pett owed Balnagown a total of over £770 sterling, he had paid only £350, yet the onus seems to have been on Balnagown to fulfil his side of the bargain rather than on Pett to pay Balnagown in full. The reasons for this become clear when other papers are examined. It appears that Balnagown was using his credit with the Crown to pay off personal debts and taxation.

In March 1668, the collector of the stent, based in Inverness, wrote to the commander of His Majesty's forces in Ross, declaring 'I have receaved from Capt Pett for what taxatione wes resting be the laird of Bellnagowne and I doo declare him not to be deficient after this day.' In September, Balnagown found that using Pett to pay off his debts was not always reliable. One of his creditors in Inverness wrote to him to reveal that 'Mr Paite did not pay the money of your two bonds conforme to your ordour.' In May 1668, Balnagown even used the state's credit to the extent of agreeing to transport deals to Edinburgh on one of the ships which would freight his timber for the navy to satisfy a debt to a creditor there. Alexander Farquhar, a merchant in Edinburgh did a deal with Balnagown that, if he provided Pett with 500 additional deals which could then be shipped to Edinburgh, Farquhar would discharge him of a debt of £257 Scots.[28] Balnagown's severe financial difficulties were the result of the imprisonment of his father in the 1650s. In 1670, he remained in such straits that he began to petition the Crown for help, persistently pointing out that it was his father's support for Charles II in 1651 that had led to his financial downfall. His father had mortgaged his lands to raise troops for the king and, on being taken prisoner, his lands were seized by Cromwell and not restored until 1660. Meanwhile, the heir, now the laird, had relied on the charity of others and had run up debts because he had had no income from his lands.[29] Although these petitions which sought a Crown pension or office were unsuccessful, Balnagown had discovered that the government was a useful source of credit and he would use this facility to the full in years to come.

Another shipment of naval timber from Ardmore was loaded at the end of September 1668. It was taken to Inverness and custom was actually paid on the wood when it was shipped out in November. The load consisted of

---

[28] NAS: GD 129/1/38, GD 129/1/6/1, GD 129/1/10/34. Stent was a land-based system of taxation.
[29] NAS: GD 129/1/44/183. There are numerous copies of this petition scattered throughout the collection.

118 'great masts', 22 logs, 89 spars and 580 deals, most of the last presumably being those destined for Alexander Farquhar.[30] Some of the timber arrived at Deptford and was being unloaded early in December. It is clear from a letter to the Navy Commissioners from Thomas Turner, an official at Deptford, that timber remained scarce, even though hostilities were over, because damages sustained in the war had emptied the stores. As soon as the ship had docked, men were put on board to see to her unloading. As a result of the Dutch war, the prices of timber and tar had risen by about 50 per cent between 1665 and 1668. Masts previously bought for £10 sterling now cost £15, so any domestic supply remained highly desirable, even if it did not reach the standards of quality and size normally demanded by the navy.[31] A report regarding the Scottish pine which arrived at Chatham about the same time revealed that it remained deficient in terms of its length. It was found to be 'very inconsiderable, there being but 155 peeces of sorry stuff'.[32]

Other reports on the Scottish timber revealed it to be substandard and the whole enterprise to have been quite costly. A memorandum of January 1669 was headed 'Ill success of our undertaking to fetch masts from Scotland'. It noted that two of the last three shiploads of timber which had arrived from Scotland had cost the state over £1,500 but 'the whole loading (so bad are the masts)' was worth only about £600. It was observed that, with the war over, going to such lengths and such expense could hardly be justified. Although it was admitted that 'circumstances were quite otherwise when this project of fetching masts from Scotland was set on foot', the writer asserted that 'the design will undoubtedly be found very chargeable and Mr Pett in Scotland blameable for encouraging it, as he did upon his first survey, when sent thither to that purpose'. It was also observed, alluding to the 1650s, that a similar scheme had been tried 'in the late times and with no better success.'[33] The basic problem, other than the obvious one of size, was that the native Scots pine was unsuitable for masts. Long, mild growing seasons and relatively mild winters in Scotland meant that the trees grew too quickly: 'the rate of growth in Scotland was often double that of Russia'

---

[30] NAS: Inverness Customs Books, E 72/11/2; NAS: Register of Deeds, RD 2/27, p. 645.
[31] *CSP Dom Addenda, 1660–85*, pp. 278–9; PRO: SP 46/137 no. 162; PRO: Memoranda and Notes, ADM 106/3520. Efforts to unload her continued into January with more obviously frustrating delays, see *CSP Dom 1668–9*, p. 151; PRO: SP 29/250, no. 55.
[32] *CSP Dom 1668–9*, p. 118; PRO: SP 29/250, no. 201. The editor of the published *Calendar* conflated 'firrewood', i.e. firwood, and 'sorry stuff' to get 'sorry firewood'. The timber was not first-rate but it was more than just a load of kindling!
[33] R. Latham (ed.), *Samuel Pepys and the Second Dutch War: Pepys' Brooke House Papers* (Navy Records Society, 1995), p. 173.

and the resulting timber was too soft. Scots pine 'needs short, hot summers and very cold winters to form sound timber of slow growth'. Such conditions were to be found in the Baltic and New England and it was from the white pine of North America that the bulk of the Royal Navy's masts came before the American Revolution.[34]

There is some evidence that Pett remained in Scotland until the end of 1669, for he wrote to one of his superiors in January 1670 stating that he had been 'detained in the Highland of Scotland getting in readiness the remainder of the masts and timber which was found out in the time of the Dutch war'. During 1669, Balnagown was still extracting masts from his woods and, as late as November of that year, he ordered the payment of oatmeal to 'the servants for careing of my masts'.[35] If the enterprise continued into 1669, it involved only clearing up the dregs and, when the Third Dutch War broke out in 1672, no further attempts were made to seek Scottish pine for masts. In spite of this, a petition by Phineas Pett, written in 1671, reveals that, although the masts were not as good as they had appeared when marked in the woods, they and all the other timber were found to be 'of good use in the navy'.[36]

Although Balnagown's trees were not very suitable for the masts of great ships, they were useful for the production of other sorts of timber, and the deals exported by Pett for the navy would have been perfectly serviceable for decks and internal partitions. Masts were a specialist commodity and did not constitute a very large proportion of the timber extracted from the woods of Ross, which appear to have continued to supply a significant amount of timber to local and regional markets during the period. Wood was supplied to Rose of Kilravock, near Cawdor, to Lord Strathnaver, the Earl of Caithness, the Earl of Sutherland, and to customers in Inverness, Findhorn, Elgin and Banff as well as closer to home. Because he owned the most northerly pinewoods in the country, his regional trade even extended as far north as Orkney. Further afield in Scotland, Balnagown's timber was purchased by merchants in Dundee, Anstruther, Leith and Edinburgh, and the Duke of Hamilton even bought substantial amounts from him in the 1670s.[37] Balnagown was obviously keen to sell as much timber as possible to extricate himself from his dire financial predicament.

The vast majority of this timber was sold in the form of deals, sawn in

---

[34] Albion, *Forests and Seapower*, pp. 30–1.
[35] *CSP Dom 1670*, p. 42; NAS: GD 129/1/44, GD 129/1/38/155.
[36] *CSP Dom 1671*, p. 374.
[37] Numerous references to this trade can be found in NAS: GD 129. See also NAS: RD 12, 14 July 1663, no. 1059.

Balnagown's sawmills at Kincardine and Ardmore. A number of tacks for those sawmills survive. Some involved the laird undertaking to supply a certain number of logs of a specified size to the mill from which the millers must supply him with a specified number of deals. If they cut any more, they might sell them freely as long as every sixth additional deal went to the laird.[38] The tacks suggest that, in the later 1660s, with the combination of naval and other demand, the woods were yielding about 1,200 large logs per annum which were sawn into 7,200 deals, giving six deals from each log. During the 1670s, the style of the contracts changed, with the millers receiving the right to fell the agreed number of trees and float them to the mill themselves in return for a money payment, or a specified number of deals, and free labour from the laird's men.[39] This level of activity probably meant that unprecedented inroads were being made into the pinewoods of Ross of Balnagown in this period.

The most successful long-distance sales of timber by Balnagown, at least in terms of a sustained trade, took place in the 1670s, although they had their origins a few years earlier. Perhaps because he spent much of his time in Edinburgh, perhaps because of the publicity gained from the naval trade or just because of the widespread market which Balnagown's woods enjoyed, a large parcel of timber was sold to two Edinburgh craftsmen in May 1667. Andrew Cassie, slater, and James Bains, wright, agreed to buy 5,000 deals, 500 spars, 500 joists and 200 unsquared trees. Both were involved in the building trade and, perhaps more significantly, both were employed by the Crown to work on the repair and upkeep of the royal palaces and castles. A great deal of timber was constantly required for this, especially for the almost complete rebuilding of Holyrood Palace which was undertaken in the 1670s.[40]

The first consignment of this timber arrived at Leith in November 1667 but, as with the naval timber, a significant proportion did not measure up to the contracted sizes.[41] Although the records for the next few years appear to be wanting, it is evident that this deficiency in the quality of the wood was not enough to deter Bains from buying more timber from Balnagown.

---

[38] NAS: GD 129/1/38/155.
[39] NAS: GD 129/1/47, GD 129/1/39, GD 129/1/46.
[40] NAS: GD 129/1/7/10; NAS: Treasury Vouchers, E 28/70/7–8; Tain Museum: 'Notes taken from papers, printed and manuscript, used in the "Balnagown Case" of 1737–1776'. 'Bains' is variously spelled Bean, Bain, Baine, Baines, Beanes, Baynes etc. The first contact, however, was perhaps that made in 1665 between Walter Ross of Invercarron and George Baine, burgess of Edinburgh, for the sale of 2,000 deals delivered to Leith by the following July, NAS: RD 12, 8 Oct. 1666, no. 1478.
[41] NAS: GD 129/1/37, GD 129/1/44/183.

In the spring of 1672, Bains wrote to Balnagown to tell him of a vessel from Kirkcaldy which was coming north to pick up more timber.[42] In July, Balnagown wrote to his kinsman, Ross of Kindace, that he had 'past a great bargain of timber... for thrie yeir' with Bains who, by this time, was styled 'his Majesties master wright'. The contract included 240 joists twenty-six feet long and ten inches square, another 240 joists twenty-four to twenty-six feet long and between seven and eight inches square, 240 roof spars between twenty and twenty-four feet long and between five and six inches square, and 3,000 deals, all of which were to be supplied over three years. In the following March, 1,900 deals and 425 'trees' were loaded aboard the *Thomas of Kirkcaldy* while another ship took on 140 spars and over 80 joists as well as 700 deals to be delivered to Bains.[43] In the contract of 1672, it was agreed that Bains would pay the originally contracted rate for any extra pieces of timber delivered over and above that specified in the contract. In June of the following year, this was formalised with a further contract in which Balnagown sold 300 joists, 300 spars and 1,000 deals to Bains and, by September of that year, these agreements had been 'compleitit and closit' and a vessel with the last of the timber was anticipated at Leith soon.[44] The deficiencies in the lengths of timber must have been rectified. Either that or the heightened demand for wood for the rebuilding of Holyrood meant that whatever timber could be acquired had to be taken and it would be paid for according to the quality of what arrived. Further shipments went out in February, May and November 1674 as the result of yet another contract between Bains and Balnagown at the beginning of that year and, as late as February 1677 and January 1678, more timber was received by Bains in accordance with the 1674 contract.[45]

As well as being involved with the royal works, Bains was also a leading procurer of timber for those works. References to payments to him for supplying timber for the rebuilding of Holyrood are peppered throughout the manuscript Treasurer's Accounts and Accounts of the Master of Works between 1667 and 1679. All in all, the Crown appears to have given Bains around £20,000 Scots for various loads of timber.[46] Surprisingly, David Ross of Balnagown appears to have received little or nothing of this money

---

[42] NAS: GD 129/1/39, GD 129/1/41, GD 120/1/46.
[43] NAS: Rose of Pitcalnie Muniments, GD 199/237; GD 129/1/10/34 (four unnumbered items from this bundle).
[44] NAS: GD 129/1/48; RD 12, 1675, no. 335.
[45] NAS: GD 129/1/10/34 (three items from this bundle); GD 129/1/44 (three items from this bundle), GD 129/1/33/129, GD 129/1/33/130, GD 129/1/34/137; RD 14, 1692, no. 178.
[46] NAS: Treasury Vouchers, E 28/70/24/2; Accounts of the Lords Commissioners of Treasury, E 26/11; Accounts of the Master of Works, E 36/30, 31, 33.

himself. Just as with Phineas Pett, Balnagown used Bains's access to government capital as a means of securing credit and paying off outstanding debt. As early as 1667, he had instructed Bains to pay £70 Scots to a shoemaker in the Canongate. In the following year, an account was made up which showed that, after freight charges had been paid, every penny of the 1667 contract, amounting to a total of £2,460 Scots, had been paid to various craftsmen and others in Edinburgh to whom Balnagown owed money.[47] It would seem that Balnagown entered into these contracts with Bains to clear some of his debt and so that he could still procure goods and services in Edinburgh without having to handle cash and run the risk of prior claims upon it by existing creditors.

In the first few months of 1672, various debts were paid to the postmaster in Burntisland and an Edinburgh merchant, among others. By June, Balnagown even owed money to Bains. The new contract of July 1672 was probably, in part, a response to this, since Balnagown agreed to 'sell' any extra timber to Bains. The main proceeds of the contract, however, were to be paid by Bains directly to a third party, one George Suttie, an Edinburgh merchant who had previously been a bailie of the burgh. The same applied to the contract of June 1673, in which Bains agreed to pay an Edinburgh lawyer and a craftsman the whole value of the timber in the bargain.[48] Balnagown was so short of money in 1673 that he even had to pay for the freight of the ships carrying the wood to Leith out of the value of the cargo of wood which they were carrying. As time went on, Bains became increasingly unwilling to act as Balnagown's credit agent. This appears to have been because the creditors thronged the dockside the moment the timber was unloaded and before Bains had received any cash for it from the king's master of works. When this occurred in April 1673, Bains 'altogether refused to accept' a bill presented by one of Balnagown's creditors.[49] In spite of this, Balnagown continued to use Bains in this way over succeeding years and the estate papers contain scores of receipted credit notes which the laird gave to his creditors who would then present them to Bains. Balnagown's debts could not be satisfied from his woods, however. In 1678, it transpired that, four years previously, Bains had lent him 2,500 merks (£1,666. 13s. 4d. Scots) for which he was able to pay only the interest and, as late as 1691, Bains was still chasing Balnagown for payment.[50]

---

[47] NAS: GD 129/1/42/173, GD 129/1/46.
[48] NAS: GD 129/1/10/34 (the 1672 contract and numerous orders by Balnagown to Bains to pay sums to creditors), GD 129/1/38; NAS: RD 2/39.
[49] NAS: GD 129/1/10/34; GD 129/1/38; GD 129/1/43.
[50] NAS: GD 129/1/44/184; GD 129/1/43/179.

A number of deeds bear witness to continuing activity by David Ross of Balnagown between the end of the wood sales to James Bains and the Union of Parliaments. In 1686 he made an arrangement with his neighbour, William Ross of Invercarron, that no fir trees should be sold independently in future by their tenants, but that they and their factors should be 'the merchands for cutting, hewing, dryving and selling', and further, that neither should undersell the other but rather arrange a cartel, fixing the price at which local wood was to be sold to outsiders. Two years later, Invercarron sold wood from Glencalvie to Patrick Dunbar of Sidderay, and in 1694 Balnagown signed a contract with Lord Strathnaver in the north to furnish him with wood at fixed prices – joists twenty-six feet long and twelve inches square, £6 Scots each; the same seven inches square, at £1. 6s. 8d.; those of twenty feet and six inches square, 16s., and so on.[51]

A new kind of exploitation was contemplated in 1705 when Balnagown signed up with William Sutherland, merchant of Elgin and William Gordon, 'late factor at Paris, now merchant in Edinburgh', to provide 'as much laying firr timber or old standing firr trees' as could be found suitable for making tar, in the space of fifteen months. Since 'young growing firr' are specifically excluded, this was probably a way to clear up the woods and make such use of them as was possible, following the hurricane of 26 November 1703, which had been one of the great destructive storms of British history.[52] The devastation was not total, however, for next year in April he was selling 600 fir logs out of Strathcarron to local merchants, and in November he negotiated to sell 12,200 fir trees to Francis Plouden in Edinburgh and two Dublin men, William Johnstoun and Christopher Jackson. They were to have 'the greatest and best trees that can be had in all the firr woods', chosen and marked by the estate, to be cut over twelve years, with permission to 'medle and intromitt [interfere] with the marble rock of Knockchirny'. The merchants planned to employ Nathaniel Jee, shipwright of Burntisland 'master builder to the Scots African Company', for one year to advise them on cutting and transporting the wood', and to make tar, build ships, barks or boats and direct the splitting of barrel staves and lathwood and the sawing of deals and planks. Unfortunately William, Lord Ross, the entailed heir to Balnagown, raised certain objections and the deed was not signed. It is unclear what was the final outcome but Ross's principle of marking the trees by the estate was noted with approval by the Earl of Breadalbane in his negotiations with an Irish partnership in 1723 (see below, p. 349).[53]

---

[51] NAS: RD 12, 1686, no. 1024; RD 14, 1691, no. 1221; RD 12 reg. 5 July 1722.
[52] NAS: RD 12, 1706, no. 1468; Daniel Defoe, *The Storm* (London, 1704).
[53] NAS: RD 14, 1707, no. 145; GD 129/1/145 Box 36 and 129/1/152 Box 37; GD 197/213.

There were later episodes of felling, especially another attempt in the 1720s to make money out of the wood for masts, by General Ross of Balnagown, which also involved Joseph Avery who was later to be concerned with the York Buildings Company (see above, p. 207). It was not sustained, or apparently very successful, despite the quality of some of the tall, straight trees and the envy of Ross's neighbour at the prices they fetched. Steven and Carlisle also relate a local tradition that until around the middle of the nineteenth century there was still a large pinewood 'in Diebidale' (probably Glencalvie), supplying both local building needs and outside sale, 'the timber being carried by ponies along a rough hill track via Strath Rusdale to the Cromarty Forth': they also mention a custom extant until the 1870s for young people leaving the Amat district to be given a 'fir chest' made from local timber to keep their possessions.[54]

It is hard to make precise judgements as to what impact the major extractions of timber would have had on the native pinewoods of Ross. The earliest maps of the area were drawn by Gordon of Straloch, presumably after Pont. One survives as a manuscript, the other is reproduced in Blaeu's *Atlas Novus*, the latter appearing to derive directly from the former. Neither is very helpful. They show a widespread scattering of woodland right across inland Ross and Sutherland, yet the tree symbols appear to have been used as much to fill in voids as to indicate the actual location of woods. As is common with these maps, where there are lots of place-names, they elbow out the trees. This is indicated most conclusively in the failure of these maps to portray the wood of Amat at all, in spite of the ample documentary evidence for its existence and extent, nor is there more than a very little woodland shown in Strath Oykell.[55] For the purposes of assessing the nature and extent of woodland, these early maps must, therefore, be discounted.

The earliest detailed map of the area is that produced by General Roy's Military Survey of Scotland in the middle of the eighteenth century. If its portrayal of woodland is to be relied upon, seventeenth- and early eighteenth-century exploitation of the woods of Strathcarron and Strath Oykell had very little impact on the extent of woodland there. The main wooded areas in the Strathcarron area, Amat and Glencalvie, were clearly indicated, swathing the slopes of Strathcarron and Glencalvie as they continued to do into the nineteenth and twentieth centuries and do to this day. It should, however, be remembered that these maps give no reliable indication as to the density of the trees, and certainly not of their age, or what species of

---

[54] Steven and Carlisle, *Native Pinewoods*, p. 209.
[55] NLS: Gordon of Straloch, 'Sutherland, Strath Okell and Strath Charron'; 'Extima Scotiae' from J. Blaeu, *Atlas Novus* (Amsterdam, 1654).

trees were to be found in the woods and what the proportions of different species were. It is clear from seventeenth-century records that the wood of Glencalvie contained a substantial number of large pines and oak (the latter only mentioned once) but Steven and Carlisle found it to be almost entirely birch by the middle of the twentieth century. Therefore, although seventeenth-century activity clearly did not destroy the wood of Glencalvie, it may well have contributed to a significant shift in the balance of species within it away from Scots pine and oak towards birch. Certainly nothing more is heard of oak in Glencalvie after 1612, so it probably went at an early date. A shift towards more birch may also have been the case for Glen Einig in Strath Oykell and, to a lesser extent, for Amat in Strathcarron. Steven and Carlisle show substantial areas of birch and of pine in the latter, which may reflect such a change in the proportion of species. In the case of Einig wood, the fact that it was found to consist predominantly of birch with only 'three groups of pine . . . with scattered trees elsewhere' would suggest that there had been significant changes in its composition since the seventeenth century.[56] The wood of 'Meulich' has not been identified beyond question as there are many *meall* hill names in the area, but it seems very likely that it was Meaghlaich in Strathcarron, which is really a southwesterly extension of the wood of Amat still visible on modern maps.[57] Glaschoille in Strath Cuileannach may also have been removed in the later seventeenth or early eighteenth century since no wood was marked there by Roy or in later maps. The name may, however, have referred, even in the seventeenth century, to a northwesterly extension of the wood of Amat into Strath Cuileannach which still existed in the fourth quarter of the nineteenth century and, to a lesser extent, when Steven and Carlisle looked at it. By and large, the most significant reduction which seems to have occurred in the extent of the woodlands of the area since the seventeenth century appears to have taken place after around 1875 when the first Ordnance Survey reached Ross. In the case of the wood of Amat, the modern Ordnance Survey shows a substantial reduction in the extent of the wood since Steven and Carlisle surveyed it in the middle of the twentieth century, although this may merely be due to the difficulties involved in classifying woodland for cartographic purposes.

One of the woods depicted in the first Ordnance Survey, at An Sgaothach at the head of Glen Alladale, still exists but was not portrayed on Roy's map at all. It consists largely of Scots pine and it reminds us that we often cannot rely on Roy for absolute accuracy. It may even be a remnant of much more

---

[56] Steven and Carlisle, *Native Pinewoods*, p. 215.
[57] W. J. Watson, *Place-names of Ross and Cromarty* (Inverness, 1904), p. 8.

extensive woods which stretched up Glen Alladale and Glenmore, linking An Sgaothach to Amat. The scattered trees portrayed on the first Ordnance Survey and revealed by Steven and Carlisle to be pines, certainly raise this possibility.

It is perhaps noteworthy that the highest proportion of pine to birch in any of these woods is to be found in the wood of Amat which, even if it is understood to extend into Strath Cuileanach, lies largely outwith the lands held by David Ross of Balnagown in the second half of the seventeenth century. It may be that the disproportionate survival of pine there can be linked to its not having been exploited by David Ross in that period. In 1677, the local people were alarmed at the approach of an English tourist because, he recorded, they feared 'that the King had sent us with orders to cut down the wood', which appears, from his description of its location, to have been the wood of Amat.[58] Were they afraid that the same fate which had befallen other woods in the area was now to be visited upon the one extensive stand of large pine trees that was left? This is supported by the fact that, in 1767, it was reported that in Glenmore, the western continuation of Strathcarron, the 'Fir is the finest . . . in Scotland. One tree rearing a straight uninterrupted trunk to the height of 30 or 40 feet' and eight feet six inches in circumference.[59]

The fact that David Ross of Balnagown was so deep in debt and apparently felling trees with no thought for the morrow would suggest that he was pursuing an entirely unsustainable policy with regard to his woods. His successors appeared to have followed the same course, at least until the 1720s. One possible explanation for the apparent decline in the proportion of pine in the woods examined here, principally to the advantage of birch, is the heavy exploitation which took place, especially for the Crown but also for local and regional markets. It may, however, be unfair on David Ross to blame him for the failure of the pine which he felled to be replaced, in spite of the fact that there is little direct evidence in the estate papers for anything resembling woodland management, except for the numerous 'convictions in wood' recorded in the Baron Court book after local people had taken timber without permission.[60] Sixteenth- and seventeenth-century references to 'gryt firr woods' and the sizes of trees necessary to provide the

---

[58] P. Hume Brown (ed.), *Tours in Scotland, 1677 and 1681, by Thomas Kirk and Ralph Thoresby* (Edinburgh, 1892), pp. 35–6. The name of the wood is left blank in the account but, since Kirk appears to have recorded that they were led up Strathcarron, Amat is the most likely candidate.

[59] D. M. Henderson, and J. H. Dickson (eds), *A Naturalist in the Highlands: James Robertson, His Life and Travels in Scotland, 1767–1771* (Edinburgh, 1994), p. 44.

[60] NAS: GD 129/2/353.

thousands of masts, joists and deals would suggest that there were thousands of trees well over 200 years old in the woods of Ross of Balnagown by the middle of the seventeenth century. Any seedlings, saplings and young trees which existed would have been ignored in the quest for such timber. The trees which were felled would have begun their lives in or before the middle of the fifteenth century and the intervening period witnessed a significant worsening in the climate which became increasingly cold and wet. Since this pine is the most northerly in Britain, its relatively marginal situation might well have meant that, by the time that Phineas Pett and James Bains came along in the 1660s and 1670s, it was well over 100 years since there had been any significant regeneration. A similar situation appears to have prevailed on the Earl of Breadalbane's lands in Glenorchy (see Chapter 13). With significantly reduced regeneration having prevailed for many years in Ross, the extraction of large quantities of mature trees would have significantly reduced the proportion of Scots pine in these woods, with birch taking its place. Grazing pressure preventing regeneration after large-scale felling is another factor which ought to be considered, yet the fact that only the proportion of pine to birch, rather than the actual extent of the woods, changed between the middle of the seventeenth century and the middle of the twentieth, would suggest that long-term climatic change has been the major contributor to the decline of pine in Ross. The pine may therefore have been existing for centuries in an unsustainable way due largely to non-anthropogenic factors. Similarly the oak in Glencalvie was among the most northerly and also among the highest in Scotland, so it too may have been particularly vulnerable to worsening climate. The intervention of man by extensive felling may merely have accelerated the rate of decline in an already precarious situation.

CHAPTER 13

The Irish and Glenorchy, 1721–40

Ever since James Lindsay's work in the 1970s, it has been cogently argued that the commercial exploitation of timber in the West Highlands, far from exacerbating the extent of woodland destruction, might rather have had a beneficial effect through the introduction of more systematic and effective management strategies.[1] Nevertheless, perceptions of the deleterious effects of large-scale external or landlord-driven enterprises in the area have a long history and still find credence today; so long as any such exploitation has taken place, there have been those who, for all number of reasons, have desired to point the finger of blame at others.

In part, this is an aspect of the reaction to the marked growth in profit-oriented commercialism among clan chiefs from the eighteenth century, often seen as responsible for the destruction of Highland culture and community, up to and including the clearances in the nineteenth century.[2] However, the commercial pressures and opportunities experienced by the clans in relation to the exploitation of natural resources in the West Highlands involved all levels of indigenous Gaelic society, not just the chiefs. Recent work makes it clear that these pressures must be placed within a wider British and even imperial context, and that they had been increasing factors in Highland history since the seventeenth century.[3]

Given that the felling of pine and oakwoods in the West Highlands was one of the most obvious aspects of the comprehensive exploitation of the area which took off in the eighteenth century, a detailed look at one such episode provides an interesting case study. There were, in such situations, always three main groups of players: the owners of the woods, often short of cash; their tenants who had utilised the timber resource for many aspects of their daily lives from time immemorial; and the 'foreign' commercial interests who provided a new dimension to woodland history from the late seventeenth century onwards. The relationships between them were complex. Each of these groups have either claimed, or had claimed for

---

[1] J. M. Lindsay, 'Charcoal iron smelting and its fuel supply; the example of Lorn furnace, Argyllshire, 1753–1876', *Journal of Historical Geography*, 1 (1975), pp. 283–98.
[2] For example, J. Hunter, *The Making of the Crofting Community* (Edinburgh, 1976).
[3] A. I. Macinnes, *Clanship, Commerce and the House of Stuart, 1603–1788* (East Linton, 1996).

them, a prominent role in maintaining and preserving the woodland resource; equally, they have each also been blamed (often by each other as well as later commentators) for destroying it.

This chapter is focused on one particular group of outsiders, a partnership of Irishmen operating in the 1720s and 1730s, about whom local feeling was explicit and unmistakably hostile. In 1725, only two years after he had entered into a contract with them, the Earl of Breadalbane expostulated about the cutting of his oakwoods in Glenorchy: 'I never was so out of humour as yesterday, I went about to see all the oak timer the Irishmen have cutt, they have not left one standing oak tree in the countrie.' The Earl was also deeply concerned for his pinewoods, which the Irishmen were beginning to cut. This sense of outrage was essentially shared by the wider community and is echoed by the Rev. Joseph MacIntyre writing for the *Statistical Account* at the end of the century:

> The higher parts of the parish abounded once with forests of the largest and best pines; but these were cut down, about 60 years ago, by a company of adventurers from Ireland, with little benefit to themselves, and less to the noble proprietor of the country.

However, he goes on to note that: 'There are still some tracts of natural firs in Glenorchy; a good deal of oak, intermixed with ash, birch and alder.' Did the 'adventurers' deserve the opprobrium that posterity heaped upon them? Was it their fault that the woods declined?[4]

The group of Irishmen in question were not the first of their countrymen to be involved in the Scottish woods, though they were certainly the most ambitious and, perhaps because of the exemplary fate of their leader, also apparently the last. Irish purchasers both of pine and oak had been operating at least from the 1660s until the Union of Parliaments in 1707 and beyond, striking bargains with woodland owners from the Solway, along the West Highland shores right round to the Dornoch Firth (see above, pp. 200–1, 248–51). They came in search of building timber and tanbark from a country increasingly short of both, and they were numerous enough to catch the anxious attention of John Spreull, well-informed Glasgow merchant, who in a mercantilist tract of 1705 proposed banning the tanbark merchants from Scottish shores in order to encourage Scots to process their own leather.[5] Nothing came of this proposal, and after 1707 the Irish remained as free as before to come and go as they pleased.

[4] NAS: Breadalbane papers, GD 112/16/11/26/1–2; *OSA*, 8, pp. 339–40.
[5] J. W. Burns (ed.), *Miscellaneous Writings of John Spreull* (Glasgow, 1882), pp. 63–4.

Our story begins with Roger Murphey, a Dublin tanner who also styled himself 'of Inniskillin'. Between November 1721 and September 1722, he entered into contracts with the following West Highland gentlemen: Sir Duncan Campbell of Lochnell, Patrick Campbell of Barcaldine, Colin Campbell of Inveresregan, James Fisher and Patrick Campbell in Inveraray, Donald Cameron of Lochiel, John Macdonell of Invergarry, John and Allan Macdonald of Morar, Ewen Cameron of Ardtornish for Allan MacLean of Inverscaddell, John MacLean of Ardgour, John Campbell of Macorne as trustee for Lord Caddell, and Donald Campbell of Losett. The contract for Breadalbane's woods in Glenorchy and Loch Etive was made in 1723.[6] Of these, the Camerons of Lochiel at least had had prior experience of the Irish market: in 1701 and 1714, Donald Cameron's father, John, had entered into contracts with a number of Irish tanners for his oakwoods in Glen Lochy and Glen Arkaig, and as early as the 1680s Sir Ewen Cameron of Lochiel had sold the Glen Lui and Glen Arkaig woods to an Irishman as part of a scheme to set up an ironworks at Achnacarry. The family's activity in this direction was sufficiently widely known for Spreull to pick out Lochiel as one who had shown the potential of native pine by selling it to the Irish. The fact that these same woods were sold again in the early 1720s suggests that insufficient activity, or at least insufficient profit, had resulted from these earlier sales.[7]

The extent of woodland encompassed by these initial contracts was, theoretically at least, considerable – from Loch Awe in the south up through Glen Kinglass, Glen Etive and Glenure, then a swathe around Glen Mallie, Glen Arkaig and Glen Lochy, through Invergarry and up into Morar and Knoydart. The trees to be cut included not only the lucrative pine and oak but many other species, including birch, alder, ash, elm and hazel. Table 13.1 and Map 13.1 indicate the woodlands sold to the Irishmen in the contracts of 1721–3.

The contracts did contain some exemptions and stipulations which were presumably designed to restrict the impact of felling. However, with the exception of the Breadalbane contract, these were fairly vague, (for example, to be 'cut and felled as should best conduce to the advantadge thereof for an after growth') or encompassed only a restriction on the number of trees to be cut, rather than how they were to be extracted, though enclosure by stake and ryce was also mentioned on one occasion. Such vagueness may be the product of a general inexperience with commercial timber enterprises or might reflect the fact that it was expected that traditional practices for

[6] Campbell of Lochnell muniments: NRA 934/58/779/30; GD 112/11/1/17.
[7] GD 1/658/1, bundle 16, ms. 1–2, 4; bundle 12, ms. 56, ms. 63; Burns (ed.), *John Spreull*, p. 31.

Table 13.1 Irish partnership's purchases in the West Highlands, 1721–3

	Proprietor	Woodland
1.	Sir Duncan Campbell of Lochnell	Ardmucknish Dalchreenacht Isle of Eriska North side of water of Kinglass (Ardmaddy, Poullich, Inverkinglass, Narrachan and Deirnafuor)
2.	Colin Campbell of Inveresregan	Kennacraig, especially the oak park of Inchketan
3.	Patrick Campbell of Barcaldine	Glenure Drielochan Lettervallin
4.	Donald Cameron of Lochiel	Glasdeir, Glacknadrochad, Clunie and Innercheagich South side of Loch Arkaig (Achnacarry to foot of water of Cairnderry) Glen Mallie, including braes of Lochiel, Conecherechan
5.	James Fisher and Patrick Campbell in Inveraray	Letters (Lochawe)
6.	John Macdonell of Invergarry	Invergarry Lochnevis in Knoydart North side of Loch Morar
7.	John and Allan Macdonald of Morar	Morar
8.	Ewen Cameron of Ardtornish (for Allan MacLean of Inverscaddell)	Inverscaddell and Conaglen
9.	Donald Campbell of Losset, baillie of Muckairn	Kilmun in Lochawe
10.	John, Earl of Breadalbane	Glenorchy (except Glen Strae)
11.	Donald Murchison, for Seaforth Estate	Letterewe
12.	John MacLean of Ardgour	? Contract probably not finalised
13.	John Campbell of Macorne, trustee of Lord Caddell	? Contract probably not finalised

woodland management would automatically be the norm and there was no need to spell them out. Unfortunately this would not be good enough in a court of law if things started to go wrong. It does also seem to be the case that a number of the proprietors (such as Ewen Cameron of Ardtornish for Allan MacLean of Inverscaddell and Donald Campbell of Losset) sold the

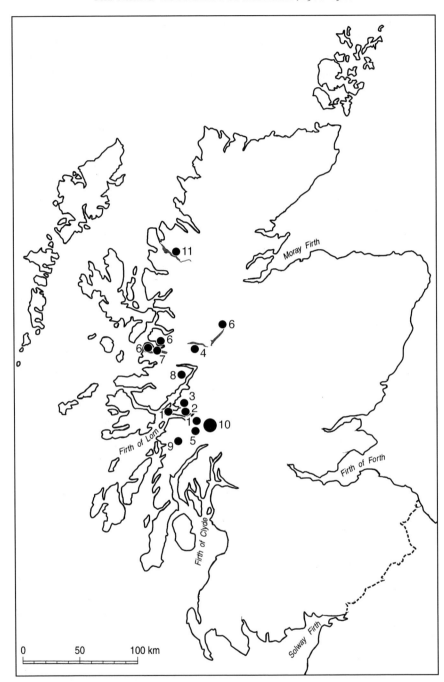

Map 13.1 Woodlands purchased by the Irish partnership in Argyll, from 1721. *Note* See Table 13.1 for explanation of the numbered estates and woods. The purchase from the Earl of Breadalbane (10) was larger than the rest.

entire timber crop without any apparent reservation to allow for regeneration. Given the sums involved (£1,000 Scots for Losset, for example, for a two-year sale; £300 sterling to Lochiel for a thirty-year sale), it is tempting to suggest that these Irish wood contracts were viewed as an extremely welcome financial fix by some Highland landowners, who had long faced increasing living costs at a time of general economic contraction.

The tanner soon found that he could not sustain these large-scale operations on his own. He subsequently entered into partnership with four other Irishmen, Captain Arthur Galbraith, John Fleming and William Kettleswell (all of Dublin), and Charles Armstrong of Mount Armstrong, Co. Kildare. These five formed the basis of what eventually became the Glenorchy Firwood Company. In tandem with all this, and apparently on the urging of Sir Alexander Murray of Stanhope who had interests in the lead at Strontian to which the York Buildings Company was attracted, he and his partners in 1725 established an ironworks at Glen Kinglass, on Campbell of Lochnell's land. Additional names were attracted to broaden the financial base: Edward Nixon of Coote Hill, Co. Cavan, merchant, and Robert Maine of Dromore, Co. Monaghan; then, by 1733, the wealthy Glasgow merchant and owner of much of Islay, Daniel Campbell of Shawfield. Glen Kinglass was a substantial concern and an important forerunner of the Bonawe furnace, founded a little to the south on Loch Etive a generation later (see above, pp. 240–7). It had a furnace and a casting house, an ore shed and a substantial charcoal shed, drawing its minerals from the haematite of Furness in Lancashire, and from the bog ores of Islay and Jura. Mostly it made pig iron, but some was cast for local use and sale as far north as Strontian; a sample discovered on site at Glen Kinglass by archaeologists was analysed as 'remarkably good iron' and the concern survived until economic depression in 1737–8 finally seems to have brought an end to its activities.[8]

As if all this was not enough, Murphey and Nixon in 1731 also joined forces with Robert Hunter, shipmaster of Irvine, to form another 'wood company' to exploit the oakwoods of Letterewe on Loch Maree, sold to them by Donald Murchison who currently controlled the Seaforth estate. Here they planned to fell 400 tons of timber and to manufacture over 2,000 barrels of bark: Murphey and Nixon quickly passed their financial interest on to Robert Hunter, who also brought in Colin Campbell of Inveresregan

---

[8] See J. M. Lindsay, 'The iron industry in the Highlands: charcoal blast furnaces', *Scottish Historical Review*, 56 (1977), pp. 56–7; J. H. Lewis, 'The charcoal-fired blast furnaces of Scotland: a review', *Proceedings of the Society of Antiquaries of Scotland*, 114 (1984), pp. 457–60.

as an additional partner, though Nixon remained as manager. However, plagued by high costs of extraction and disappointing quality of wood, and entangled in debt and cash flow problems, the partnership seems to have abandoned Wester Ross, also in 1737–8.[9]

The history of the Irish adventurers does not make elevated reading: Murphey was hanged in 1732 for murdering his servant; Nixon was constantly harangued by his partners in both the Glen Kinglass and the Letterewe Company for mismanagement; Galbraith, the most responsible and dedicated of the group, was nevertheless sued by the chamberlain of Glenorchy, Hugh Campbell, for non-payment for timber. Yet murder, incompetence and indebtedness, together with the apparent disapprobation of the local community, by no means prove that they wrecked the woods under their control. The remainder of this chapter focuses on their operations in the Breadalbane estates in Glenorchy.

The documentary evidence in some respects provokes more questions than it answers, when trying to assess where the initiatives came from and the extent to which these timber operations were in any sense sustainable. In the first instance it was not only outsiders who had identified the commercial potential of the woods. Much activity prior to the 1720s involved local, or at least Scottish, merchants and craftsmen. One of the most important players in Argyll was James Fisher, the provost of Inveraray, to whom a number of the Glenorchy woods had already been sold, although the Earl of Breadalbane was certainly not convinced of the expertise of the local contractors in his woods.

Landowners like Breadalbane may have looked to Murphey and his colleagues to provide an opportunity to move onto a much bigger, and more lucrative, commercial stage. On the other hand, having a primary responsibility to maintain such natural resources for the use of future generations both of their own families and those of their tenants, they should have been aware of the need to safeguard the long-term future. Nevertheless they, like us, were not immune to the vagaries of fortune and if an individual family hit a difficult patch financially, then decisions might well be taken which could lead to an asset which might otherwise have been carefully husbanded being sold off for the highest price. In the case of a number of the proprietors who sold their woods to Murphey, their association with Jacobitism suggested a lack of access to lucrative political patronage and favour.

---

[9] Lochnell: NRA 934/58/779, various; 934/55/750 and 752 various. A sale of timber seems to have taken place in 1737 but thereafter correspondence related only to the clearing of accounts: 934/55/750/10, 20; 934/55/752/19, 34.

However, a fine line had to be drawn between these undoubted short-term pressures and the needs of the communities who lived on the estates over the longer term. As some of the contracts indicate, the activities of the Irishmen were seen as destructive of more than just the timber. The construction of an ironworks in the area implied the production of charcoal, 'whereby they must have the privilege of digging pitts in the earth or how up earth and burn heaps of wood in these pitts whereby not only the ground but the grass will be considerably damnified'.[10]

But such considerations were only caveats to the sale process. Murphey arrived in Scotland some time in 1721 and proceeded to approach landowners or their representatives with offers to buy their woods. In addition to the contracts that Murphey negotiated himself, he also managed to inveigle his way into existing contracts, including one currently held by Fisher (who had died by 1722), Patrick Campbell in Inveraray, Sir James Campbell of Ardkinglas and Robert Fisher, merchant in Glenorchy.[11]

There is no doubt, however, that the lease of the Glenorchy woods was a major prize for the Irish partnership, the largest and most expensive woodland sale they were able to negotiate. They had their eye, first, on the extensive pinewoods of the barony, which they intended for building timber and masts in the Irish market; next, on the oak around Loch Awe, no doubt mainly for tanbark and charcoal; and lastly, on a variety of broadleaf trees in the woods towards the head of Loch Etive, also useful for bark and charcoal. It was the pine that was the prize, and a survey carried out by the estate at about this time valued it at £8,900 (presumably a gross valuation in sterling, but it is the relative values rather than the sum which is significant).

The woods were scattered in seven different localities (see Map 13.2).

Table 13.2 Values of Glenorchy and Loch Etive pinewoods, ca. 1721

Locality	Valuation (£ sterling)
Glen Strae	1,200
'The two Larigs'	1,000
Invergaunan	1,200
Doire Darach	1,200
Glen Fuar	1,000
Crannach	2,700
'The wood of Gualinliguish'	600
Total	8,900

[10] Lochnell: NRA 934/58/779/28.
[11] See NAS: GD 112/10/1/3/12, and GD 112/16/11/1/1–2 for examples of these earlier contracts.

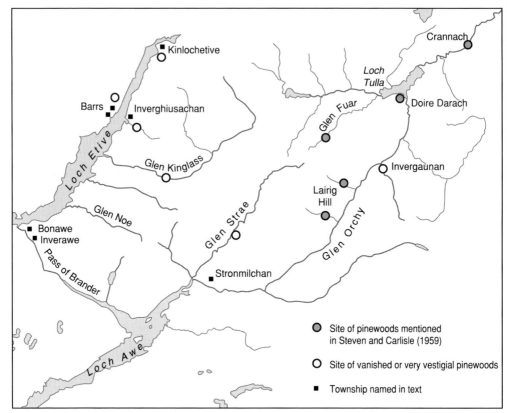

Map 13.2 Glenorchy and Loch Etive pinewoods.

The last named cannot now be identified, but it was the smallest. There was no pinewood left in Glen Strae or Invergaunan when Steven and Carlisle made their survey before 1959, but two small remnants in Allt Coire Bhiscair and Allt Broighleachan represented 'the two Larigs'. Doire Darach on Loch Tulla is still on the map, as is Crannach on the Water of Tulla to the east and, more uncertainly, a few trees at Glen Fuar to the west.[12] The attraction was that the trees were evidently large and valuable; the problem, that they were in diverse and difficult localities.

The process whereby Murphey acquired rights to the Earl of Breadalbane's woods in Glenorchy, as opposed to the other contracts which were completed by the end of 1722, was complex. The estate managers certainly did not rush into the final agreement. In the first instance it is clear that Breadalbane wished to sell a set number of the best trees, but Murphey

---

[12] NAS: GD 112/16/11/1/11; H. M. Steven and A. Carlisle, *The Native Pinewoods of Scotland* (Edinburgh, 1959).

insisted on taking the whole with reservations on the young wood which, he claimed 'will be near to every second tree'. The estate was also hoping to raise £1,000 sterling from the sale, a large sum, and more than the Irishman was initially prepared to offer. It was admitted that 'the great wood lies very far from the sea and an ill road to bring it there whether by land or water', a common problem for these West Highland woods, which may, in fact, often have prevented their earlier large-scale exploitation. It was also stated elsewhere that the price for both bark and timber was falling in the early 1720s, though the fact that the Irish wished to invest so much time and money in these woods suggests either that they were incompetent businessmen or else that prices were still, or were soon expected to be, comparatively buoyant in Dublin, where the main market for their building timber lay.[13]

To begin with, negotiations moved briskly. Breadalbane indicated to his chamberlain, Campbell of Barcaldine, what he thought of the conditions Murphey had set out at the end of 1721. He reluctantly approved of the reservation of all trees under twenty-four inches at breast height, remarking that 'they will be too small for his work, and will grow to me'. It is interesting to note that the offer which Provost Fisher had made for the Glenorchy woods in February 1721 mentioned reserving trees of only twelve inches at breast height. Breadalbane's reaction to that offer was not concerned with whether or not twelve inches was a sensible figure, but with the hassle involved in having to measure all the trees, which is why he continued to press for the cutting of a specific number. Interestingly, he noted at that time that this had been the method used in a recent sale of Ross of Balnagown's woods in the north.[14]

With regard to Murphey's offer, what did concern the Earl was the amount of time (twelve years) which the Irishman wanted to complete the cutting and carrying off of the timber, for the good reason that 'the longer they are about it the young very small timber suffers by the going daily through it, cutting the old to fall on the young, and drawing and leading horses through the wood daily'. The Earl was also canny enough to realise (and this was part of his objection to the length of time the Irishman proposed to take over the contract) that he needed to pay 'a good substantial honest man' to watch over the cutting for its entire duration, despite the fact that the trees to be cut would be marked. He was still very keen, believing that he had a considerable number of mixed-age trees scattered

---

[13] NAS: GD 112/16/11/1/8, 6; Campbell of Barcaldine papers, GD 170/271/2.
[14] NAS: GD 112/16/11/1/3; in 1707, Balnagown had sold 12,200 pines 'to be chosen and marked' by the estate, GD 297/213/. See above, p. 335.

in different woods over a large area, to bring Murphey to a set number to be cut down. However, given a parlous financial problem, he was anxious for the sale, urging Barcaldine and his other estate managers to bring negotiations to a satisfactory conclusion because the affairs of the family 'do very need and require such help'.[15] He ordered that, should Murphey insist on buying the woods as a whole, an immediate inspection was to take place to find out exactly how many trees there were above twenty-four inches circumference at chest height. One can only presume, from what happened later, that this inspection either did not take place, or was not done properly.

All of the above related to the considerable pinewoods of Glenorchy itself, but Murphey was also interested in the oakwoods of the barony which had hitherto been sold to Fisher and others. Again, the Earl showed himself to be both personally aware of the condition of his woods and educated in the specifics of wood management. Because oak, unlike pine, springs from coppice stools, it was good practice to fence it after felling to protect the regrowth from animals. Breadalbane knew this, but remarked that some of the Lochawe oakwoods would prove difficult to fence because 'it is scattered wood'.[16]

Part of the problem in negotiating the contract with Murphey was the fact that he was not in the country for long, a fact that he seems to have used to try to put pressure on the cautious Breadalbane. The latter intended to make a final answer in the following spring, when Murphey was expected back in the country. In the meantime, the Earl set about sorting out the rather unsatisfactory situation (from his point of view) relating to his outstanding wood contracts. In particular, he was tired of the excuses presented by William Campbell of Glenfalloch's son, who had taken a contract worth 2,700 merks (around £900 Scots) on the woods of Innis Chonain, an island in Loch Awe. Apparently Glenfalloch now alleged 'that the bark will not separate, because of the birds' dung on the top of the tree', a story that Breadalbane regarded as 'nonsense, that may wither the top of the tree and no more. It is because they are knavish and ignorant, for in England great old timber, the bark is best, but there is more pains in managing it.' He was determined, therefore, to allow no reduction in the price.[17] The birds in question were perhaps cormorants, the droppings of which can even kill trees in the long run. Herons, the other possibility, make large untidy nests in the tree tops but contain their droppings within them.

---

[15] NAS: GD 112/16/11/4.
[16] NAS: GD 112/16/11/1/7.
[17] NAS: GD 112/16/11/1/15.

Perhaps in anticipation of the finalisation of a contract with Murphey, and in recognition of the fact that that contract would doubtless require a considerable degree of oversight on the part of the estate, Breadalbane planned the recruitment of a suitable person to look after his Glenorchy pinewoods. The qualifications were to be an understanding of 'woods, young and old, the projecting part of them' (presumably he meant the marketing aspects), as well as grazing and the selling of cattle, which was still a fundamental part of the Highland economy. However, the Earl admitted that finding one man with such skills in the area was extremely unlikely, because 'our people are generally selfish and have not the language'. He even went so far as to say that 'no Highland man [has] a notion of woods' and it was proposed that a search be made for a suitable 'good mettled young man' in the Loch Lomond area. Unfortunately there is no evidence that he found one.[18]

There was no further reference to Murphey in 1722 and this, together with the fact that a new roup of an oakwood in the barony of Glenorchy (at Kilbride in Glenaray) was conducted in February 1723, suggests that, for some reason, negotiations had stalled. The roup was won by William Campbell, brother of Colin Campbell of Glenfalloch, who offered the sum of £20 sterling, £5 over the minimum asking price.[19] However, the following month, one of Murphey's new partners, Edward Nixon, wrote to Barcaldine, having had a look at the Glenorchy pinewoods himself. This letter indicates that negotiations were now continuing, Murphey being left to finish them off. The Irish partners seem to have thought a lot of the Glenorchy pinewoods, and Captain Galbraith described them as 'the best in Britain' in December 1723, after the contract had been agreed.[20] This might suggest that it had been Breadalbane who had gone cold on the deal in 1722. However, though the specifics of the contract agreed in 1723 seem to represent something of a victory for the Irishmen over the Earl, it came at the price of agreeing to his original demand for £1,000 sterling.

The Breadalbane contract of September 1723 was exceptional in the detail contained within it and may reflect a long-standing interest in, and understanding of, woodland management on the part of that landowner which may not have been entirely the norm within the West Highlands. As well as exempting from the contract the pinewoods in Glen Strae, it was expressly determined that no pine tree 'of which the circumference at the distance of three foot from the ground does not exceed twenty four inches round' was

---

[18] NAS: GD 112/16/11/19.
[19] NAS: GD 112/16/11/1/20.
[20] NAS: GD 112/16/11/1/10; GD 170/3067.

to be touched. Sensibly, such trees were to be 'marked out by indifferent persons to be nominated and appointed by both the said parties'. However, this was, of course, a victory for Murphey, who had originally suggested this exemption, over Breadalbane's desire to specify a set number of trees. In addition to the pinewoods, the oakwoods of the barony were sold, and all the birch, alder, hazel, ash, oak, rowan and sallow growing in four townships at the top of Loch Etive. All the trees were to be cut:

> according to the common and reasonable custom in such cases, and at the proper times and seasons and in such manner as the same may spring and that within the space of fifteen years, and in like manner to draw off and carry away the timber so cutt within the space of five years after the expiration of the time allowed for cutting of the same.

It would appear, once again, that Murphey had won out against Breadalbane's objections, even managing to extend his original offer to cut down the trees in twelve years to fifteen.

Additional regulations were made for the cutting of the oakwoods, which was to be completed within two years and done in haggs. The mention of haggs within the Highland context is rare and therefore extremely interesting. However, it is not the first time that a Glenorchy wood contract mentioned the practice of dividing woods into divisions for cutting, a common enough practice elsewhere, primarily associated with coppicing. In 1669 Breadalbane's father (when he was still laird of Glenorchy) sold some of the Glen Etive woods to a local entrepreneur, under fairly stringent stipulations for the cutting down and removal of the trees. Twenty of the youngest and straightest trees were reserved in all cases. All the timber cut each year was to be removed that same year, doubtless to facilitate the process of regeneration, though the disturbance caused by taking timber out, presumably by horse, might still be problematic. In addition, the contractor was 'to cut clean over as much of the wood as he is to cut' and that was to be done 'close by the ground with sufficent hague-men'.[21] It may be that in this 1669 contract, Glenorchy had been keen to set down what was regarded as best practice, though whether this was learned from Lowland timber contracts or was based on English examples is impossible to tell. Either would be possible, since the family was involved in politics at a British level. Though the 1723 contract did contain much detail, it may have been the case that Breadalbane felt that the Irish already knew what best practice was and that he did not therefore need to spell it out fully. He

[21] NAS: GD 112/16/11/2/6.

may also have been under some considerable financial pressure to finalise the sale. He undoubtedly lived to regret the terms of the contract, which was drawn up and signed by estate commissioners appointed to act while he was in the south. Carelessness in legal detail was perhaps a penalty of absenteeism.

The terms of the contract were also concerned with the processing of the felled timber. Clearly Breadalbane envisaged a major industrial exploitation of his timber resource and he therefore agreed with the Irishmen a number of other important conditions essential to the successful exploitation of the woods. These included grazing for their horses and permission to:

> Erect and build saw milns and to remove them to proper places if needfull, draw watter courses, dig saw-pits, build store houses and other houses necessary for themselves and workmen and inclose timber yards, cutt fearns and cast divotts necessary for their said houses and work, as also liberty of running the waters straight [i.e. canalising for floating] betwixt the woods of Cranich and Lochtolly [Loch Tulla], but still with as little prejudice as possible to the grounds.

They were also entitled to:

> Cutt, strip, squair and carry away by themselves and others employed by them both by land and watter the timber hereby sold from the severall woods to Bonnask [Bonawe] and to cutt, coall, cord and carry away all the said woods and under woods proper for such use.

Such conditions, involving the exploitation of woods from one end of Glenorchy to the other, imply a range of processing activity, presumably designed to take advantage of prevailing market conditions for timber, bark and charcoal. However, Breadalbane foresaw that such activities were likely to be viewed with suspicion (or perhaps opportunism) by his tenants when he offered his protection to Murphey and his men 'from stealth and violence', and to assist them in recovering stolen goods. The resistance of the local tenantry to commercial timber schemes was a recurring theme in the history of the Highland woods.

Murphey, like many possessed of a good idea, was nevertheless soon experiencing serious cash-flow problems. Within six months of negotiating his many contracts of 1721–2, he was impelled to pass over half of them to Captain Arthur Galbraith, to whom the tanner owed a considerable sum of money. In the following months three additional partners (Armstrong,

Fleming and Kettleswell) were brought in to share the burden and expense of the enterprise in Scotland, attesting to the fact that this was not only a major operation, but an attractive one (see p. 345 above). By this time it also included the construction of an ironworks in Glen Kinglass, to be kept going by all the wood contracts, except Breadalbane which was not yet negotiated. William Kettleswell, the newest recruit, was to take care of the ironworks once it was built and to attend to the company's business in North Britain and England. Bristol was mentioned, presumably as an outlet to sell bar iron.[22]

Unfortunately all the Irishmen except for Galbraith seem to have been already up to their necks in debt and the poor Captain found it difficult to get things moving for the clamour of creditors. Nevertheless, some progress was certainly made, including the construction of the furnace in Glen Kinglass. This soon resulted in the addition of another two partners – Edward Nixon and Robert Maine – as Murphey desperately sought to shore up his finances, given the inevitability of large-scale outgoings long preceding any profit in a venture of this type. By 1725 he was in such debt that he had, temporarily at least, to turn over his entire quarter share in the Glen Kinglass ironworks to Nixon and Maine until his debts were paid.[23]

Galbraith seems to have been the most reliable member of the Irish company; he was certainly the most active. In November 1725 he made arrangements with John Smith of Doune near Stirling, and the local factor, Hugh Campbell of Stronmilchan, to deal with the timber coming down the river. Campbell was to cut, fell, cross-cut and then haul to various streams all the wood belonging to Galbraith in Glenorchy. Interestingly, it is clear that Bonawe was already the choice of venue as the point at which the Glenorchy wood was to be gathered and secured above the water mark, having been floated down from as far afield as Crannach. From there it would be transported on ships. It was intended (but, sadly, we do not know if they ever did it) to build an 'engine' (essentially a large netting device) near Bonawe to stop the timber floating into the salt water. All the timber was to be cross-cut into whatever size was deemed best for selling. Galbraith was, however, to indicate which pieces of timber were to be left intact for masts, a very valuable commodity at this time. That this was a major investment is also indicated by the fact that Smith and Campbell were to be given the option to take up tacks of land, presumably in the area already held by Galbraith, and timber yards at Bonawe were to be found.[24]

---

[22] Lochnell: NRA 934/58/779/30, 25.
[23] Lochnell: NRA 934/58/779/24.
[24] Lochnell: NRA 934/58/779/7.

Unfortunately, and as ever, the whole procedure ran quickly into debt, though we can perhaps also glimpse some essential problems with the whole enterprise in letters written by Galbraith to Hugh Campbell, primarily in order to reassure the latter and also to urge him to continue timber operations, without which no profit could obviously be made. The hand-to-mouth nature of these operations is fully revealed in a letter of January 1725 wherein Galbraith expresses the concern that provisions currently *en route* for Bonawe would be seized by creditors for Kettleswell, one of his partners. The straits in which the Captain found himself are eloquently expressed in his exhortation: 'dear Hugh be sure you keep private my goeing to Ireland and whatever you doe get all the timber in your power as soon as possible down to bunaw and be sure there is noething in my power that I shall spare to make you amends for it.' Equally worrying was the apparent need for pitching axes at Glen Kinglass 'for the people are idle there and want of axes'.

Another letter, probably written at the end of 1725, indicates that the situation had certainly not improved. Though ships from Dublin were promised, the quality of the timber (and therefore the lucrativeness of the sale) was in some doubt, not least in relation to the cost of bringing it down the river. Galbraith was quick to blame 'the idleness of villains' (possibly meaning his own countrymen but also local workmen) but the biggest problem with making money out of Highland timber was its remoteness from markets. Within a month Hugh Campbell was suing Galbraith and the rest for debts totalling £43. 2s. 3d. for wood which he had already floated down. Galbraith urged Campbell to continue to send wood down (for without it, he could not sell any to pay off the debt) but unfortunately Hugh died six years later before it was settled.[25] This story was typical.

However, our assessment of the role of these Irish entrepreneurs in the history of these West Highland woods does not rest solely on the financial shenanigans associated with the partnership, however problematic they may have been for the individuals involved. The accusation levelled by Breadalbane and in the *Statistical Account* rests on a far longer-term danger from the Irishmen's activities: the devastation of the woods themselves. The detail contained within the Breadalbane archive permits us to examine the state of the woods over the longer term, together with the management practices engaged upon by his estate officers to deal with the perceived destruction caused during their exploitation. Since the worst accusations came from the Earl himself, it is useful to examine them and then make

---

[25] Lochnell: NRA 934/58/779/8a, 11, 8h.

some assessment of any action subsequently taken to alleviate the damage, as well as noting any incidental information on the state of the woods in the following years. We have already noted the terms of the contract agreed between the Earl and Murphey in 1723. In theory, and despite the fact that they reflected a capitulation to the Irishman's demands, these stipulations, and the preparations made by Breadalbane to ensure sufficient oversight of the contract, were surely sensible. We have every reason to believe that Breadalbane sincerely expected that they would preserve his woods.

The first queries about the contract from the estate did not, in fact, relate to problems with the woods, but rested on the perennial problem of the financial robustness of the Irishmen's activities. By late 1724, Breadalbane's estate managers, at Galbraith's instance, were investigating the failure of Edward Nixon to pay £100 sterling to the Captain, which was obviously required by the latter to fulfil his own responsibilities. It was also alleged that the management of the woods was not being attended to, both aspects having been temporarily devolved on Nixon by Galbraith as part of the constant renegotiation of the Irish debts among themselves. However, it soon transpired that in the previous March, Nixon and an associate, John Cathcart, had managed to pay £400 sterling owed as the initial payment for the woods of Glenorchy. This may not have helped Galbraith, but it presumably satisfied Breadalbane, for the time being at least. However, the Earl's managers were as keen as Galbraith himself to see the latter restored to active involvement in the manufacturing of the woods, a recognition of the Captain's diligence and effectiveness, presumably in contrast with his colleagues. The issue was not helped by the fact that Nixon was currently in Ireland (thereby perhaps proving the point), though in the first instance it was proposed that Galbraith should follow him there to try to resolve the problem 'in a friendly manner'.[26]

At the same time, perhaps belatedly, the estate began to look more closely at the economics of the whole enterprise. They based their calculations on an offer by Glasgow merchants made some years before, when pine at Bonawe could sell for between eighteen and twenty shillings per tun. A tun was reckoned at forty feet long and twelve inches square, which would require three to four trees. Therefore 30,000 trees would equate to 7,500 tuns, fetching a price of £7,500 sterling. The problem, as the estate also acknowledged, was the cost of getting the timber down to Bonawe, though it was alleged that the task was not totally 'impracticable' and 'proper hands will make it easy'. It was reckoned that twenty men 'who are good cutters' would be required for twelve weeks to work the timber in the woods. It

[26] NAS: GD 112/16/11/1/25.

would then be rough-squared and sent on its journey downstream, a feat described in the memorandum:

> The same men puts it and leads it to the adjacent burns in the woods, which burns when a spate is, doeth carry the timber to Lochtollie [Loch Tulla], which Loch is the head of the water of Urchie [Orchy] and carries the timber to Lochow [Loch Awe]. There two boats and men gather the timber and float it to the Brander. There it runs down the water of Awe to Bona [Bonawe], where two boats must gather it shore, to lay to be shipped.

It was reckoned that the expenses associated with the cutting, squaring and floating amounted to only about £1,500, leaving at least £5,500 as profit. If this was extraordinarily optimistic, it was also extraordinarily tempting. It was also reckoned that all this could be achieved in three years, not the fifteen stipulated in the contract. However, while it was easy enough for the estate to calculate the profit which it hoped would come from the timber of upper Glenorchy, further correspondence makes it clear that the system of oversight which underpinned the successful execution of the Irish contract from the estate's point of view had still not taken place. Admittedly Galbraith and his partners had not yet cut all the pine (they had presumably started with the oakwoods on Lochaweside). Nevertheless, Barcaldine insisted in August 1724 that it would be necessary, when the Irish did begin operations, 'that you apoynt some proper person to marke out the timber that's reserved'.[27] Given what happened next, it is unlikely that the estate ever got round to marking out the reservations or appointing anyone to any kind of on-the-spot supervision before Breadalbane arrived in person. He was appalled by what he found.

In July and August 1725, the Earl, who usually resided either in London or on Lochtayside, spent 'four days together thro mountain bogs and rocks', partly on foot and partly on horse, to inspect the progress of the woodland operations on his ancient western estates. He was incandescent at the destruction:

> The countrie looks like a desolation, and will be more, for as there is already not one oak tree in the countrie for any use, so there will not be ere they are done one fir tree, to help a house, nor the mans to the minister here, nor a bridge on a burn in the countrie wherof there are a great many.

[27] NAS: GD 112/16/11/1/23, 24.

Firstly, whereas the intention had been only to sell 'woods or bushes of timber', the Irish had gone three or four miles to cut an isolated tree: thus, presumably, depriving the community of useful pollards. Secondly, the oak was cut so badly, three or four feet above the ground but then peeled and stripped to the ground, so that it 'must die, and nothing could spring'. Then they had ignored the proper season for cutting, 'so that no growth can come this year, and before next year the root and stock will be dead'. In cutting pine, they had picked out 'all the fir trees in all different places' (the phrasing suggests that much of the wood was mixed pine and oak) and trampled the young growth: 'they looke like run awaymen, taking advantage of the easiest and so spoyling the wood'. Already they had cut more than half the pine, in breach of contract, and what was left was the most remote from water carriage 'and in the hills of little value'.

He wanted to sue the Irishmen for breach of contract, but he admitted 'without telling lawyers or any other' that the estate had made one serious mistake in respect of the pines:

> The 24 inches is a very small measure for in a wholl wood the smallest tree I could see, I measured; and it was above 30 inches, they are very tall trees, and there will not be any trees under 24 inches, any where in the woods; where by my notione of that measure I thought there would have been thousands, there will be nothing, and very little good for any thing of young growth.

This could not have been literally true, for elsewhere at Crannach many of the trees were said to be young, and at Loch Tulla, under the stipulated size. Nevertheless, the Earl's observation does seem to indicate a large-scale failure of most of the trees to regenerate in the recent past. He concludes:

> I would not for double the value have the bargain hold, it is ane everlasting ruine to us, and a 15 year slaverie, and the tennents gets nothing by them. And we may easily now make double that monie of it.[28]

So Breadalbane was extremely keen to establish that Galbraith and Murphey had contravened the regulations for good management and had abused the contract. Unfortunately for him, much of the regulation for cutting relied on the vague phrase that they were supposed to cut down the trees 'according to the common and reasonable custom in such cases and

---

[28] NAS: GD 112/16/11/26/1–2.

at the proper times and seasons and in such manner as the same may spring and that within the space of fifteen years'. The estate was claiming that the Irishmen's activities rendered the wood 'altogether common and no growth can ever get up', but it was certainly too early to establish that this was indeed the case. In the event, the Glenorchy Firwood Company continued in occupation of their lease until 1736 at least, and there is evidence of an 'overseer of the fir timber att Bunaw' at that time and as late as 1740.[29]

All in all, this seems to have been a classic case of shutting the stable door once the horse had bolted. Ultimately, responsibility for ensuring proper management of the woods rested with the landowner and (particularly in this case, since he was effectively absentee) his officers on the ground. Breadalbane, who had, belatedly, gone personally on foot through his woods, seems to have immediately ordered that many of them, including those presumably recently cut by the Irish, be enclosed. But what is remarkable is the fact that no such stipulation had been included in the recent contract. It also seems to have been the case that the flurry of legal activity initiated by Breadalbane at this time to look into the Irish contract was primarily concerned with the possibility of exacting the financial penalties which would accrue if the contract was deemed to have been broken. Indeed, one might surmise, in the light of the memorandum summarising the potential profit which the estate believed could be made from the as-yet-untouched pinewoods of upper Glenorchy, that the real rationale behind the Earl's ill-humour was to have the Irish contract declared null and void through their alleged bad conduct, pocket the fine and then reap the profits from selling the pine timber direct himself. Certainly it is stated explicitly in one of the legal documents that 'it should be liesome and lawful to the said earl . . . to sell and dispose upon what remains of the said wood uncut as they shall think fit in the same manner as if the said contract had never been entered into'.[30] However, despite all this legal activity, the contract continued.

Some three years later, a survey was done, part of the purpose of which was to ascertain 'how the Irish manage their affairs in that country [Glenorchy]', undoubtedly with a view to assessing the damage still deemed to have been inflicted by them on both the pine and the oakwoods. Though it was asserted that one of the oakwoods had 'little or no appearance of growth by the havock made there by the Irish in cutting of it' and the pinewoods at Loch Tulla were similarly noted as having been subject to

---

[29] NAS: Campbell of Dunure papers, GD 170/271/9; Loch Etive trading company, RH 4/93/2/5/4. fo. 107.
[30] NAS: GD 112/16/11/1/27, 28; GD 112/16/10/1/5.

abuse, most of the other woods appear by then to have been well enclosed and in good condition. With regard to those that were not, the blame was laid at the door of 'ill-keeping', as a result of having been grazed when the trees were young and before the Irish had made their appearance in Glenorchy. It was also noted, in relation to the pinewoods of Doire Darach at Loch Tulla, which the Irish were currently engaged in cutting:

> There are a great many reserves [trees of under twenty-four inches] which probably will be cut down if not timeously prevented and by any information could be had the trees reserved by the contract were not marked out by neutral persons as thereby appointed.[31]

The estate managers really had no-one to blame but themselves if this wood contract was not proceeding to their master's satisfaction.

A further survey was conducted some twenty years later, in 1744, in order to establish when the woods could next be cut. According to that survey, although the Glenorchy woods (presumably meaning the remaining pinewoods) should be left till last, they would then 'be fully ripe', hardly a description likely to be applied to them if they had all suffered irredeemable damage in the 1720s.[32]

Ultimately, it is difficult to avoid the conclusion that the Irishmen merited the epithets of 'adventurers' and 'run awaymen'. Their attitude compares very unfavourably with the management regime instituted some thirty years later by the Lorn Furnace Company of English ironmasters. The Irish apparently intended to take out as much timber as possible without consideration for the longer-term future and in disregard for the principles of sustainable woodland management as practised elsewhere in contemporary Scotland. However, our sources, which largely reflect the views of the estate with little direct input from men like Galbraith, undoubtedly represent an extremely biased viewpoint. It is clear that the outrage expressed in 1725, and which has been part of the history of the Glenorchy woodlands ever since, was real, even though the estate-owner at that time also wanted to make more profit out of the woods by voiding the contract. It is also clearly the case that the estate itself was extremely remiss in not taking sensible precautions, particularly in terms of oversight, both before the contract was concluded and certainly thereafter. Breadalbane himself seems to have had both sufficient know-how and the determination to contain the damage through proper measures; however, the fact that he was so rarely in Argyll

---

[31] NAS: GD 112/16/10/2/1.
[32] NAS: GD 112/16/10/2/2.

to oversee the situation and issue appropriate instructions lies at the root of the problem. One certainly has the impression that the estate's managers in Argyll had insufficient expertise in the management of the woodlands. Barcaldine, who acted for Breadalbane in the initial stages of the sale of the Glenorchy woods, freely admitted that he had 'no more skill about firr woods than a child'.[33] Although he repeatedly pressed for expertise to be brought in, it never was. It is always easy, and perhaps more romantic, to blame outsiders; sadly, the more mundane reality seems to be that responsibility lies a lot closer to home.

How much long-term damage was done to the woods in the years of the Irish operations, irrespective of whether the blame was theirs or the estate's? It will be recalled how, in the 1790s, the local minister described the higher parts of his parish as having sixty years before abounded 'with forests of the largest and best firs', now all gone, though there were still some 'tracts of natural firs' in the parish and extensive deciduous woods, including oaks (see above, p. 341). This is clear evidence that there was a large, serious and observable decline in the pinewoods, and the five small remnants described by Steven and Carlisle are unrecognisable as the finest pinewoods in Britain, as Captain Galbraith had called them. Probably the long-term damage to the oak, however, was much less serious and general than the Earl had thought it would be, and subsequent enclosure and better estate care largely repaired it. Certainly when the Lorn Furnace Company arrived in the 1750s to use the oak they had no complaints.

One final observation is in order. There were other pinewoods in the immediate area which have completely gone, and in which the Irish were in no way involved. The partnership was explicitly excluded from the pinewood in Glen Strae, where the Gaelic place-names Inbhir nan-giubhas and Allt nan Giubhas indicate its former location, but where Steven and Carlisle could find nothing more than a few trees along the river and in a birchwood nearby. The local minister in 1843 explained that here the 'last remains' of the forest had been cut down thirty-five years earlier and used to build the parish church: it sounds as if by then it had held nothing but a few large pines, probably another case of failing regeneration.[34] Further west around the top of Loch Etive, the name Allt Ghiusachan indicates another lost pinewood, perhaps already gone by the time the Irish arrived, as they were mentioned only as cutting deciduous trees in the township of Inverghiusachan. At Barrs on the opposite bank, where the Irish also cut only deciduous trees, and at Allt Mheuranby near Kinlochetive,

---

[33] NAS: GD 112/16/11/8.
[34] NSA, 7 (Argyllshire), p. 92.

where again they had the same rights, Steven and Carlisle found a few trees of native pine that also seemed to indicate former lost woods. Traces of native pine have recently been found in Glen Kinglass itself.

Some light on the circumstances in this area comes from a memorandum of 1726 from the tacksman of Glennoe, former woodkeeper of Loch Etive, recently dismissed and replaced by another at the behest of the estate commissioners. It reveals a state bordering on anarchy, but involving local people and not the Irish:

> Glenno does not deny but he has frequently convertit to his own use some of the earle's timber in Lochetive par'larly at the tyme that the woods of both Barrs and Glenno were acutting and in place of justifying himself for what he has done he is hartily sory that he did not more since where he slew his tens other slew thair thousands par'larly Inveraw who as its notterly known who cutt doune the earle's green woods at Lochetive for their constant fyre wood.

He went on to say that more pine and oak had been carried from Loch Etive under pretence of repairing the mills of Netherlorn than all those mills were worth, and that the pine 'of the contraverted spott in Glenkellen' had been cut by the miller of Netherlorn and Robert Campbell of Kentrae, 'and that for any thing that ever you know'.[35] It was not a good defence of his stewardship of the woods of Loch Etive, but it was very revealing.

Also illuminating is the evidence of the minister of Glenorchy in 1843, who observed the deforestation of his parish with dismay. Not so long ago, he said:

> The greater part of our moors and valleys and the sides of our mountains, midway to their summits were clothed with trees of various kinds. The braes were clothed with a dense and magnificent forest, partly of oak, birch, ash and alder, but chiefly of pine.

He described the fate of the pine, mainly blaming the Irish, and said that the hills were still partially clothed with oak coppice, birch, aspen, ash, elm and holly, 'but these generally speaking are rapidly disappearing, and our mountains and valleys and straths have become comparatively naked and bare'. Plainly he considered deforestation a continuing, even an accelerating, problem, not one that lay only in the past. Taken alongside other observations

---

[35] NAS: GD 112/16/11/1/29.

on the modern prevalence of black-faced sheep in his parish, it is easy to propose the underlying cause.[36]

In other words, the depredations of the Irish were not alone in leading to the contraction or extirpation of pinewoods in this part of Argyll – overgrazing, poorly regulated exploitation by the tenants and probably climate change hampering regeneration must all come into the equation. The exploits of Roger Murphey and his friends certainly do not figure among the instances where the introduction of commercial exploitation led to better woodland management, but neither should they bear all the blame for the fate of the pinewoods in the barony of Glenorchy.

---

[36] NSA, 7 (Argyllshire), p. 92. Since this chapter was completed, an attractive illustrated book on the woods of the area has appeared: P. Wormell, *Pinewoods of the Black Mount* (Skipton, 2003).

CHAPTER 14

The MacDonald woods on Skye, 1720–1920

The value of the woods of Sleat, the southernmost of Skye's many peninsulas, has long been recognised. As early as 1463, John, Earl of Ross and Lord of the Isles, with the consent of his council, granted Sleat to his natural brother Celestine and entailed it to his heirs male by 'Finvola', the daughter of Lachlan MacLean of Duart, in return for the service of a ship of eighteen oars.[1] The charter includes the usual formulaic list of things pertaining to the lands, including *silvis* (woods). As well as this standard form, the charter includes the telling item *quercis* (oaks). In the middle of the fifteenth century, the value of extensive tracts of oak woodland was clearly understood. The timber could be used for building and repairing galleys and the bark was a source of tannin for conditioning leather. Six years later, in 1469, Sleat was regranted by the Lord of the Isles to Celestine's brother Hugh, the founder of the MacDonalds of Sleat.[2] In the third quarter of the seventeenth century, Sleat was described as exceeding 'anie part of the whole [of Skye], as to its woods' and, in 1695, Martin Martin wrote that 'There are several coppices of wood scattered up and down the isle. The largest [is] called Letter-hurr' (Leitir Fura).[3] (See Map 14.1).

Before the eighteenth century, due to the lack of surviving documentary sources, it is difficult to discern much about how the woods on the MacDonald estates on Skye were treated – how they were exploited and how they were managed. The early charters also mentioned *virgultis*, which can mean anything from thickets to coppiced trees. It is thus possible that management of the oak and other woods was at least contemplated as early as the fifteenth century. Not until the nineteenth century does any attempt appear to have been made to make commercial profit from naturally

---

[1] J. and R. W. Munro (eds), *The Acts of the Lords of the Isles, 1336–1493* (Scottish History Society, Edinburgh, 1986), pp. 126–8.
[2] *Ibid.*, pp. 152–5, 303–5.
[3] A Mitchell (ed.), *Geographical Collections Relating to Scotland made by Walter Macfarlane* (Scottish History Society, Edinburgh, 1906), 2, p. 221; M. Martin, *A Description of the Western Islands of Scotland, c.1695* (edn Glasgow, 1884), p. 142.

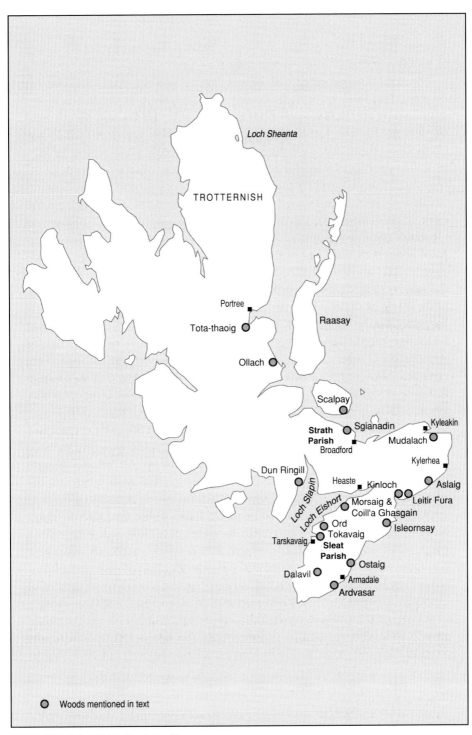

Map 14.1 Woodlands on Skye.

growing timber but it is clear that, many years earlier, some care was taken to preserve it and to utilise it as a resource.

The earliest surviving records of woodkeepers on the MacDonald estates on Skye date from 1720. At this point, the lands were in the hands of the government, forfeited as a result of the support by Sir Donald MacDonald of Sleat for the 1715 Jacobite Rising. This is the first year from which any financial accounts for the estates survive but there is every reason to suppose that paying men to look after the woods was a practice of some years' standing. It seems unlikely that the commissioners for the forfeited estates would have introduced such a new practice. Indeed, the accounts from the early 1720s describe the payments made to the woodkeepers of Sleat and Ollach as their 'usual salary'.[4] Although factors' accounts do not survive in good sequences until the nineteenth century, those which do all include payments to woodkeepers. As late as 1805, the whole of Sleat was still under the supervision of one woodkeeper and, during the eighteenth century, other tracts of natural woodland elsewhere on Skye had their own woodkeepers: Ollach, Tota-thaoig and Uilder in Trotternish, and Mudalach in Strath (acquired by the MacDonalds in 1751 and with a woodkeeper in 1720 if not before).[5] After ca. 1750, no payments to woodkeepers in Trotternish are recorded. Sleat and the eastern end of Strath possessed the only substantial tracts of woodland and, consequently, they were the areas upon which attention was bestowed.

When John Walker visited Skye in 1764 to report on the natural resources of the island to the Commissioners of Annexed Estates, he noted that the treeless aspect was not due to any defect 'either in soil or climate', as various remnant woods witnessed to the contrary. He mentioned an old wood of over 300 acres on the edge of the moss above the kirk of Portree, 'extremely sparse' but composed of birch, hazel, alder, bird cherry ('hagberry') and 'water elder'. There were 'no good trees' although it grew on a good woodland soil, and he judged that it would make an admirable plantation were it to be enclosed and the gaps planted up with 'good forest trees'. He also mentioned a tract of 150 acres on the north side of the Bay of Oransay (the woods of Duisdale, near Isleornsay in Sleat) with the remains of very large trees of ash, birch, alder, rowan and hazel, with 'all the young growth miserably stunted by the grazing of the cattle', and he found near MacKinnon's old castle (probably Dun Ringill on the shores of Loch Slapin)

---

[4] NAS: E 656/24/2 and 4.
[5] Clan Donald Lands Trust, MacDonald Estate Papers (MEP): 927/3, 3810; University of Glasgow Business Records Centre: Papers of J. MacLean WS, UGD 37/2/2; NAS: E 645/13/1 and 3.

'about 200 acres of coppice, chiefly of ash and birch', in the same sad condition, 'all open and so eat down by the cattle, that not a plant of them has been suffered to arrive to a tree'.[6]

By contrast, Walker apparently found in parts of Sleat some signs of forest management. He admired first the fruit trees in the garden at Armadale, and the plantations there with many ash trees as large and vigorous as any in Scotland. He went on:

> The wood of Dunscaich [Ord/Tokavaig] in Slait is also very considerable and thriving, consisting of birch, oack, ash, alder, rowan, holly, hazel and grey willow. Here I measured an alder which was 7 feet in circumference at the height of 4 feet above the ground.[7]

The tone of his comments indicates that he saw a wood well cared for, although he does not mention whether it was actually enclosed or not.

By the second decade of the nineteenth century, the woods of Sleat had been divided in two for the purposes of supervision – the east side (probably the woods of Kinloch, Leitir Fura and Aslaig) and the west side (the woods of Ord/Tokavaig, Coill' a Ghasgain, Morsaig and Dalavil). By the next decade, Dalavil had its own woodkeeper. In the later 1830s, the number of woodkeepers burgeoned, probably as a result of an estate survey of 1829 which recommended more assiduous protection of the woods.[8] The woods of Dalavil, Ord, Aslaig and Kinloch, in Sleat, and Scalpay and Sgianadin, and Mudalach, in Strath, were each given a woodkeeper. From 1850, however, enthusiasm for the maintenance of natural woodland appears to have waned and only three woodkeepers are recorded: one for Ord, one for Scalpay and Sgianadin, the other based at Kylerhea (probably with responsibility for Aslaig, Mudalach, Leitir Fura and Kinloch). By 1881, only one woodkeeper remained, at Kylerhea, and he died in 1886[9] (Fig. 14.1).

Some of the chronological gaps in supervision may have been filled by ground officers taking over responsibility for the woods or by those who held tacks of the lands on which wood grew. A series of tacks survives from 1734 which demonstrates the landlord's desire to see his woods properly looked after. Dalavil, Leitir Fura, Tokavaig and Ollach were all set in tack in April and May that year. The tacksman of Dalavil bound himself, 'to do his

---

[6] M. M. McKay (ed.), *The Rev. Dr. John Walker's Report on the Hebrides of 1764 and 1771* (Edinburgh, 1980), p. 205.
[7] Ibid., pp. 204–5.
[8] MEP: 5913, 135–6.
[9] MEP: 5900, 5902, 5905, 3847/1, 3852/1.

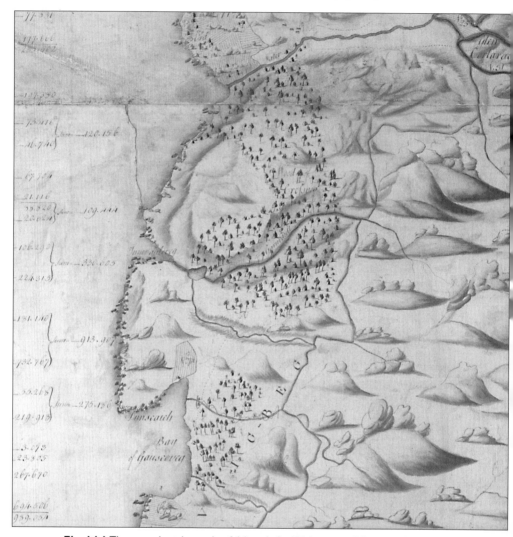

Fig. 14.1 The unenclosed woods of (above) Ord/Tokavaig and (opposite) Dalavil in Sleat as shown on an estate plan of 1763. Both cover a similar area today and were at least partially enclosed in the early nineteenth century. The burn winding from Loch a' Ghlinne to the sea at Dalavil was straightened and deepened in the nineteenth century. From Mathew Stobie's 'Plan of ... part of Slate on the Island of Skye', by kind permission of Clan Donald Lands Trust.

utmost endeavour to preserve any wood or trees growing on the ground of his possession for his master's use and not to pasture or feed any goat thereon'.[10] Similarly, the tacksman of Leitir Fura agreed 'not to pasture or feed

[10] MEP: 4274/6.

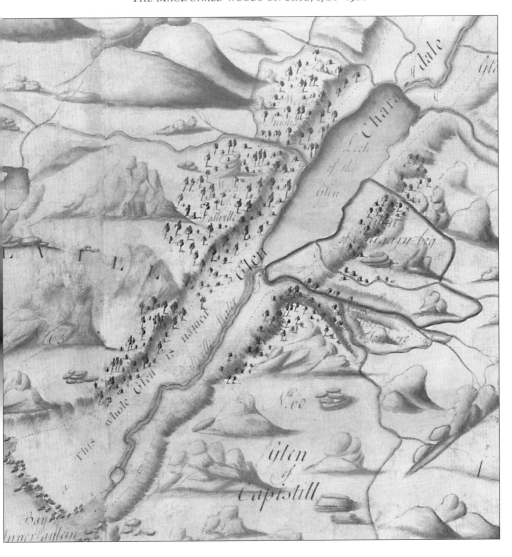

goats on the ground of his possession but to do his utmost endeavour for preserving the wood and timber growing thereon' and the same clause was included in the tack of Tokavaig.[11] The tack of Ollach – to Archibald MacQueen and John Nicolson – formally appointed MacQueen as woodkeeper for the duration of the lease.[12] All four leases were to last for eleven years and, in that period, there is no evidence for any of these tenancies having been under the charge of a separate woodkeeper. After ca. 1800, a

[11] MEP: 4274/3 and 5.
[12] MEP: 276/4.

standard set of printed estate regulations was appended to every lease and the lessees were bound to abide by them. The woods on tenancies were explicitly reserved to the landlord: 'All grounds under timber, whether of natural growth or planted, and whether copse wood or more fully grown, are reserved to the Proprietor, without compensation, and with liberty [to the landlord] to inclose the same.'[13]

A developing regime of protection and preservation can thus be discerned. From 1720, and before, there were woodkeepers responsible for at least some of the natural woods on the estate. On his appointment in 1734, the woodkeeper of Ollach was commanded 'not to suffer or allow the same [the wood] to be cutt or destroyd by any persons whatever without my [Lord MacDonald's] licence in writing'. From that year too, responsibility for some areas was handed to the tacksmen of the wooded tenancies. The regulations introduced at the beginning of the nineteenth century went further in their explicit reservation of all timber to the landlord. This may have led to the increased importance and prevalence of woodkeepers on some parts of the estate in that century. In spite of that, however, every tacksman and tenant was to be 'accountable for the whole growing timber and wood of every kind upon his possession and, in case any tenant by himself or his servants shall be found guilty of cutting, peeling or destroying any wood' he would be 'prosecuted without distinction with the utmost rigour of the law'. All muirburn within one mile of woods and plantations was strictly prohibited and the keeping of goats anywhere on the estate was completely banned.[14]

The introduction of a more strict set of regulations was a result of the appointment of commissioners to oversee the running of the estate by Alexander Wentworth, Lord MacDonald, in 1797, because, he stated, 'I generally reside at a great distance from my estates.'[15] They included Sir John Sinclair of Ulbster and were given full power to run the estate. The regulations which they drew up continued to be applied long after their role as commissioners ceased in 1824.[16] By the 1860s, the regulations had been adapted for, in 1863, a lease of Dalavil contained new terms relating to woodland. The woods were still reserved to the landlord who, in addition, 'shall at all times have free access thereto and the power of removing timber therefrom without any claim at the instance of the tenant for surface

---

[13] See e.g., MEP: 4289/4.
[14] *Ibidem*.
[15] NAS: RD 3/276, 548–50. I am grateful to Margaret MacDonald, archivist at the Clan Donald Visitor Centre, for this reference.
[16] University of Glasgow Business Records Centre: UGD 37/2/9/18.

damage'. The tenant was also bound to 'uphold the fences round these plantations' although the landlord would pay half the expense of any repairs.[17] In 1873, a similar lease of Ord and Morsaig reserved all the wood to the landlord, although it granted the tenant the right to graze sheep and cattle in the woods. The landlord was also allowed to enclose woodland and to have 'all accesses and facilities requisite for the purpose of thinning, pruning, cutting and manufacturing' the wood without being liable for any damage done by the exercise of such powers.[18] This may be directly related to plans for the commercial exploitation of the woods.

How well these regulations were kept is, however, unknown, for there are very few surviving records of cases of prosecution for their breach. In July 1801, the commissioners insisted on the proper enforcement of the rule against keeping goats, although there are no recorded instances of eviction for failing to adhere to it. Four years later, the chamberlain wrote to the commissioners telling them of tenants' abuse of the woods of Dalavil, Ord, Tokavaig and Leitir Fura. The woodkeeper, Angus MacDonald, was dismissed for having 'paid no attention to that duty' and indeed having 'winked at the depredations'. The chamberlain had information that a number of tenants might be convicted of abuse of the woods and recommended that Lord MacDonald should prosecute, warning that 'if there is no notice taken of it, his Lordship should desist in laying out money in rearing of wood and planting . . . as I am thoroughly convinced when he is stript of all his forest woods the next sacrifice will be his planting'.[19] In 1806, the commissioners insisted that 'considerable depredations' committed in the plantation at Portree and other woods should be investigated and the perpetrators brought before the sheriff. They went on to make clear their 'determination to punish delinquency of this kind in the most exemplary manner'.[20] In the following year, investigations were being made about tenants taking oak from Dalavil without permission. The chamberlain had it intimated in the parish kirk that 'any person or persons that will be found guilty in trespassing upon the woods will be prosecuted according to law'.[21]

In January 1811, a Justice of the Peace court at Broadford heard charges against 'sundry' persons for cutting and barking trees at Sgianadin, Scalpay, Aslaig, Leitir Fura and Kinloch. They had been summoned by the

---

[17] MEP: 4325/3.
[18] MEP: 4331/2.
[19] J. Macpherson to estate commissioners, November 1805, MEP: 927/3.
[20] MEP: 4180/14, 4212/3/1.
[21] MEP: 1054/2.

chamberlain of the estates, the ground officer of Sleat and the grass-keeper in Leitir Fura. Some were fined between two guineas and £5 each while others were remitted to the chamberlain so that he could make of them whatever example he wished.[22] Three years later, Alexander Nicolson in Tokavaig was evicted for taking wood without permission to repair his fire-damaged house.[23]

Perhaps this rigour, apparent in the estate records after ca. 1800, was due to the relatively efficient running of the estate by the commissioners. In 1818, five tenants of Lord MacDonald who lived on Scalpay were prosecuted by James MacLeod, laird of the adjacent island of Raasay, for taking wood from his estate.[24] Perhaps the woods of Scalpay and Sgianadin were monitored so well that it was easier to go into the next estate for timber. Payments are recorded in estate accounts for apprehending woodcutters in 1823 and the expenses for prosecuting others were paid in 1831–2, demonstrating that some vigilance continued after the commissioners ceased to operate in 1824.[25] There are no further records of action against tenants for abuse of the woods until the 1880s but this may be due to the relative lack of survival of estate correspondence in the intervening period.[26]

As well as the imposition and enforcement of regulations, some positive measures, other than the employment of woodkeepers, were taken for the maintenance and preservation of the woods. As early as 1763, the young landowner, Sir James MacDonald, paid a visit to his estates on Skye and North Uist. A Lowland map-maker, Matthew Stobie, was commissioned to survey the estates, and Sir James expressed the wish that his lands should be run for the good of those who lived on them as well as that of the landowner. The 'many useful improvements which his fruitful genius suggested and his active spirit promoted', as the inscription on his monument in the parish kirk of Sleat has it, were lost to the estate with his death in Rome in 1766. When in Skye in 1763, the young Sir James had looked over his lands and appears to have taken a particular interest in the woods. In a letter to his lawyer in Edinburgh, MacKenzie of Delvine, he wrote:

> I find it necessary to inclose some places on account of oak woods which are entirely consumed by the cattle but which if properly taken

---

[22] MEP: 5609/1.
[23] MEP: 424/1.
[24] NAS: SC 32/5/1.
[25] MEP: 5884, 5897.
[26] e.g. MEP: 2546/1. The two periods for which estate records are most full coincide with the regime of the commissioners in the early nineteenth century and the factorship of Alexander MacDonald in the later 1870s and 1880s.

care of may be worth in twenty years as much as the estate. No one who has not seen it can have any idea of the destruction which is everywhere visible in these woods. Peter Hench whose indignation has been raised to the highest pitch upon this as well as upon many other subjects has devised a method for the salvation of them which I shall immediately adopt. You will be so good to send me immediately to Inverness for this purpose six hedge bills and two felling axes.[27]

Unfortunately, there is no evidence to suggest that this work was carried out. The tools would have been sent to Skye and tidying up of some of the woods may have been attempted. As far as enclosure is concerned, however, an estate survey of ca. 1800 appears to show that no areas of woodland had yet been enclosed. It recommended that a number of the woods should be enclosed, and a map which accompanies the survey shows where the surveyor felt some of these enclosures should go.[28] There appear to have been a number of moves made towards the enclosure of various tracts of woodland in the first decade of the nineteenth century. In 1802 a 'Set of the estates of Skye' (a plan for their reorganisation) reserved the woods of Dalavil, Tokavaig, Leitir Fura and Aslaig to the proprietor and recommended that they should be 'enclosed as soon as possible'.[29] Nothing seems to have resulted from this but, in 1805, the chamberlain of the estate viewed the woods of Sleat and recommended the cutting and enclosure of the oak of Coille Dalavil. A letter from the chamberlain to one of the commissioners in January 1810 shows that this had been carried out. The oak at Dalavil had been sold to an Edinburgh tanner in 1809 and the wood was enclosed, to allow the stools to send up shoots without disturbance from livestock and deer.[30] The fact that, in December 1809, repairs were made to the dyke at Dalavil might suggest that enclosures had been erected prior to sale and extraction and that any breaches caused by the removal of timber and bark were being made good.[31]

Much of that impressive enclosure dyke still stands (see Plate 9b and c). It is built almost entirely of dry stone with a vertical, flat outer face over three feet high. The inner face varies from that height to nothing at all in places where natural features, such as outcropping stone, make a free-standing dyke unnecessary. It is topped with a turf coping into which

---

[27] NLS: Delvine Papers, 1309, fos 228–9. I am grateful to D. W. Stiùbhart for this reference.
[28] NAS: RH 2/8/24, 29–30, 40, 46; NAS: RHP 5999/27.
[29] MEP: 4389/3.
[30] MEP: 1624/1.
[31] MEP: 3707/2, 'Reports by the gardener at Armadale to the estate commissioners', 1809–10.

branches would probably have been stuck to make the effective height of the barrier greater. In 1801, birch brushwood was certainly used on the estate to increase the effectiveness of physical barriers round the plantations at Armadale and, in January 1810, brushwood was being cut at Dalavil.[32]

In the Spring of 1810, presumably with Coille Dalavil already enclosed, further enclosures were proposed. Planning reached such a stage that, by the end of March, the lines of the proposed dykes had been laid out 'by two skillfull men' and an estimate of the cost submitted to one of Lord MacDonald's commissioners for enclosing the woods at Ord/Tokavaig and Kinloch/Leitir Fura. The estimate gives a good description of the sort of dyke which was built at Dalavil: 'a double stone dyke with two rows of feal', that is, a dyke with two stone faces and two thicknesses of turf on top.[33] Ordnance Survey maps of Ord/Tokavaig, both first edition and modern, suggest that the wood was enclosed or that the existing dykes round the grazings of the neighbouring townships were added to so that most of the wood was protected from livestock. That this was the case is supported by an estate survey of 1851 which, in dealing with pasture at Ord/Tokavaig, mentioned 'the woods which are grazed', implying that some were not. There is no evidence that any enclosure went ahead at Kinloch and Leitir Fura. The dyke there would have been longer; it was estimated as more expensive per unit of length because the ground slopes up from the woods which run down to the sea. The same estate survey of 1851 also noted the continuing need to enclose the woods of Kyleakin, Leitir Fura and Aslaig.[34]

There can be no doubt that the woods on the MacDonald estates on Skye would always have been exploited like any other natural resource. The earliest surviving reference to such use comes from the last decade of the seventeenth century in Martin Martin's *Description of the Western Islands of Scotland*. Martin, a native of Skye, wrote of a sacred well at Loch Siant (probably Loch Sheanta, north-east of Digg in Trotternish) near which was a small copse. He stated that 'none of the natives dare venture to cut the least branch of it, for fear of some signal judgement to follow upon it'.[35] As well as revealing something of folk belief on seventeenth-century Skye, it suggests that the people were accustomed to taking timber from other woods on the island and that this particular grove of trees was exceptional in remaining untouched. Martin also mentioned that it was traditional to fell timber 'in time of the decrease of the moon', that the steam from hazel

---

[32] MEP: 686/3; MEP: 3707/2.
[33] MEP: 1273/1, 1283/10.
[34] MEP: 3738.
[35] Martin, *Description of the Western Islands*, p. 141.

sticks put in boiling water was used to bring out a sweat, and that a rod of oak a few inches long, cut before the middle of May, twisted, boiled and dried, was used to preserve yeast.[36]

As has been discussed above, many tenants simply helped themselves to what the woods had to offer and records of action against them reveal what sort of things they took. In 1811, a number of tenants fell foul of the chamberlain for barking birch and oak trees, taking young trees for hoops (for barrels and buckets), cutting sticks for repairs to boats, for caschroms (footploughs) and for roof cabers.[37] Tenants could, however, obtain such things by legitimate means, through the estate employees with responsibility for the woods. In 1773, Samuel Johnson recorded that birch and oak bark were used for tanning leather.[38] This was probably normally procured legitimately from the estate. In 1745, five men were paid for taking hides from Kyleakin and Kylerhea to Isleornsay in Sleat.[39] Since both Kyleakin and Kylerhea lay close to oak and birch woods, this could be a record of tanned hides being brought to a brogue-maker or to an important harbour for export. In 1747, the factor wrote, in a letter to MacKenzie of Delvine, that one of the baillie's perquisites was 'a pair of shose from every tenant that barked leather'.[40] This is suggestive of at least some regulation of the tanning of leather and thus of the use of oak and birch bark.

The estate accounts provide only four instances of sales of bark during the nineteenth century, although the amounts involved were, by the standards of small sales to tenants, quite large. During 1859, £59. 7s. 3d. worth of bark and wood were sold at Kinloch – probably comprising a number of small sales – and, in 1860, Christopher MacKinnon, the woodkeeper of Kinloch, Aslaig, Leitir Fura and Mudalach, sold £9. 4s. 6d. worth of bark.[41] It is possible that other small sales of 'wood' included bark without mentioning it, but the four instances mentioned above are the only ones in which bark is specified as having been sold.

The amount of detail in annual accounts tended to vary from year to year. Almost every account during the nineteenth century recorded a small amount of income from sales of wood to tenants. In most years it is not easy to calculate how much came from the natural woodland and how much

---

[36] *Ibid.*, pp. 174, 189, 192. His account of when to fell in relation to the moon was the opposite of recorded practice elsewhere in Scotland but the same as in Denmark.
[37] MEP: 5609/1.
[38] Samuel Johnson, *A Journey to the Western Islands of Scotland* (Penguin Classics edn, London, 1984), p. 68.
[39] NAS: CS 96/4261.
[40] NLS: Delvine Papers, 1306, fos 46–7.
[41] MEP: 3840/7, 3841/2.

from plantations. The location of the sale, the purpose for which the wood was sold, the type of timber and the name of the person conducting it, if details of these are provided, can be of some help with this. In the accounts for 1839–40 Christopher MacKinnon sold £1 worth of wood, and Alexander Kennedy, woodkeeper at Kinloch, sold 19s. worth. It should be noted that this compares to total sales of £51. 5s. 6d. in that year, the bulk of which probably came from the coniferous plantations.[42] Larch from the plantations at Armadale and Ostaig was sold in considerable quantities for boat-building. The deciduous wood was not sold in trifling quantities, it was just cheap. In 1853–4, ninety-six hazel cabers for roofing were sold for 4s., which was only $^1/_2$d. each.[43]

Separate lists and accounts of wood being supplied to tenants survive for the years between 1878 and 1885. In 1878–9, there were seventeen different instances of wood being 'given' to tenants. This would perhaps suggest that some tenants did not have to pay, or that wood for certain purposes, such as house repairs, was given out free of charge. Wood for rafters, cabers and boat timbers as well as unspecified 'sticks' and 'wood' were among the items given out. Similarly, the accounts of sales reveal that numerous people took small amounts of wood, both from the plantations and from the natural woods.[44]

How was this wood obtained? It does not seem to have been a process of turning up at a public sale. If one wanted wood, one applied to the factor. As noted above, in 1734, the woodkeeper of Ollach was ordered not to allow anyone to take wood without the permission of the landlord. In the landlord's absence, the factor took his place. Before the late nineteenth century, it is possible to see only by default that permission to get wood was required in that it becomes apparent when people took wood without it. In 1882, one Malcolm Macpherson wrote to the factor in Portree seeking wood to repair his house 'twice blown down by the late violent gales'. He asked that the factor 'would be so good as order Christopher MacKinnon' to give him wood for repairing it since the family were 'living in the barn and not at all comfortable'.[45] Similarly, there is the pitiful story of the widow of John MacInnes in Heaste. She had three children to support, her last rent had been paid by a benefactor and, she wrote, with 'a little help from a cousin of my husband in Australia . . . I was able to weather through'. Her house had collapsed during the winter while she was in it and her brother and

---

[42] MEP: 5902.
[43] MEP: 5906.
[44] MEP: 3759, MEP: 5635/1–13, MEP: 4044–5.
[45] MEP: 2380/1.

sister-in-law had rebuilt it for her but they needed wood for the roof. She was therefore writing to the factor in April 1884 asking to be allowed to take some wood from Ord, just across Loch Eishort.[46] The same system appears to have applied to plantation wood. In May 1882, one Kenneth MacRae of Broadford asked the factor for 'a line stating that I'm to get ... wood' at Armadale for repairing a boat.[47] It was not only the ordinary tenants who needed to apply to the factor. In 1883, Duncan MacInnes, ground officer of Sleat, wrote to Alexander MacDonald at Portree concerning commercial sale of birch and the general state of the woods. In a postscript he wrote, 'Kindly let me know if I can have a few stabs [stobs] from Dalville for repairing my fence between Gillin and Stonefield – horses which are becoming troublesome. Of course I will be glad to pay for the stabs'.[48]

During the eighteenth century, and again at the beginning of the twentieth century, the woodland was of use to the tenants because of the shelter it provided for livestock. The beautiful estate plans drawn up by Matthew Stobie in the 1760s, although they do portray woodland, treated it as part of the pasture of the estate.[49] Although goats were to be excluded from the woods, they could provide shelter in winter and grazing for sheep and cattle. The reservation of the woods to their proprietor in the early nineteenth century, combined with the enclosure of some of them, must have reduced this but some pieces of woodland would have continued to be grazed throughout the period. The fact that the woods of Trotternish appear not to have been looked after as such after ca. 1750 would suggest that, as far as the estate was concerned, they came to be regarded primarily as sheltered pasture for the tenants' stock.

As well as the tenants, the estate itself needed wood. The charter of the Lord of the Isles, giving Sleat to Celestine of the Isles in 1463, demonstrates the perceived value of oak to the estate as early as the fifteenth century. Having readily available timber of one's own certainly made many ordinary estate jobs a good deal cheaper. It would seem, however, that the MacDonalds of Sleat did not have enough wood for all their needs. In the middle of the sixteenth century, Donald MacDonald of Sleat was served with an interdict preventing him from taking timber from the MacKenzie lands of Kintail for building galleys. By the 1730s, the owners of Sleat seem to have come to an arrangement by which timber from the mainland could

[46] MEP: 2863/6.
[47] MEP: 2457/14.
[48] MEP: 2726/6.
[49] MEP: MAP/CD/325, 'Plan of ... [*damaged*] part of Slate on the Island of Skye and County of Inverness', Matthew Stobie, 1763; NAS: RHP 22110, 'Strath and Sleat from Loch Daal to Callich Point', Mathew Stobie, 1766.

be procured legally. In 1733, the factor's accounts record payments for sending the 'birline' (galley) to the wood of Knoydart and a tack of the mills of North Uist from 1746 allowed the holder to take a boat-load of timber from the woods on the mainland. The origins and nature of these rights are not clear, although MacDonald of Sleat may have had servitude rights in the woods of Knoydart.[50]

The estate accounts show that, throughout the eighteenth century, large building timber had to be bought in. Unlike many mainland Highland estates, Skye had no native Scots pine. The earliest record of plantation, in 1703, relates to Sir Donald MacDonald's attempts to rectify this by buying in enough seed to grow 'eight or ten thousand firrs', but even this, and further efforts at planting during the eighteenth century, appears to have been insufficient to meet demand.[51]

As a result of the relatively unspecific nature of the accounts, it is often difficult to be certain, when timber is referred to, whether it was coming from natural or planted woods. In 1733, payments were recorded to certain estate servants 'when sent to the wood' and, in 1745, one Alexander Robertson, a wright, was paid £4. 6s. 8d. Scots 'for his pains in the woods'. One of the uses to which this wood was put was the making of new stocks into which 'the fidler's son' was put soon after they were made.[52] It is clear from the fact that plantations were mentioned in the accounts and portrayed on maps of the period that they had reached maturity by the second half of the eighteenth century.[53]

One of the few early instances of unambiguous evidence for timber being taken from natural woodland for estate purposes dates from 1749. One Neil MacLean was paid £26. 13s. 4d. Scots for taking a boat-load of cabers from the wood of Ord for the kirk of Portree.[54] In 1750, a boat-load of timber was sent from Sleat for the mills of Portree and Romisdale in Trotternish. Ash was commonly used for millwheel axles but it is not possible to be sure whether this would have come from a natural or planted source. There are and were large quantities of natural ash at Ord and Leitir Fura. Walker in 1764 had admired the size and quality of the ash in the planted Armadale woods, and in 1773 it was also noted by Dr Johnson that there

---

[50] A. Nicolson, *A History of Skye* (Glasgow, 1930), p. 100; NAS: Court of Session CS 96/4261; MEP: 4280/7.
[51] NLS: Delvine Papers, 1307 fo. 205.
[52] NAS: RH 4/90/11/3–4 (Microfilm); CS 96/4261.
[53] NAS: CS 96/4261; MEP: 3807; M. MacKenzie, 'The South part of Sky (*sic*) Island and the adjacent Main of Scotland', from *A Maritim Survey of Ireland and the West of Great Britain* (published in 1776 but surveyed 1751–7).
[54] MEP: 3907/10.

were 'tall ash trees' in the plantations there. In 1780 and 1831, it was those ash trees from Armadale which were used for repairing estate mills.[55] In 1839, however, natural wood was taken from Aslaig and Kinloch for a mill in Sleat.[56]

During the nineteenth century, most of the wood used for estate purposes probably came from plantations – although it is not clear whether the boat-loads of timber from Scalpay for 'palings' around the Portree plantations came from natural or planted wood there.[57] Many items, such as wheelbarrows and carts and even road bridges were made from estate timber. In the 1850s, a sawmill was built at Portree to process timber from the plantation there and, ten years later, another was built at Armadale.[58]

As noted above, brushwood, probably birch, was used to heighten dykes around enclosed woods, both planted and natural. In 1810, brushwood at Dalavil was cut in January and May and, in 1824, £50 was paid to Mr MacKinnon of Corry, near Broadford, for cutting brushwood at Kinloch. There is no record of what the large amounts of twigs which must have been produced by this work were used for, but enhancing enclosures would probably have been one purpose to which they were put. It is likely that a 'loghouse' at Dalavil in 1809 would have been made with the natural wood available there.[59] It was probably built to store some of the oak timber left behind after the bark had been removed by an Edinburgh tanner. At the same time, wood was cut for a 'rustic bridge' for the policies at Armadale.

Then there was the hope that commercial exploitation of the woods would provide much-needed income, quite apart from estate use. The yield from the semi-natural woods was overshadowed by the yield from the plantations, but with the financial pressures of a debt-ridden estate being what they were, money was welcome from any available source.

The oak at Coille Dalavil was the first of the semi-natural woods to be exploited for substantial commercial gain. After a suggestion from the chamberlain of the estates in 1805, the oakwood there was advertised for sale some time before January 1808 when an offer of £400 was received from a Dumbarton timber merchant.[60] Perhaps it was because of the rising price of oak bark that this sale did not go through and, instead, a sale of the

---

[55] MEP: 3810, 5897; Mackay (ed.), *The Rev. Dr. John Walker's Report*, p. 205; Johnson, *Journey to the Western Islands*, p. 67; J. MacDonald, *General View of the Agriculture of the Hebrides* (Edinburgh, 1811), p. 339.
[56] MEP: 5902.
[57] MEP: 5895, 5902.
[58] MEP: 5904, 3841/5 and 7.
[59] MEP: 3707/2.
[60] MEP: 1084.

oak of Coille Dalavil by public roup was held in Edinburgh in March 1809 at the upset price of £500. It was bought by John Swan, an Edinburgh tanner, for £635 and seems to have been cut soon after.[61] At or before the time of this sale, the wood was enclosed and, in spite of breaches of the articles of roup by the buyer and disagreements about the way in which the oak had been cut, it was reported early in the following year that 'the shoots are very strong and thriving'. John Blackadder's report on the estate of 1811 mentioned the same wood to be 'in a very thriving state'.[62]

In 1810, there may have been commercial sales of ash. Reports of the work done on the gardens, woods and plantations noted that, in January, ash at Ord and Leitir Fura was measured.[63] Although no note of a sale survives, the accounts are not complete for that year. Five years later, expenses were paid to men 'marking the ash wood at Letterfure previous to sale'. This time, a sale definitely went ahead and it raised £39. 1s. but the purchaser's name was not recorded.[64]

There are no records of further commercial sales from the woods until the late 1850s. In 1858, £75. 3s. 6d. was taken from what may have been the sale of semi-natural wood on Scalpay.[65] Over the following two years, substantial amounts of bark were sold from Dalavil and Kinloch. Presumably due to competition from foreign imports, and since about fifty years had elapsed since Dalavil had last been cut, the price obtained for the bark was low and only £320. 18s. was raised from all these sales.[66]

There are signs that, in the 1870s, plans were afoot to undertake more serious commercial exploitation of the estate's semi-natural woodlands. In 1873, a lease of Ord and Morsaig stipulated that, as far as the woods were concerned, the proprietor would have the right to 'all accesses and facilities requisite for the purpose of thinning, pruning, cutting and manufacturing the same, without being liable to the tenant in any damage which may be occasioned by the exercise of such powers'.[67] In the following year, £115 was spent on the deepening and widening of a channel from Loch a' Ghlinne, at Dalavil, to the sea. This cut had been made, as part of a scheme of drainage projects around the estate, in the early 1830s to drain the land between the head of the loch and the sea for cultivation. In 1874, the width of most of the 1,872-yard cut was increased from about twelve to twenty-seven feet

---

[61] MEP: 927/3, 4226/11, 5910; MacDonald, *Agriculture of the Hebrides*, p. 360.
[62] MEP: 4226/11 1624/1–4, 1269/5, 5912.
[63] MEP: 3707/2.
[64] MEP: 3822.
[65] MEP: 3840/4.
[66] MEP: 3840/7, 3841/1–2.
[67] MEP: 4331/2.

and its depth was increased from three to eight feet.[68] Coille Dalavil lies at the side of the loch and, if timber were to be shipped from the coast, it would have been far easier to float it to the sea by the loch and the cut than to drag it overland for well over a mile of undulating terrain. Only along the last 200 yards or so of the watercourse does it revert to a natural burn, but with a high tide and a reasonable flow of water which could have been provided by a sluice at the end of the loch – a local blacksmith was paid £15. 11s. 8d. for work on the cut – this would probably not have presented too much of an obstacle.

On 8 August 1879, J. and R. Cooper, Glasgow timber merchants, offered to buy oak and birch at Kyleakin (Mudalach) and Kinloch. This firm had already dealt with the MacDonald estate, having paid over £700 for plantation timber from Armadale in the 1860s.[69] At first, Lord MacDonald declined the offer but by the end of September it had been provisionally accepted by the factor, on the condition that Lord MacDonald agreed, and a cheque for £450 was banked. In November, Lord MacDonald having failed to agree, J. and R. Cooper requested the return of their cheque but the sale appears finally to have been completed, since on 31 January 1880 J. and R. Cooper were sent a receipt for the original amount.[70]

The woods of Kinloch and Mudalach having been opened up again for commercial gain, Ord, Morsaig and Dalavil were next. At Dalavil, the oak which had been coppiced in 1809 and again in 1859 appears to have been cleared, at least from the lower ground close to the shore of the loch. Beech, possibly planted as early as 1810 among the coppiced oak, was allowed to grow in its place and it may have been with a view to extracting this larger timber that the cut from the loch to the sea had been widened in 1874. It was the birch, there and at Ord and Morsaig, however, which was to be sold during the 1880s.

Advertisements for this birch must have been published in the early months of 1880 and, in May, offers began to come in.[71] A small quantity was bought by the Raasay estate, along with much larger amounts of larch and spruce, in that month.[72] The offer which was accepted for substantial quantities of birch came from Wylie and Company of Ardersier near Inverness but final arrangements for the sale were not made until September 1881. In that month, Wylie agreed to buy upwards of 2,000 tons of birch at

[68] MEP: 4029/1, 3823/3, 5897, 5898.
[69] MEP: 3841/8 and 10.
[70] MEP: 2067; 2090/1–2.
[71] e.g. MEP: 4039/2.
[72] MEP: 4039/1.

6s. 6d. per ton to be cut between October and March 'when there is no sap in the wood'. They expressed a wish to go ahead as soon as possible and, to that end, security for payment was secured from a Glasgow insurance firm by the middle of October.[73]

The work began at Ord and, by January 1882, in spite of bad weather, 'the cutting and dragging of the birch ... [was] well advanced'. The weather had, however, slowed the work sufficiently to prevent them from getting any cargoes ready for shipment by the beginning of February – the winter was described by Wylie as 'one continual (*sic*) westerly gale'.[74] By this time, Wylie's 'friends at Paisley' were 'pressing for delivery'. It is likely that, since the timber was being shipped to Paisley, it was intended to be made into thread bobbins at the Coats mills there. Unfortunately, the records of the Coats company are not explicit about the origin of the wood which they bought, so it is impossible to be sure.[75]

Although the birch at Ord was still to be shipped out by the beginning of March, those engaged in cutting it felt that over 100 tons could be loaded by the 11th. It was at this time that the birch at Dalavil began to be cut. A further hitch prevented the shipment of any birch until April when the SS *Wallace* made two separate visits to load at Ord, soon to be followed by two other ships, the *Jane and Mary*, which took some birch from Kinloch, and the *Dolphin*, which was to load at Dalavil.[76] All of their cargoes were shipped directly to Paisley. The amounts of birch must have been fairly substantial since shipping carried on through the summer and into September.[77] Cutting had begun again by October 1882 and, during that season, it was planned to take birch from Ord, Morsaig and Dalavil. Cutting at Morsaig was 'well ahead' by early November, with a yield of 150 tons expected.[78] By January, further cargoes had been shipped from Ord and Dalavil in the *Bee* and the *Dolphin* and cutting appears to have continued throughout that season, with shipping going on during the summer of 1883.[79]

Wylie intended to begin cutting birch once more in the autumn of 1883. A family bereavement caused some delay and, in November, the firm wrote to the factor of the MacDonald estate informing him of this. In stating their

---

[73] MEP: 2324/1, 4–7.
[74] MEP: 2372/1, 2386/2/1, 2406/8/1.
[75] The records of J. and P. Coats Ltd are housed at the University of Glasgow's Business Records Centre: UGD 199.
[76] MEP: 2406/8/1, 2438/9 and 15.
[77] MEP: 2565/4 and 6, 2497/3, 2451/2 and 3.
[78] MEP: 2528/23, 2451/11.
[79] MEP: 2568/3, 2697/7.

readiness to begin that season's work, they wrote, 'at the same time, should you think that in the interests of Lord MacDonald it would be better that another had the wood we would forego any claim we may have'.[80] There is no further record of Wylie of Ardersier in the MacDonald muniments and one can only assume that, hoping for a better price from another buyer, the factor took them up on their offer of withdrawal. Perhaps Wylie had made the offer in the knowledge that what was left was not worth having – a 'half load' already cut at Coill' a Ghasgain was not collected.[81]

Wylie's ambitious plan to extract over 2,000 tons of birch from the woods of Sleat was never realised. Only about 800 tons in all appear to have been shipped. Birch was not a very lucrative commodity and the income received from these sales was probably just over £200.[82] It should be stated, however, that these two estimates of the actual tonnage and money may be too low. For example, although the *Wallace* is known to have loaded twice in April at Ord, as mentioned above, a record of the tonnage survives for only one of those visits. Nearly half of the tonnage recorded came from Ord. The spaces in the wood which this extraction opened up are clearly visible on the 1901 revision of the Ordnance Survey, originally surveyed in 1876. Although the wood has, over the last century, begun to encroach back into these gaps, it is only with recent exclosure that appreciable regeneration has begun to take place.[83]

As far as the commercial side of timber is concerned, much more money was to be made from the sale of larch and spruce from the plantations at Armadale and Ostaig. While Wylie and Company were shipping wood from the west side of Sleat, J. and R. Cooper of Glasgow paid over £2,000 for thousands of mature conifers.[84] It was to this company, after the activities of Wylie of Ardersier had ceased, that the estate turned for the sale of the birch. Less than a week after the offer by Wylie to withdraw, J. and R. Cooper were invited to look over the birch.[85] Nothing resulted from this approach and nearly eighteen months elapsed before anything more was done. In April 1885, presumably in response to an advertisement, Robert Hutton, a Paisley timber merchant, made an enquiry about the birch of Sleat. He arrived in the middle of May and, after having looked over the

[80] MEP: 2726/3.
[81] MEP: 2726/6.
[82] MEP: 3849/1, 3850/1, 5635/14, 2528/30.
[83] Ordnance Survey 1:10560 Isle of Skye [Inverness-shire], sheet 51, 1st edn, surveyed 1876 (published 1881), revised 1901 (published 1903); Ordnance Survey 1:25000, Pathfinder Series, no. 219, North Sound of Sleat (1974).
[84] MEP: 3849/1, 3850/1, 3851/1.
[85] MEP: 2726/5.

woods of Dalavil, Ord 'and some other places', he wrote to Alexander MacDonald, the factor. His verdict was that 'all the birch that was at all convenient for shipping had been cut by former purchasers'. He opined that what was left could be sold only if the estate were to 'contract with men in the various localities to do all the work to the putting it on board the vessels'. Only in such circumstances did he feel that he could make an offer.[86] Presumably the factor felt that the profitability of this would be too small to be worth pursuing.

In September 1885 the interest of J. and R. Cooper was again stimulated and their representative, Mr Smith, came to look over the birchwoods. No purchase was made, however, in spite of the estate's view that there was a 'good deal of birch' at Dalavil. Another Glasgow timber merchant, Archibald Campbell, made an enquiry, stating that he wanted the wood for bobbins. In October, after having looked over the woods, he too felt that there were insufficient quantities of wood at the size he wanted to justify purchase.[87]

This was the last serious attempt in the nineteenth century to exploit the natural woodland for commercial profit. The MacDonald estates remained deeply in debt. North Uist and parts of the Skye estates had been sold off. None of these sales, however, was sufficient to bring the family back into the black and, by 1918, sale of the whole remaining estate was contemplated.[88] Desperation is reflected in a letter from the estate office in Portree, written in December that year. Along with attempts to find a market for the plantation timber, realising cash from the natural woods was being considered. The writer stated that:

> There is a good deal of birch in the south end of the property which I think we should endeavour to dispose of by every means possible and endeavour to prevent sale of the property becoming necessary. That wood is of no value whatever to us and is simply cumbering good hill grazing. I would like permission to approach the Coats firm to see if they would purchase it for making bobbins for their thread works. I understand they are in the habit of purchasing wood standing for this purpose... I have made a careful list of all the timber on the ground ... and if the question of cutting is to be gone into we should get a report from an absolutely independent man of undoubted standing ... Large cutting would of course interfere with the amenity

[86] MEP: 2959/10, 2970/1–4, 2980.
[87] MEP: 3032/8–10, 3049/2–3.
[88] MEP: 3664/1.

to some extent but if we are to save the Estate for the family it is worth considering.[89]

The 'independent man' they found was David Keir, wood manager on the Atholl estates in Perthshire. On 10 March 1919, he submitted a report on the woods which stated that the birch left at Ord was 'only scrub and of no commercial value whatever at the present time and it would be useless wasting money, time and labour trying to improve it. It is useful as a shelter for stock, which is all it will ever be good for.'[90] The hired expert thus disagreed with the panic-driven views of the estate office concerning their timber and its utility in relation to stock, and so Sleat was rescued from being denuded of its natural woodland.

Long-standing notions about a historically recent, drastic decline in woodland cover die hard. In the middle of the eighteenth century, the Rev. John Walker wrote that Skye was 'very destitute of trees yet formerly it has been filled with wood, of which there are still some considerable remains'.[91] Recently, some of the woods of Sleat have been described as 'decaying'. A fuller report, written in 1972, described the situation in the following terms:

> The tree cover... although probably at one time extensive, is now sparse ... the cumulative effects of past mismanagement has had such an effect with the production of blanket bog that the waterlogged mor humus no longer constitutes a suitable seedbed for the establishment of birch seedlings.[92]

This report went on to conjecture that, before about 1600, Sleat was almost entirely covered in trees and that these were cleared, first as a result of 'demand for charcoal for smelting during the seventeenth century' and later by an increase in grazing pressure which accelerated with the introduction of sheep after the middle of the eighteenth century. Clearance of woodlands for sheep all over the Highlands 'by either felling or burning the existing woodland' was assumed and the author suggested that 'research

---

[89] MEP: 3364/5.
[90] MEP: 3783.
[91] Mackay (ed.), *The Rev. Dr. John Walker's Report*, p. 204.
[92] 'Gleann Meadhonach, Sleat, Isle of Skye: an archaeological survey for the Clan Donald Lands Trust', *Dualchas*, Skye and Lochalsh District Council Museum Service (unpublished report in possession of Clan Donald Lands Trust, 1994); R. A. Pellew, 'An ecological appraisal of the proposed wildlife reserve, Ostaig Estate, Sleat', (unpublished report in possession of Clan Donald Lands Trust, 1972), p. 8. I am grateful to the Archivist and the Ranger Service at the Clan Donald Visitor Centre for permitting access to these sources.

into the mode of woodland clearance' should be carried out.[93] No evidence was provided for any of these assertions.

The evidence from maps and place-names would tend to suggest that the extent of woodland today in Sleat and Strath differs very little from that shown either on Matthew Stobie's maps of the 1760s or on the first edition of the Ordnance Survey of the 1870s. Over the last 200 years, limited localised decline has undoubtedly occurred, largely because of the detrimental effects of grazing on the ability of the woods to regenerate. Only at Mudalach does woodland decline appear to have been of greater significance, a phenomenon which may be linked to its more northerly exposure. A judgement based on historical evidence would tend to suggest that the vast majority of woodland decline which has occurred on Skye took place many centuries ago.

In the light of the evidence presented here, the notion that there was extensive woodland clearance on Skye in the seventeenth and eighteenth centuries should be rejected. There is certainly nothing to suggest that commercial iron smelting had any impact. Even if it had taken place, it has been shown that those who used West Highland woods for charcoal to smelt iron were not silvicultural vandals but careful managers of what was an essential resource to their operations which involved a great deal of capital investment and were sited close to sources of wood rather than ore. To build an ironworks and then destroy the source of fuel would have been commercial madness.[94] That the MacDonald estate employed people to look after its woods from the early eighteenth century, if not before, does not suggest any enthusiasm for their large-scale removal. The woods of Sleat and Strath were managed as a resource which could satisfy many needs. They were valuable to the estate itself and to its tenants for the timber and bark which they produced and for the sheltered grazing they afforded to livestock. Occasionally they could also be exploited for modest financial gain. Had more effort been devoted to the management of the woods once their commercial potential was discovered, then they might have been of greater commercial significance. That this was not done was probably the result of a combination of lack of interest from a succession of absentee landlords and the greater profitability of other sources of income. In the first half of the eighteenth century, the principal source of cash, other than

---

[93] Pellew, 'An ecological appraisal', pp. 13–14.
[94] Chapter 9 above, and J. M. Lindsay, 'Charcoal iron smelting and its fuel supply; the example of Lorn furnace, Argyllshire, 1753–1876' *Journal of Historical Geography*, 1 (1975), pp. 283–98 and 'The iron industry in the Highlands: charcoal blast furnaces', *Scottish Historical Review*, 56 (1977), pp. 49–63.

rents, was the sale of black cattle. As that century wore on, the manufacture of kelp, especially on the shores around North Uist, took over and lasted as the most important source of finance into the third decade of the nineteenth century. After the market for kelp had collapsed, those who ran the estate appear to have concentrated on exploiting its sporting potential, building shooting lodges and renting them out to wealthy tourists. Forestry remained a secondary concern. Even in the realm of the timber trade itself, more effort was expended on the commercial potential of forestry plantation around Armadale and Ostaig, principally of larch and spruce, than on the exploitation of the native woods. Until the last quarter of the nineteenth century, only occasionally were there attempts to raise substantial sums from the long-distance sale of the latter. When large-scale felling with no consideration for the future of the woodlands was contemplated, both in the 1880s and after the Great War, it was thwarted and the woods were preserved by the limitation of the resource itself as a commercial commodity.

CHAPTER 15

Conclusion

Were the native woods of Scotland managed sustainably? At first sight, obviously not, as they suffered a crashing decline. From covering half or more of the land surface 5,000 years ago, they were in steep retreat from the Bronze Age onwards, much of the land was open when the Romans appeared and the next 1,500 years saw further loss. No doubt the graph, if we had one, would fluctuate, but the Middle Ages end with a shortage of large timber, the rise of imports and start of legislative concern about felling and planting. By 1750, we suggest (in an upward revision of previous estimates), that the extent of Scotland under wood was a little less than 10 per cent. The percentage under ancient semi-natural wood then fell to about 3 per cent around 1900 and is now about 1 per cent. If one wished to think in terms of right and wrong, it might appear appropriate to quote the famous aphorism of Aldo Leopold: 'a thing is right when it tends to preserve the integrity, stability and beauty of the biotic community. It is wrong when it tends otherwise.'[1] In this light, the destruction of the biotic community that covered half Scotland may seem a great wrong, an offence against nature, a parallel to the many instances of deforestation taking place across the world: many from Fraser Darling onwards have seen it so.

Against this, one must also see that much of the decline was either caused by or assisted by climate change of an entirely natural kind, and that it is now recognised that biotic communities seldom tend naturally towards stability ('the balance of nature'), as Leopold thought, but towards alteration. It is largely fruitless at this distance of time to speculate upon exactly how much of the collapse of the forests in prehistory was natural and how much man-made, but the onset of marked climatic oceanicity around the start of the Bronze Age would have severely changed the forests irrespective of any human intervention. The fact that so many of the remnants of pine and oak in peat bogs throughout Scotland are about 4,000 years old is witness to this; in these climatic conditions, peat must have become the new natural land cover of much of the country.

[1] A. Leopold, *A Sand County Almanac and Sketches Here and There* (New York, 1949), pp. 224–5.

To say that, however, is not to deny that human intervention at an early date may have helped the impact of the climate change, including the formation of peat, in certain places. As time passed human impact alone became quite capable of fundamentally altering the face of the land. In the Iron Age, as Richard Tipping has suggested, in parts of the Lowlands the efforts of farmers brought about 'by far the most substantial anthropogenic alteration of the landscape to have taken place, its scale apparently exceeding later clearance episodes during the historic period'.[2] A relatively simple, non-market society, using the first true ploughs, could completely transform one biotic community, a wooded landscape, into another totally different, a set of farms. For many centuries such human endeavours were described as progress, and we would be hypocritical indeed to deny that our own material wealth rests today, ultimately, on efforts like these. Only a society that has already altered a great many biotic communities in its favour can now afford the luxury of lamenting the alteration of more.

In the Middle Ages, the influence of humanity continued to be crucial, but both at that period and on into the early modern age after 1500, when our study begins in proper detail, it is clear that climate also continued to be a major factor in woodland history. The twelfth and thirteenth centuries were marked by a warm spell that came to an end around 1300, to be succeeded by a phase of colder, wetter and windier weather of varying intensity, which reached its nadir between about 1570 and 1740 and did not clearly improve until the middle of the nineteenth century. Since then amelioration has become more and more obvious, greatly accelerated in the late twentieth century by global warming, apparently with human causes.

There are many indications of woods disappearing within historic time without a record of human interference. Some years ago, H. L. Edlin drew attention to the frequency of tree place-names at high elevations, on Rannoch Moor and in the Grampians: oak at 1,000 feet, alder at 1,200 feet, willow at 1,750 feet, birch at 2,000 feet, pine at 2,200 feet, and rowan possibly higher:

> The existence, in former times, of woods at these levels has seldom been questioned; but it has been customary to assign them to a remote, prehistoric past. Study of place names suggests that they were still flourishing not so long ago – a few generations, or a few hundred years back, at the most.[3]

---

[2] Royal Commission on the Ancient and Historical Monuments of Scotland, *Eastern Dumfriesshire, an Archaeological Landscape* (Edinburgh, 1997), p. 20.
[3] H. L. Edlin, 'Place names as a guide to former forest cover in the Grampians', *Scottish Forestry*, 13 (1959), pp. 63–7. Dr Simon Taylor notes, however, that occasionally Edlin

Such high woods were probably neither large nor dense: frequently the names suggest a small wood or a scatter of trees standing out as a distinguishing feature on the hill. Unrecorded human activity such as fire or grazing may indeed have played some part in their demise, but a drop in the tree line across Central Europe of some 200 metres between the thirteenth and the seventeenth century was a recognised effect of climatic change,[4] and it is altogether probable that it affected the Scottish Highlands as well. In the seventeenth and earlier eighteenth centuries, examples of documented woodland decline at lower altitudes in the west of Scotland in particular often appear entirely natural, as at Little Loch Broom and Loch na Sealga (see above, pp. 53, 58–9), or exacerbated by felling in circumstances where trees already had problems of regeneration, as in Glenorchy (see above p. 358). In short, the last time many of the Scottish hills saw trees upon them was a few thousand years ago, or at best several hundred years ago, in climate and soil conditions different from now. This needs to be appreciated when new native woods are planned in the uplands.

Leaving to one side, then, the large but unquantifiable contribution of natural causes to the deforestation of Scotland, how sustainably were the remaining semi-natural woods managed in the centuries after 1500? The verdict depends on time and place, and on what they were managed for. Almost all woods before the Victorian period, apart from a few on inaccessible slopes or in gorges, were used as pasture and shelter for animals. In the nineteenth century for the first time, there came to be some from which domestic stock were permanently excluded in the interests of timber or sport. Even woods in the eighteenth century devoted to the market production of charcoal and tanbark were only enclosed for a set number of years, characteristically seven or eight, in a rotation of twenty or twenty-five.

If all woods were pasture, it is certainly open to question what the term 'wood pasture' can mean in an historical, as opposed to an ecological, sense. In this book we have used it as a convenient term to describe a wood with a thin canopy of 20 per cent cover or less, and one that was historically either an enclosed deer-park in the Lowlands, as at Cadzow or Dalkeith, or one that got left out of enclosure in the uplands, like Glen Finglas. It is important, though, for modern conservationists to recognise that all our

---

mistakes the Gaelic for a single tree for that denoting several trees (e.g. on p. 65 he translates Meall à Chaoruinn as 'Hill of the Rowans' not 'Hill of the Rowan', and on p. 66 Beinn à Chaoruinn is similarly 'of the Rowan'). An additional point is that pine tree names in particular may allude to the presence of bog wood, not living wood.

[4] H. H. Lamb, 'Climate and landscape in the British Isles', in S. R. J. Woodell (ed.), *The English Landscape, Past, Present and Future* (Oxford, 1985), p. 155.

ancient semi-natural woods, without any exception, had domestic stock in them on a regular basis. It follows from that, whether Vera's theories of the grove-like nature of the original woods are correct or not, that woods in historic time were likely to have been full of glades which animals kept open by grazing, and where the grass grew best. There are numerous references to fields and even to cultivation in the woods: it is best to think of them as typically mosaics of wood, meadows and bogs.

Was the grazing regime practised in these woods sustainable? The survival of some woods in some places might be considered itself as limited positive evidence, but there is no indication from documents that peasant farmers themselves ever had the preservation of a wood as a conscious priority or even an aim of management. Had they owned the woods themselves, instead of being tenants with little security of tenure, they might, of course, have devised different practices. Souming was certainly a method of sharing grazing within a community, presumably by relating it to some kind of carrying capacity of the land as a whole, but as it was stated as the number of beasts allowable per tenant, it was probably too inflexible to take account either of growing population, or of a deteriorating environment due to climatic or other factors. Rural population was rising in our period in the sixteenth and early seventeenth centuries, and again from some point in the eighteenth century until around 1840. Climate was probably at its coldest and wettest from the late sixteenth century through the seventeenth century, but did not decisively improve until after 1850. In these circumstances the woods were bound to come under continuous pressure unless there was a positive incentive otherwise.

Some of the animals kept before the nineteenth century by upland farmers, notably the cattle and the horses, would, unless their numbers became too great, serve to keep the woodland open, and by their heavy feet and tearing grazing break up mossy or heathery ground, allowing a measure of regeneration that would otherwise have been choked by vegetation. Others were less benign: keeping goats became common especially in the seventeenth and eighteenth centuries, until suppressed by the landowners because of observed damage to the woods. Herding by children gave an opportunity to spread the pressure, but whether the community would herd carefully or not would probably depend partly on the abundance of wood in relation to the perceived grazing needs of the flock and partly on the vigilance of the landowners. Most eighteenth-century landowners believed that the farming activities of the tenants were often harmful to the wood, sometimes deliberately so. One vivid instance was when the factor of Glen Finglas in 1707 told the Earl of Moray that he suspected the farmers were cutting the trees and killing the deer in a concerted attempt to get rid of both, and so improve their holdings. A wood may survive only because a family took their long-term stewardship seriously and resisted this sort of pressure for longer than most, but landowners, like everyone else, had their own priorities and short-term needs.

Throughout Scotland, if wood was plentiful in relation to need, it might be destroyed, used extravagantly, or denied the chance to regenerate by repeated muirburn and grazing, usually with the tacit agreement of the estate. As peat was normally plentiful, and coal in places available, there was relatively little need for peasants to keep an area of woodland for fuel, as in New England, France, Germany or Denmark. There is little recorded evidence for cutting branches for 'leaf hay', as over much of Europe or in Sweden and Norway, but when animals were brought down from the shielings in late summer they were, in areas like Lochaber and Argyll, allowed to browse on the leaves and branches of small trees and scrub, apparently over wide areas (see pp. 113–15). If this grazing was properly regulated, it might even have been sustainable over a period: we have no evidence either way. Such scrub woods no longer exist. Subsequent changes in the use of hill pastures with different objectives would in any case have destroyed them. That even extensive mature woods with acknowledged crucial functions for shelter and pasture could die out, probably due to overgrazing exacerbated by climate change, is demonstrated by examples in the late eighteenth and early nineteenth centuries from Wester Ross and northern Sutherland (see above, pp. 119–21).

Woods were seldom, however, maintained just to succour stock. They were multi-functional, and as many of the products (timber, charcoal, tanbark) came to yield direct and growing profit to the landowners, they were subjected to increasingly direct estate intervention in how they were managed. It was the expanding industrial economy, notably the ironworks and the tanning industry that provided, through the market, exactly the incentive to conserve. As long as there was money to be made, management was intended to hit a certain level of sustainability. It took some forms that were age-old, but applied in the eighteenth and early nineteenth centuries with more vigour and in more places, notably by enclosing the whole wood, by enclosure of 'haggs' or felling coups, and cutting in rotations that became more systematic as time passed. It also included detailed specifications on how a wood was to be cut, what trees were to be left as standards, how regrowth was to be treated, how gaps were to be filled and when and how grazing was to be allowed in after a felling.

A level of management aiming at sustainability was higher and better understood for broadleaf woods than for pinewoods, but it was hardly thought worth applying to upland birchwoods. Oakwoods enjoyed the full coppice routine similar to much of England and parts of Europe, and one that spread in time from the Lowlands into the Highlands of Argyll, Stirlingshire and Perthshire. It was eventually as widespread and well developed as any south of the Border, though for reasons of climate and soil seldom so consistently profitable. Certainly in Scotland, as E. J. T. Collins describes in England, the later eighteenth and early nineteenth centuries were a golden age of traditional coppice forestry, 'when in terms of output and productivity, and standards of woodmanship', management 'reached its apogee'.[5]

The management of pinewoods, on the other hand, was usually restricted to imposing a lower limit to the girth at which trees could be cut, in order to give young ones a chance to grow on. Little seems to have been understood before the nineteenth century of the conditions under which the woods might regenerate, except that it was generally beyond their own shade. If a pinewood failed to regenerate after a felling, as it often did in the west, no attempt was made to scarify the ground or to limit grazing: as far as is known, Highland lairds invariably took no action, but wrung their hands and deplored the loss of a capital asset. Sustainability in the management of a pinewood remained an accident dependent on climate and location.

The nineteenth century after ca. 1830 differed in many respects from all that had gone before. Firstly, damage to unprotected upland woods, wood pastures and scrub of all sorts, was much more intense than before, due primarily to the spread of commercial sheep farming: the increase in grazing pressure on the hill by a factor of four to eight times (see above, p. 69) was fatal to most montane scrub except on inaccessible ledges, and overgrazing combined with the 'improvement' of pasture by burning apparently wiped out the stock of sub-montane scrub on which shieling animals had in some places formerly fed in late summer. This grazing pressure also brought natural regeneration to a halt in most woods that were open to sheep and cattle, whether originally left out of enclosure or laid open to animals again after the profits of coppice began to fail. After half a century of profitable management, coppice oakwoods suffered increasing neglect, conversion to game covert and replanting with larch or Norway spruce. Birchwoods found an external market for the first time, but were not treated sustainably: they were usually seen as a crop that could be taken off before the ground was turned to pasture or replanted with something more commercial.

[5] E. J. T. Collins, 'The wood-fuel economy of eighteenth century England', in S. Cavaciocchi (ed.), *L'uomo e la foresta secc. XIII–XVIII* (Prato, 1996), p. 1110.

Pinewoods fared better, at least in the sense that their silvicultural needs were increasingly understood and large-scale nurseries established to replenish stock and extend the planted area, though often with disregard of the provenance. German seed was quickly found to be unsuitable for Scottish conditions, but exotic conifers of other species (larch, Norway spruce, Corsican pine) were introduced into some of the ancient woods with little feeling for 'authenticity' (for this term, see pp. 7–8 above). Still, natural regeneration continued to be seen as an excellent way to perpetuate a wood of Scots pine.

If the calculations in Chapter 3 are correct, the area of the ancient semi-natural woods of Scotland fell by one-half or two-thirds in the nineteenth century, partly by extirpation, partly by conversion to plantation. Nothing of this could now be attributed to climate change: it was entirely due to human destruction and alteration. In the course of the twentieth century it dropped by a further two-thirds, mainly due to afforestation but also due to further conversion to farmland.

With this in mind, we can consider again the categories of management outlined in Chapter 1. Within our period, the wildwood, if it had existed so late, finally came to an end except, perhaps, as vestiges on inaccessible slopes. The creation of the semi-natural wood had evolved over many centuries before 1500: although plantations began early, as late as 1800 at least four-fifths of the existing woods were still semi-natural, though only about one half by 1900. Today, with 17 per cent of Scotland afforested, only about 1 per cent is ancient semi-natural woodland.

The way the woods were managed in the eighteenth century was more likely to be sustainable in the commercial sector than in the subsistence sector. In the latter, the wood would remain at the mercy of peasant grazing pressures that were apparently not regulated with the survival of the wood in mind, and could easily bring about its degradation and demise, particularly in a situation of population expansion. In the commercial sector, the rewards from timber, charcoal and tanbark were large enough to ensure that attention was paid to the ongoing production of wood on the site.

Nevertheless, in a typical coppice regime of the time, biodiversity and authenticity were modified to a degree. The structure and character of many of the west-coast and Perthshire oakwoods were changed by rigorously weeding out other tree species and, more radically, by planting up gaps in the wood or extending them beyond the existing boundaries: often the acorns used were not of local provenance and may not always have been of the same species. It is also likely enough, as Susan Barker found at Coniston Water in the Lake District, that a long period of removal of wood products accompanied by spells of grazing led to nutrient impoverishment of much of the area: at Coniston, the most biodiverse areas were the inaccessible gorges that transected the woods, because these had suffered less from the axe and the tooth.[6] The minister of Clunie in Perthshire was noting diminishing returns from the coppice woods in 1792, blaming uncontrolled grazing by cattle and poor management of the stools, and also observing that ground which produces trees, like that which produces other exhaustive crops, 'must, in a certain number of years, become wasted and fatigued, and consequently must require a certain period of repose'.[7] Nevertheless, the systems of coppice management used in Scotland (and England) were more benign than those used in parts of Germany, where felling was followed by burning the underbrush, turning the surface between the stumps and taking a one-year catch crop of rye before the trees grew up again, a cauterisation of the woodland floor and its flora that had no parallel in Britain, except beneath the charcoal and bloomery stances themselves.[8]

In the nineteenth century, grazing pressures became crushingly heavy over all the hillside; markets first encouraged and then wiped out the commercial use of wood produce from enclosed woods; and forestry developed as a much more intrusive science than the old woodmanship. The management of woods was transformed. In the uplands, much was effectively abandoned to sheepwalk or used as a one-off resource before conversion into hill pasture or plantation. Many of the ancient semi-natural pinewoods had a particularly dramatic history when for the first time they went through an episode of clear-fell in the three decades before 1830, though they proved remarkably resilient and mostly returned to cover the same ground as before, their ecosystem basically intact. Nevertheless, aspects of their biodiversity suffered, as demonstrated by the extinction or near-extinction of the goshawk, the great spotted woodpecker, the red squirrel and the pine marten. The capercaillie had gone before (in the 1770s) for reasons that are obscure and probably not primarily related to the management of the woods. That all the missing birds and mammal species subsequently either found their own way back or were reintroduced testifies to the basic health of the ecosystem, but certain less conspicuous invertebrates and some of the flora may have suffered permanent damage. Some of the ants are very patchily distributed now (Map 15.1 and Fig. 15.1), and

---

[6] S. Baker, 'The history of the Coniston woodlands, Cumbria, UK', in K. J. Kirby and C. Watkins (eds), *The Ecological History of European Forests* (Wallingford, 1998), pp. 167–84.
[7] Quoted in C. Dingwall, 'Coppice management in Highland Perthshire', in T. C. Smout (ed.), *Scottish Woodland History* (Edinburgh, 1997), p. 163.
[8] E. Westermann, 'Central European forestry and mining industries in the early modern period', in Cavaciocchi, *L'uomo e la foresta*, p. 940.

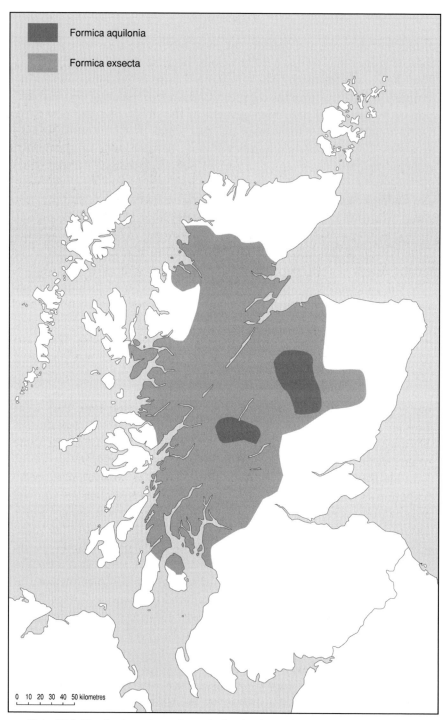

Map 15.1 Distribution of wood ants in Scotland. Crown copyright. All rights reserved. Forestry Commission.

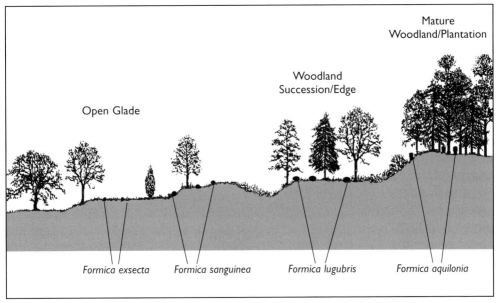

Fig. 15.1 Niche preferences of wood ants. Each species has slightly different requirements, and only a functioning woodland ecosystem can provide them all.
Reproduced by kind permission of Jonathan Hughes.

much of the distinctive pinewood flora common in Scandinavia in similar woods is rare and isolated in Scotland. Plants like *Linnaea borealis* and *Moneses uniflora* are known to be susceptible to clear-fell, and may have been more plentiful when an informal continuous-cover forestry was practised.

In the broadleaf woods, management under modern pressure either became more radical or was abandoned altogether. Coppicing was gradually abandoned before 1914, and the trees allowed to grow on as high forest. This would lead to changes in biodiversity and losses among light-loving plants of the woodland floor and the insects that depended on them, like fritillary and chequered skipper butterflies. The increase in shade would be especially significant in woods where former glades had been planted up earlier in the tanbark and charcoal boom. Many broadleaf woods were now given over to game coverts, with laurel, snowberry or rhododendron introduced, smothering the original flora with no concomitant gain. This was often combined with planting non-native trees, beech, larch, sycamore and Norway spruce, and Scots pine in the Lowlands, not necessarily with biodiversity loss (some birds like goldcrests, coal tits and bramblings would benefit) but certainly at a cost to the authenticity and natural feel of the wood. More radical still was any decision that all the broadleaves be cut out and the ground replanted with larch, pine and Norway spruce. The great age of Sitka spruce and lodgepole pine as commercial trees lay in the future, though they were being raised and studied in arboreta and their good qualities for the forester beginning to be noted before the First World War.

So the period ends with sustainability in the management of semi-natural woods very heavily in retreat, many such woods having been given over to moorland and rough pasture (occasionally to arable), or having suffered the total or partial replacement of their native trees by exotic ones. By 1911 there were already isolated voices heard in praise of the ecological and historic interest of the ancient woods, but more influential and numerous were the voices of the radical forestry lobby that would replace them with something capable of producing a fast-growing crop of timber and therefore supposedly more profitable. The future looked bleak, but it was to become bleaker still before it improved.

It is also worth enquiring how Scottish woodland history compares in its broad perspective to that of two of our nearest neighbours, Ireland to the west and Denmark to the east. There are a lot of similarities. Both attained a woodland maximum 5,000 to 6,000 years ago. Ireland clearly suffered like Scotland the consequences of increased oceanicity at the onset of the Bronze Age, as the Atlantic fringe became much less hospitable to tree growth.[9] Estimates of their forest cover in the sixteenth century, however rough, make it clear that, as in Scotland, it had very greatly reduced by the end of the Middle Ages: the estimate for Denmark is 20 to 25 per cent afforested, for Ireland about 12 to 13 per cent or less.[10] By the second half of the eighteenth century it is 8 to 10 per cent for Denmark (much as for Scotland) but perhaps under 2 per cent for Ireland.[11] In all three countries a plantation forestry movement gathered pace from the eighteenth century, progressing furthest in the nineteenth century in Denmark, and in the twentieth century resulting in a substantial increase in forest cover in all three countries. Little of this in Ireland could be described as ancient semi-natural woodland by the twentieth century. In Denmark, apart from one or

---

[9] For general studies, see J. R. Pilcher and S. Mac an tSaoir, *Woods, Trees and Forests in Ireland* (Dublin, 1995); E. Neeson, *A History of Irish Forestry* (Dublin, 1991); E. McCracken, *The Irish Woods Since Tudor Times: their Distribution and Exploitation* (Newton Abbot, 1971); B. Fritzbøger, *Kulturskoven: Dansk Skovbrug fra Oldtid til Nytid* (Copenhagen, 1994).

[10] T. Kjærgaard, *The Danish Revolution, 1500–1800: an Ecohistorical Interpretation* (Cambridge, 1994), p. 20; McCracken, *Irish Woods*, p. 15. The extent of Irish woods has been much disputed; Rackham would put the figure ca. 1600 as low as 3 per cent, but Nicholls has strongly disputed the minimalist view. O. Rackham, *The History of the Countryside* (London, 1986), pp. 112–16; K. Nicholls, 'Woodland cover in pre-modern Ireland', in P. J. Duffy, D. Edwards and E. FitzPatrick (eds), *Gaelic Ireland c.1250–c.1650: Land, Lordship and Settlement* (Dublin, 2001), pp. 181–206.

[11] Kjærgaard, *Danish Revolution*, p. 20; McCracken, *Irish Woods*, p. 15.

two woods like Rold Skov in Jutland, none of the old woods were comparable in authenticity and naturalness to the best of the Scottish pine or oak woods, though probably Denmark still held at least as much wood of ancient semi-natural origin as Scotland.

While the broad profiles there have many similarities, the three countries also demonstrate significant differences. In Ireland the near-total destruction of woodland was brought about, at least in the seventeenth century, by the outsider, in this case by the landowners of the invading British colonial power. The woods were cleared away partly for military reasons (the dread of the forest remaining as a refuge for guerrilla 'wood-kernes'), partly as a response to market opportunity for timber and iron.[12] Of about 150 recorded Irish ironworks, a very few survived for a century or longer, implying sustainable use in their vicinity until the nineteenth century.[13] Most were very short lived, suggesting low margins of profitability that could not run to sustainable practices, and a wish by proprietors to consider, as Sir William Petty in Co. Kerry put it in 1672, 'how we may destroy the woods... how we may engage the iron works to take them off presently'.[14] But, as Rackham states, 'Irish trees grow again, just as English ones do': they were not allowed to do so by the farmers as well as by the landlords, due to the extraordinary growth of Irish rural population and their animals between the early seventeenth century and the mid-nineteenth, and to the fact that abundant peat meant wood was not needed as fuel.

Fraser Darling, writing of his interpretation of Scottish woodland history, believed that outsiders were the villains of the piece, and linked this to a general point about the development of an external market in resource use:

> Man does not seem to extirpate a feature of his environment as long as that natural resource is concerned only with man's everyday life: but as soon as he looks upon it as having some value for export – that he can live by selling it to some distant market – there is real danger.[15]

The history of Ireland might be read as giving at least some support to those generalisations, but the history of Scotland does not. Outsiders

---

[12] T. C. Smout, 'Energy rich, energy poor: Scotland, Ireland and Iceland 1600–1800', in D. Dickson and C. Ó Gráda (eds), *Refiguring Ireland: Essays in Honour of L. M. Cullen* (Dublin, 2003), pp. 32–4.
[13] Rackham, *History of the Countryside*, p. 116.
[14] T. C. Barnard, 'Sir William Petty as Kerry ironmaster', in *Proceedings of the Royal Irish Academy*, 82 C (1982), p. 31.
[15] F. F. Darling, *Natural History in the Highlands and Islands* (London, 1947), p. 59.

involved in exploiting the Scottish woods such as the Irish and English adventurers in the pinewoods, were almost never successful or persistent: the most successful, the Hull merchants at Glenmore towards the close of the eighteenth century, superficially caused devastation but actually left behind them a site that could and did grow another native wood. Those who involved themselves in the oakwoods as ironmasters, when they were successful like the Cumbrian smelters in Argyll after 1750, introduced superior methods of management that were widely copied in the Highlands. But most Scottish woods were left to the management skills and inclinations of Scotsmen, who, when they scented a market generally took better care of their woods than before. The biggest threat to the Scottish woods came in the nineteenth century, not from exports but from a new intensity of imports, which, by destroying the market that had been there before, also demolished much of the previous incentive to treating the woods sustainably.

The comparison with Denmark is also instructive, not least because the evidence for much of the detail of wood management (this does not survive for Ireland) includes many close parallels to Scotland. For example, the Danes, like the seventeenth-century Highland Scots, believed that trees were best cut with attention to the moon, though they seemed to hold this belief for longer. They used their woods extensively for grazing, so that, as in Scotland, the distinction between wood and wood pasture is hard to make before the nineteenth century. The best grass hay meadows in Denmark were found in a mosaic among the woods, and animals were only banned from woodland when modern winter feed such as turnips became available. The Danes made little or no use of leaf hay after the end of the Middle Ages (again paralleling Scotland, though we have little information about Scottish medieval foddering), though their neighbours to the north and south continued to do so.[16]

In other respects, however, Danish and Scottish practice was fundamentally different. Though both societies since Neolithic times had used wattle as a major element in building form, coppicing in Denmark remained until the nineteenth century relatively casual and informal, with little of the organised cutting of enclosed haggs in rotation, systematically leaving standards, which characterised England and came to characterise much of Scotland. The pattern of ownership in Danish woods precluded this kind of development. Most Danish woods were described in medieval and early modern law as 'common woods', in which the lords had the right to take the mature, high-stemmed oak and beech trees (*overskov*) and the peasants all the other trees and bushes (*underskov*). Except in a minority of private

---

[16] Fritzbøger, *Kulturskoven*, pp. 125–52, 156–7.

woods where these rules did not apply, no-one had enough control over enough ground to instigate the kind of coppice regimes common in Britain. Cutting for wattle remained much more haphazard and was only permitted in winter.[17]

A system of split ownership rights could only work well when woodland was plentiful. It became a recipe for conflict and chaos when wood grew scarce, as it did in the course of the seventeenth and eighteenth centuries. As the price of wood fuel rose (there was little good quality peat and no coal in Denmark), peasants took care that their young *underskov* did not become the lord's high *overskov*, and the lords in return accused the peasants of 'destroying' the wood, though in fact there is little evidence of the area under common wood declining as drastically as the lords alleged. Rather, less grew to high forest.[18]

The upshot was the Danish forest law of 1805, by which the lords secured ownership of the entire wood in return for compensation to the peasants, who were given land in lieu of the forfeited grazing and rights to *underskov*. In the short run this led to a reduction in the forest area, as woodland that had consisted before primarily of *underskov* was converted to grazing or arable. In a longer perspective it enabled modern forestry, inspired both by German and English example, to take hold.[19] Danish tanbark became a major competitor to domestic British supplies, before both were supplanted by modern substitutes.

The legal contrast with Scotland could hardly be starker. Here, the landowner was absolute, and the tenants had few rights. This put formal responsibility for stewardship of the woods squarely on the landowner. If he behaved with good judgement, as many did, the credit is his; if with poor judgement, like the Earl of Breadalbane in his failure to monitor and control the Irish partnership, the blame is likewise his. Tenants, being deprived of the ultimate control, are likewise absolved from the ultimate responsibility. Such a system, for better or worse, enabled new courses to be consciously charted and decisively followed, whereas the Danish regime tended towards drift and the continuation of custom.

In practice, however, management of a wood in Scotland, as in Denmark and elsewhere, often had to be a compromise, with recognition of common interest on both sides. The peasants might complain that a landlord shut them out of a wood, but the latter knew he must provide access to alternative timber and grazing or endure a reduction of rent. The lord might complain

---

[17] *Ibid.*, pp. 37, 153–7, 313–30.
[18] *Ibid.*, pp. 39–60.
[19] *Ibid.*, pp. 270–92.

of peasants stealing and cutting in his property, but the tenants knew they had to come to an accommodation with forester and factor and rub along together. Danger to the continued existence of an ancient semi-natural wood might indeed come about through unilateral action by the landlord, if he felled it through pressing financial need and failed to replant, or if he allowed the peasants' animals to eat the regrowth through carelessness or indifference, or if he sensed a more profitable use of the land from rising rents when the external market rewarded the sheep or cattle grazier. In a different way, a semi-natural wood might perish if the landlord was an enthusiastic nineteenth-century or twentieth-century forester, persuaded to erase the present native trees and replace them with exotics. More often in earlier centuries, though, woods perished by consent, when misuse went unrecognised by either side until it was too late, as in Strathnaver in the late eighteenth century, or when both sides could see no merit in leaving the ground under trees, as when sheep-farmers cleared out the scrub of Lochaber and Argyll in the early nineteenth century. Woodland management has usually been driven by the exigencies of people making a living in the short run, though sometimes its consequences might be modified by accidents of location, or the vision of a family owner with a sense of stewardship for more than one generation. Unless the short-term pressures were compelling, the woods would no doubt benefit by community and landlord awareness that timber and shelter would be needed in the future as it was today, for all the normal processes of husbandry; but even this cautionary sense was eroded when imports became everywhere cheap and accessible in the nineteenth century. It is hard to detect any other ethos, any land ethic or environmental consciousness of respect for the woods. If such a thing did exist historically, it is impossible to see that it prevailed against materialistic and utilitarian forces.

Things may be different in the future. To ensure the integrity of our ancient native woods they will have either to yield an income directly to their owners in their present condition, to attract public funds for their maintenance, or to fall into the hands of charities able to keep them in perpetuity. Charities, however, are only able to own a few, public funds are under competing pressures and native wood produce is presently worth little. An informed altruism, a new environmental consciousness by the public that will swell the coffers of the charities, put pressure on the government to pay, and be shared by owners themselves may, after all, be the best hope. The old native woods are too precious to lose, but it will take people who care about the history and biodiversity that they contain, to make a strong effort of will and money to save them. Understanding is the foundation for action. If we have shown in our book how the native woods are both nature and culture, their story entangled with ours just like that of historic buildings and archaeological monuments, and just as worthy of care, then we shall rest content.

Bibliography

For a list of archives consulted, please see Acknowledgements, p. vii.

Abbreviations

APS = *Acts of the Parliaments of Scotland*, in 12 volumes (Edinburgh, 1844–1875).

CSP Dom = *Calendar of State Papers (Domestic)*, volumes covering the period 1603–9 (HMSO, 1867–1972).

DNB = *Dictionary of National Biography*, ed. L. Stephen and S. Lee in 63 volumes (London, 1885–1905).

NSA = *New Statistical Account of Scotland* [ed. J. Gordon], in 15 volumes (Edinburgh, 1845).

OSA = J. Sinclair (ed.), *Statistical Account of Scotland* (Edinburgh, 1791–9) in 21 volumes; reissued and rearranged by county in 20 volumes by D. J. Withrington and I. R. Grant (Wakefield, 1983). Our references are to the first edition.

RMS [*Registrum Magni Sigilli*] = *The Register of the Great Seal of Scotland* (first volume issued 1814, subsequent volumes reprinted Edinburgh, 1984).

RPCS = *The Register of the Privy Council of Scotland*, three series of volumes (HMSO, 1877–1967).

TRHAS = *Prize Essays and Transactions of the Highland and Agricultural Society*, later *Transactions of the Royal Highland and Agricultural Society*, in several series.

TRSAS = *Transactions of the Royal Scottish Arboricultural Society*, continuation of *TSAS*.

TSAS = *Transactions of the Scottish Arboricultural Society*, continued as *TRSAS*.

Other Printed Works Consulted

[Anon.], (n.d.), *The Case of the Tanners of Ireland Briefly Represented* (n.p.), National Library of Ireland, P.4034.

[Anon.], (1913), 'Prices of home timber', *TRSAS*, 27, p. 223.

[Anon.], (n.d., but 1914), *Poacher's Guide* (n.p.): a copy in St Andrews University Library.

[Anon.], (ed.) (1997), 'Ane fair pallice of greine tymber', *Reforesting Scotland*, 16, pp. 33–4.

Aalen, F. H. A. (ed.) (1996), *Landscape Study and Management*, Dublin.

Adam, R. J. (ed.), (1960), *Home's Survey of Assynt*, Scottish History Society, Edinburgh.

Adam, R. J. (ed.) (1972), *Papers on Sutherland Estate Management, 1802–1816*, Scottish History Society, Edinburgh.

Adam, R. J. (ed.) (1991), *The Calendar of Fearn*, Scottish History Society, Edinburgh.

Albion, R. G. (1926), *Forests and Seapower: the Timber Problem of the Royal Navy, 1652–1862*, Cambridge, MA.

Aldhous, J. R. (ed.) (1995), *Our Pinewood Heritage*, Farnham.

Anderson, M. L. (1967), *A History of Scottish Forestry*, 2 vols, Edinburgh.

Anson, P. F. (1930), *Fishing Boats and Fisher Folk on the East Coast of Scotland*, London.

Armet, C. M. (ed.) (1953), *Kirkcudbright Sheriff Court Deeds, 1676–1700*, 2 vols, Edinburgh.

Armit, I. and Ralston, I. (2003), 'The coming of iron', in T. C. Smout (ed.), *People and Woods: a History*, Edinburgh, pp. 40–59.

Atterson, J. and Ross, I. (2002), 'Man and woodlands', in C. Gimingham (ed.), *The Ecology, Land Use and Conservation of the Cairngorms*, Chichester, pp. 120–9.

Austad, I. (1989), 'Tree pollarding in western Norway', in H. H. Birks *et al.* (ed.), *The Cultural Landscape, Past, Present and Future*, Cambridge, pp. 31–45.

Badenoch, C. O. (1994–5), 'Border woodlands I – Berwickshire' and 'Border woodlands II – Roxburghshire', *History of the Berwickshire Naturalists Club*, 46, pp. 115–22 and 272–86.

Baillie, M. and Stell, G. (1993), 'The Great Hall and roof of Darnaway Castle, Moray', in W. D. H. Sellar (ed.), *Moray: Province and People*, Scottish Society for Northern Studies, 163–86.

Baldwin, J. (ed.) (1994), *Peoples and Settlement in North-west Ross*, Edinburgh.

Baker, S. (1998), 'The history of the Coniston woodlands, Cumbria, UK', in K. J. Kirby and C. Watkins (eds), *The Ecological History of European Forests*, Wallingford, pp. 167–84.

Bangor-Jones, M. (2002), 'Native woodland management in Sutherland: the documentary evidence', *Scottish Woodland History Discussion Group Notes*, 7, pp. 1–5.

Barber, K. E., Dumayne, L. and Stoneham, R. (1993), 'Climatic change and human impact during the late Holocene in northern Britain', in F. M. Chambers (ed.), *Climate Change and Human Impact on the Landscape*, London, pp. 226–36.

Barlow, Professor (1841), 'On the strength of the fir in the Forest of Mar', *TRHAS*, 13, p. 122.

Barnard, T. C. (1982), 'Sir William Petty as Kerry Ironmaster', in *Proceedings of the Royal Irish Academy*, 82 C, pp. 1–32.

Barrow, G. W. S. (1999), 'The background to medieval Rothiemurchus', in T. C. Smout and R. A. Lambert (eds), *Rothiemurchus: Nature and People on a Highland Estate, 1500–2000*, pp. 1–6.

Bartholomew, A., Malcolm, D. C., and Nixon, C. J. (2001), 'The Scots pine population at Glen Loyne, Inverness-shire: present conditions and regenerative capacity', *Scottish Forestry*, 55, pp. 141–8.

Baxter, R. (1878), 'On the best kinds of wood for charcoal and processes of charring', *TSAS*, 8, pp. 246–9.

Bayfield, N. G. and Conroy, J. W. H. (eds) (1995), *Cairngorms Assets Review*, Cairngorm Partnership, n.p.

Bechman, R. (1990), *Trees and Man: the Forest in the Middle Ages* (trans. K. Dunham), New York.

Bennet, K. D. (1994), 'Post-glacial dynamics of pine', in J. R. Aldhous (ed.), *Our Pinewood Heritage*, Farnham, pp. 23–39.

Bennet, K. D. (1996), 'Late Quaternary vegetation dynamics of the Cairngorms', *Botanical Journal of Scotland*, 48, pp. 51–63.

Bil, A. (1990), *The Shieling, 1600–1840*, Edinburgh.

Birks, H. H., et al. (ed.) (1989), *The Cultural Landscape, Past, Present and Future*, Cambridge.

Blackwood, R. C. (1955), 'An estate's forest history – Eliock, Dumfriesshire', *Scottish Forestry*, 9, pp. 13–21.

Blaeu, J. (1654), *Theatrum Orbis Terrarum Sine Atlas Novus*, 5, Amsterdam.

Boutcher, W. (1775), *A Treatise on Forest Trees*, Edinburgh.

Bowman, J. E. (1986), *The Highlands and Islands, a Nineteenth-century Tour*, Gloucester.

Breeze, D. J. (1992), 'The great myth of Caledon', *Scottish Forestry*, 46 (1992), pp. 331–5, reprinted in T. C. Smout (ed.), *Scottish Woodland History*, Edinburgh, 1997, pp. 47–51.

Breeze, D. J. (1996), *Roman Scotland*, London.

Brown, J. (edn 1882), *The Forester*, Edinburgh.

Brown, P. Hume (ed.) (1891), *Early Travellers in Scotland*, Edinburgh.

Brown, P. Hume (ed.) (1892), *Tours in Scotland, 1677 and 1681, by Thomas Kirk and Ralph Thoresby*, Edinburgh.

Brown, P. Hume (ed.) (1893), *Scotland Before 1800 from Contemporary Documents*, Edinburgh.

Buchanan, W. (1723), *An Inquiry into the Genealogy and Present State of Ancient Scottish Surnames*, Edinburgh.

Buckland, P. C. (2003), 'Holocene woodland history: a palaeoentomological perspective', paper given to conference at Sheffield Hallam University, 'Working and walking in the footsteps of ghosts', May/June 2003.

Bunce, R. G. H. and Jeffers, J. N. R. (eds) (1977), *Native Pinewoods of Scotland*, Cambridge.

Bunting, M. J. (1996), 'The development of heathland in Orkney', *Holocene*, 6, pp. 193–212.

Burns, J. W. (ed.) (1882), *Miscellaneous Writings of John Spreull*, Glasgow.

Burt, E. (1754), *Letters from a Gentleman in the North of Scotland*, 2 vols, London.

Camden, W. (edn 1806), *Britannia: or a Chorographical Description of the Flourishing Kingdoms of England, Scotland and Ireland*, London.

Campbell, C., Tipping, R. and Cowley, D. (2002), 'Continuity and stability in past upland land uses in the western Cheviot Hills, southern Scotland', *Landscape History*, 24, pp. 111–19.

Carlisle, A. (1977), 'Impact of man on the native pinewoods', in R. G. H. Bunce and J. N. R. Jeffers (eds), *Native Pinewoods of Scotland*, Cambridge, pp. 70–7.

Carmichael, A. (1884), 'Grazing and agrestic customs of the Outer Hebrides', appendix to *The Report of the Royal Commissioners of Inquiry on the Condition of Crofters and Cottars in the Highlands of Scotland* (Parliamentary Papers, 1884), 1, pp. 451–82.
Carmichael, A. (1928), *Carmina Gaedelica*, 2 vols, Edinburgh.
Caseldine, C. and Hatton, J. (1993), 'The development of high moorland on Dartmoor: fire and the influence of Mesolithic activity on vegetation change', in F. M. Chambers (ed.), *Climate Change and Human Impact on the Landscape*, London, pp. 119–31.
Cavaciocchi, S. (ed.) (1996), *L'uomo e la foresta secc. XIII–XVIII*, Prato.
Chalmers, G. (edn 1887), *Caledonia, or a History and Topographical Account of North Britain from the Most Ancient to the Present Times*, Paisley.
Chambers, F. M. (ed.) (1993), *Climate Change and Human Impact on the Landscape*, London.
Cheape, H. (1993), 'Woodlands on the Clanranald estates', in T. C. Smout (ed.), *Scotland since Prehistory: Natural Change and Human Impact*, Aberdeen, pp. 50–63.
Cheape, H. (1995), '"A few summer shielings or cots": material culture and buildings of the '45', *Clan Donald Magazine*, 13 (1995), p. 68.
Clapham, J. H. (1939), *An Economic History of Modern Britain*, 2 vols, Cambridge.
Clarkson, L. A. (1971), *The Pre-Industrial Economy in England, 1500–1750*, London.
Clough, M. (1994), 'Early fishery and forestry development on the Cromartie estate of Coigach, 1660–1746', in J. Baldwin (ed.), *Peoples and Settlement in North-west Ross*, Edinburgh, pp. 230–43.
Cochran, L. E. (1985), *Scottish Trade with Ireland in the Eighteenth Century*, Edinburgh.
Cockburn, H. (1888), *Circuit Journeys*, Edinburgh.
Collins, E. J. T. (1996), 'The wood-fuel economy of eighteenth century England', in S. Cavaciocchi (ed.), *L'uomo e la foresta secc. XIII–XVIII*, Prato, pp. 1097–121.
Coppins, A., Coppins, B. and Quelch, P. (2002), 'Atlantic hazelwoods: some observations on the ecology of this neglected habitat from a lichenological perspective', *British Wildlife*, 14, pp. 17–26.
Corbet, G. B. (ed.) (1998), *The Nature of Fife*, Edinburgh.
Cordiner, C. (1780), *Antiquities and Scenery of the North of Scotland*, Banff.
Crawford, B. E. (1995), *Earl and Mormaer: Norse-Pictish Relationships in Northern Scotland*, Rosemarkie.
Crawford, R. and Imlah, M. (eds) (2000), *The New Penguin Book of Scottish Verse*, Harmondsworth.
Crawford, R. M. M. (1997), 'Oceanicity and the ecological disadvantages of warm winters', *Botanical Journal of Scotland*, 49, pp. 205–21.
Crawford, R. M. M. (2000), 'Ecological hazards of oceanic environments', *New Phytologist*, 147, pp. 257–81.
Cregeen, E. R. (ed.) (1964), *Argyll Estate Instructions: Mull, Morvern, Tiree, 1771–1805*, Scottish History Society, Edinburgh.

Crone, A. and Mills, C. M. (2002), 'Seeing the wood and the trees: dendrochronological studies in Scotland', *Scottish Woodland History Discussion Group Notes*, 7, pp. 14–22.

Crone, A. and Watson, F. (2003), 'Sufficiency to scarcity: medieval Scotland 500–1600', in T. C. Smout (ed.), *People and Woods in Scotland: a History*, Edinburgh, pp. 60–81.

Crozier, J. D. (1910), 'Sitka spruce as a tree for hill planting and general afforestation', *TRSAS*, 23, pp. 7–16.

Cruikshank, G. (ed.) (1988), *A Sense of Place: Studies in Scottish Local History*, Edinburgh.

Cullen, L. (1968), *Anglo-Irish Trade 1600–1800*, Manchester.

Cunliffe, B. W. (1985), 'Man and landscape in Britain, 6000 BC–AD 400', in S. R. J. Woodell (ed.), *The English Landscape, Past, Present and Future*, Oxford, pp. 54–67.

Cunningham, I. C. (2001), *The Nation Survey'd: Timothy Pont's Maps of Scotland*, East Linton.

Daly, H. E. (1991), *Steady State Economics*, Washington, DC.

Darling, F. F. (1947), *Natural History in the Highlands and Islands*, London.

Darling, F. F. (1949), 'History of the Scottish forests', reprinted in *Reforesting Scotland*, 7 (1992), pp. 25–7.

Darling, F. F. (1956), *Pelican in the Wilderness*, London.

Darling, F. F. (1968), 'The ecology of land use in the Highlands and Islands', in D. S. Thomson and I. Grimble (eds), *The Future of the Highlands*, London, pp. 27–56.

Darwin, T. (1994), 'Sacred trees in Scottish folklore', *Reforesting Scotland*, 10, pp. 8–10.

Davies, I., Walker, B., Pendlebury, J. (2002), *Timber Cladding in Scotland*, Edinburgh.

Defoe, D. (1704), *The Storm*, London.

Department of the Environment (1994), *Biodiversity: the UK Action Plan: Summary Report*, London.

Dickson, D. and Ó Gráda, C. (eds) (2003), *Refiguring Ireland: Essays in honour of L. M. Cullen*, Dublin.

Dickson, J. H. (1993), 'Scottish woodlands: their ancient past and precarious future', *Scottish Forestry*, 47, pp. 73–8.

Dingwall, C. (1997), 'Coppice management in Highland Perthshire', in T. C. Smout, *Scottish Woodland History*, Edinburgh, pp. 162–71.

Ditchburn, D. (1990), 'A note on Scandinavian trade with Scotland in the later Middle Ages', in G. G. Simpson (ed.), *Scotland and Scandinavia 800–1800*, Edinburgh, pp. 73–85.

Dixon, G. A. (1975), 'William Lorimer on forestry in the Central Highlands in the early 1760s', *Scottish Forestry*, 29, pp. 191–210.

Dixon, G. A. (1976), 'Forestry in Strathspey in the 1760s', *Scottish Forestry*, 30, pp. 38–60.

Dixon, J. H. (1886), *Gairloch in North-west Ross-shire*, Edinburgh.

Dixon, P. (2003), *Puir Labourers and Busy Husbandmen*, Edinburgh.

Dodgshon, R. A. (1981), *Land and Society in Early Scotland*, Oxford.

Dodgshon, R. A. (2004), 'The Scottish Highlands before and after the clearances: an ecological perspective', in I. D. Whyte and A. Winchester (eds), *Upland Landscapes* (Society for Landscape Studies).

Donaldson, J. (1794), *General View of the Agriculture of the County of Nairn*, London.

Dougall, M. and Dickson, J. (1997), 'Old managed oaks in the Glasgow area', in T. C. Smout (ed.), *Scottish Woodland History*, Edinburgh, pp. 76–85.

Dow, F. D. (1979), *Cromwellian Scotland, 1651–1660*, Edinburgh.

Dudley, N., Stolton, S. and Jeanrenaud, J.-P. (1993), *Towards a Definition of Forest Quality, a WWF Colloquium*, Godalming.

Dudley, N. (1996), 'Authenticity as a means of measuring forest quality', *Biodiversity Letters*, 3, pp. 6–9.

Duffy, P. J., Edwards, D. and FitzPatrick, E. (eds) (2001), *Gaelic Ireland c.1250 – c.1650: Land, Lordship and Settlement*, Dublin.

Dumayne, L. (1993), 'Iron Age and Roman vegetation clearance in northern Britain: further evidence, *Botanical Journal of Scotland*, 46, pp. 385–92.

Dunlop, B. M. S. (1997), 'The woods of Strathspey in the nineteenth and twentieth centuries', in T. C. Smout (ed.), *Scottish Woodland History*, Edinburgh, pp. 176–89.

Dunlop, J. (1978), *The British Fisheries Society, 1786–1893*, Edinburgh.

Edlin, H. L. (1955), 'Coppice-with-standards and the oak tanning bark trade in the Scottish Highlands', *Scottish Forestry*, 9, pp. 145–8.

Edlin, H. L. (1959), 'Place names as a guide to former forest cover in the Grampians', *Scottish Forestry*, 13, pp. 63–7.

Edward, R. (edn 1883), *The County of Angus, 1678*, Edinburgh.

Englund, C. (1996), 'Woodland management on the forfeited estate of Perth, 1746 to 1784', unpublished University of St Andrews MA dissertation.

Erskine, J. F. (1795), *General View of the Agriculture of the County of Clackmannan*, Edinburgh.

Fell, A. (1908), *The Early Iron Industry of Furness and District*, Ulverston.

Fenton, A. (1976), *Scottish Country Life*, Edinburgh.

Fenton, J. (1997), 'Native woods in the Highlands: thoughts and observations', *Scottish Forestry*, 51, pp. 160–4.

Fife, H. (1994), *Warriors and Guardians: Native Highland Trees*, Glendaruel.

Firth, C. H. (ed.) (1895), *Scotland and the Commonwealth*, Scottish History Society, Edinburgh.

Foot, D. (2003), 'The twentieth century: forestry takes off', in T. C. Smout (ed.), *People and Woods in Scotland: a History*, Edinburgh, pp. 158–94.

Foster, S. and Smout, T. C. (eds) (1994), *The History of Soils and Field Systems*, Aberdeen.

Fowler, J. (2002), *Landscape and Lives: the Scottish Forest Through the Ages*, Edinburgh.

Fraser, W. (1883), *The Chiefs of Grant*, 3 vols, Edinburgh.

Fritzbøger, B. (1994), *Kulturskoven: Dansk Skovbrug fra Oldtid til Nytid*, Copenhagen.

Fuller, R. J. and Warren, M. S. (1995), 'Management for biodiversity in British woodlands – striking a balance', *British Wildlife*, 7, pp. 26–37.

Gibson, A. J. S. and Smout, T. C. (1995), *Prices, Food and Wages in Scotland, 1550–1780*, Cambridge.

Gilbert, J. M. (1979), *Hunting and Hunting Reserves in Medieval Scotland*, Edinburgh.
Gilchrist, A. (1872), 'On the pruning of oak coppice', *TRHAS*, 4th series, 4, pp. 197–204.
Gilchrist, A. (1874), 'On the treatment and management of oak coppice in Scotland', *TRHAS*, 4th series, 6, pp. 118–32.
Gilchrist, A. (1876), 'On natural coppice wood, of other species than oak', *TRHAS*, 4th series, 8, pp. 210–19.
Gilpin, W. (ed. T. D. Lauder, 1834), *Remarks on Forest Scenery and Other Woodland Views*, 2 vols, Edinburgh.
Gimingham, C. (ed.) (2002), *The Ecology, Land Use and Conservation of the Cairngorms*, Chichester.
Gladstone, J. (1961), 'The natural woodlands of Galloway and Nithsdale', *Forestry*, 34, 174–80.
Goodare, J. (1999), *State and Society in Early Modern Scotland*, Oxford.
Gordon, G. P. (1911), 'Primitive woodland and plantation types in Scotland, in *TRSAS*, 24 (1911), pp. 153–77.
Grant, A. (1984), *Independence and Nationhood: Scotland 1306–1469*, London.
Grant, E. (edn 1978), *Memoirs of a Highland Lady*, London.
Grant, E. (1991) (ed. P. Pelly and A. Tod), *The Highland Lady in Ireland: Journals, 1840–50*, Edinburgh.
Grant, E. (1994), *Abernethy Forest: its People and its Past*, Arkleton Trust, Nethy Bridge.
Grant, I. D. (1978), 'Landlords and land management in north-eastern Scotland, 1750–1850', unpublished University of Edinburgh Ph.D. thesis, 2 vols.
Grant, I. F. (1930), *The Social and Economic Development of Scotland before 1603*, Edinburgh.
Grant, I. F. (edn. 1995), *Highland Folk Ways*, Edinburgh.
Grigor, J. (1839), 'Report on the native pine forests of Scotland', *TRHAS*, 2nd series, 12, pp. 122–31.
GUARD (1996), 'Scottish Bloomeries Project: Interim Report'.
Gunn, W. F. (1885), 'The woods, forests and forestry of Ross-shire', *TRHAS*, 4th series, 17, pp. 133–202.
Hæggström, C.-A. (2003), 'Report on Scottish wood pastures visited 17–20 March 2003', unpublished paper for the Woodland Trust.
Haldane, A. R. B. (1952), *The Drove Roads of Scotland*, Edinburgh.
Hall, W. T. (1954), 'Utilisation and exploitation technique in Scottish forestry', *Scottish Forestry*, 8, pp. 176–90.
Hambler, C. and Speight, M. R. (1995), 'Biodiversity conservation in Britain: science replacing tradition', *British Wildlife*, 6, pp. 137–47.
Hamilton, H. (1963), *An Economic History of Scotland in the Eighteenth Century*, Oxford.
Hannigan, K. and Nolan, W. (1994), *Wicklow, History and Society: Interdisciplinary Essays on the History of an Irish County*, Dublin.
Hay, G. D. (1973), 'The cruck-building at Corrimony, Inverness-shire', *Scottish Studies*, 17, pp. 127–33.
Henderson, D. M. and Dickson, J. H. (eds) (1994), *A Naturalist in the Highlands: James Robertson, His Life and Travels in Scotland, 1767–1771*, Edinburgh.

Henderson, J. (1812), *General View of the Agriculture of the County of Sutherland*, London.
Highland, A. (1989), 'Gateside Mills: the Scottish bobbin and shuttle trade in the British and international setting, 1860–1960', unpublished University of St Andrews Ph.D. thesis.
Hodges, H. (1964), *Artifacts: an Introduction to Early Materials and Technology*, London.
Holdgate, M. (1995), 'How can development be sustainable?', *Royal Society of Edinburgh Journal*, 143, p. 25.
House, S. and Dingwall, C. (2002), '"A Nation of Planters": introducing the new trees, 1650–1900', in T. C. Smout (ed.), *People and Woods in Scotland: a History*, Edinburgh, pp. 128–57.
Humphrey, J. and Quine, C. (2001), 'Sitka spruce plantations in Scotland: friend or foe to biodiversity?', in *Alien Species: Friends or Foes?, Glasgow Naturalist*, 23 supplement, pp. 66–76.
Hunter, J. (1976), *The Making of the Crofting Community*, Edinburgh.
Hunter, J. (1995), *On the Other Side of Sorrow: Nature and People in the Scottish Highlands*, Edinburgh.
Hunter, T. (1883), *Woods, Forests and Estates of Perthshire*, Perth.
Hutchison, R. (1870), 'On the economic uses and comparative value of different descriptions of timber grown in Scotland', *TSAS*, 5, pp. 109–13.
Hutton, J. (1878), 'On the woods and plantations of the Mackintosh estate in Brae Lochaber', *TSAS*, 8, pp. 233–40.
Imrie, J. and Dunbar, J. G. (1982), *Accounts of the Masters of Work, for Building and Repairing Royal Palaces and Castles 1616–1649*, Edinburgh.
Innes, C. (1861), 'Some account of early planting in Scotland', *TRHAS*, new series, 9, pp. 40–53.
Innes, C. (ed.) (1855), *The Black Book of Taymouth*, Bannatyne Club, Edinburgh.
Innes, C. (ed.) (1859), *The Book of the Thanes of Cawdor*, Spalding Club, Aberdeen.
Jacobs, M. (1991), *The Green Economy*, London.
Jenkins, D. (2003), *Of Partridges and Peacocks* [Aboyne].
Johnson, Samuel (edn 1984), *A Journey to the Western Islands of Scotland*, Penguin Classics, London.
Johnston, J. L. and Balharry, D. (2001), *Beinn Eighe: the Mountain above the Wood*, Edinburgh.
Kirby, K. J. (2003), 'What might a British forest-landscape driven by large herbivores look like?', *English Nature Research Report*, 530.
Kirby, K. J. and Watkins, C. (eds) (1998), *The Ecological History of European Forests*, Wallingford.
Kitchener, A. C. (1998), 'Extinctions, introductions and colonisations of Scottish mammals and birds since the last Ice Age', in R. A. Lambert (ed.), *Species History in Scotland*, Edinburgh.
Kitchener, A. C. and Conroy, J. (1996), 'The history of the beaver in Scotland and the case for its reintroduction', *British Wildlife*, 7, pp. 156–61.

Kjærgaard, T. (1994), *The Danish Revolution, 1500–1800: an Ecohistorical Interpretation*, Cambridge.

Lamb, H. H. (1985), 'Climate and landscape in the British Isles', in S. R. J. Woodell (ed.), *The English Landscape, Past, Present and Future*, Oxford, pp. 148–67.

Lambert, R. A. (1997), 'Preserving a remnant of the old natural forest: the joint venture of the NTS and RSFS in the mid-1930s', *Scottish Forestry*, 51, pp. 31–3.

Lambert, R. A. (ed.) (1998), *Species History in Scotland*, Edinburgh.

Lambert, R. A. (1999), 'In search of wilderness, nature and sport: the visitor to Rothiemurchus, 1780–1820', in T. C. Smout and R. A. Lambert, *Rothiemurchus: Nature and People on a Highland Estate, 1500–2000*, Dalkeith, pp. 32–59.

Langhelle, S. I. (1999), 'The timber export from the Tysvær area in the sixteenth and seventeenth century', in [S. I. Langhelle] (ed.), *Timber and Trade*, Localhistorisk Stiftelse, Tysvær, Norway, Fagrapport nr. 1, pp. 24–36.

[Langhelle, S. I.] (ed.) (1999), *Timber and Trade*, Localhistorisk Stiftelse, Tysvær, Norway, Fagrapport nr. 1.

Latham, R. (ed.) (1995), *Samuel Pepys and the Second Dutch War: Pepys' Brooke House Papers*, Navy Records Society.

Latham, R. and Matthews, W. (eds) (1970–83), *The Diary of Samuel Pepys*, London.

Lee, U. (2002), 'Native timber construction: Strathspey's unique history', *Scottish Woodland History Discussion Group Notes*, 7 (2002), pp. 23–9.

Leopold, A. (1949), *A Sand County Almanac and Sketches Here and There*, New York.

Lewis, J. H. (1984), 'The charcoal-fired blast furnaces of Scotland: a review', *Proceedings of the Society of Antiquaries of Scotland*, 114, pp. 445–63.

Lillehammer, A. (1986), 'The Scottish-Norwegian timber trade in the Stavanger area in the sixteenth and seventeenth centuries', in T. C. Smout (ed.), *Scotland and Europe, 1200–1850*, Edinburgh, pp. 97–111.

Lillehammer, A. (1990), 'Boards, beams and barrel-hoops: contacts between Scotland and the Stavanger area in the seventeenth century', in G. G. Simpson (ed.), *Scotland and Scandinavia 800–1800*, Edinburgh, pp. 100–6.

Lindsay, I. G. and Cosh, M. (1973), *Inveraray and the Dukes of Argyll*, Edinburgh.

Lindsay, J. M. (1974), 'The use of woodland in Argyllshire and Perthshire between 1650 and 1850', unpublished University of Edinburgh Ph.D. thesis.

Lindsay, J. M. (1975), 'Some aspects of timber supply in the Highlands, 1700–1850', *Scottish Studies*, 19, pp. 39–53.

Lindsay, J. M. (1975), 'Charcoal iron smelting and its fuel supply; the example of Lorn Furnace, Argyllshire, 1753–1876', *Journal of Historical Geography*, 1, pp. 283–98.

Lindsay, J. M. (1977), 'The iron industry in the Highlands: charcoal blast furnaces', *Scottish Historical Review*, 56, pp. 49–63.

Lindsay, J. M. (1980), 'The commercial use of woodland and coppice management', in M. L. Parry and T. R. Slater (eds), *The Making of the Scottish Countryside*, London.

Linnard, W. (2000), *Welsh Woods and Forests*, Llandysul.

Loeber, R. (1994), 'Settlers' utilisation of the natural resources', in K. Hannigan and W. Nolan, *Wicklow, History and Society: Interdisciplinary Essays on the History of an Irish County*, Dublin, pp. 267–304.

Louden, J. C. (1842), *An Encyclopaedia of Trees and Shrubs*, Edinburgh.
Lovat, Lord and Stirling of Keir, Captain (1911), *Afforestation in Scotland: a Forest Survey of Glen Mor*, TRSAS, 25.
Lythe, S. G. E. (1960), *The Economy of Scotland in its European Setting, 1550–1625*, Edinburgh.
Mabey, R. (1996), *Flora Britannica*, London.
Macadam, W. I. (1886–7), 'Notes on the ancient iron industry of Scotland', *Proceedings of the Society of Antiquaries of Scotland*, new series 9, pp. 89–131.
McArthur, M. M. (ed.) (1936), *Survey of Lochtayside, 1796*, Scottish History Society, Edinburgh.
M'Corquodale, W. (1868), 'On the conversion of coppice land into a more remunerative crop', *TSAS*, 4, pp. 47–50.
McCracken, E. (1971), *The Irish Woods since Tudor Times: their Distribution and Exploitation*, Newton Abbot.
MacDonald, A. R. (1997), 'That valuable branch of the common good: the Perth plantation', *Scottish Forestry*, 51, pp. 34–9.
MacDonald, A. R. (1999), 'The first English ironworks in Scotland? The "forge" at Canonbie, Dumfriesshire', *Transactions of the Dumfriesshire and Galloway Natural History and Antiquarian Society*, 73, pp. 209–21.
MacDonald, J. (1811), *General View of the Agriculture of the Hebrides*, Edinburgh.
MacDonald, N. M. (1972), *The Clan Ranald of Knoydart and Glengarry*, n.p.
Macdougall, N. (1989), *James IV*, Edinburgh.
Macgill, W. (1909), *Old Ross-shire and Scotland as seen in the Tain and Balnagown Documents*, 2 vols, Inverness.
Macinnes, A. I. (1996), *Clanship, Commerce and the House of Stuart, 1603–1788*, East Linton.
McKay, M. M. (ed.) (1980), *The Rev. Dr. John Walker's Report on the Hebrides of 1764 and 1771*, Edinburgh.
MacKenzie, D. F. (1884), 'Scottish timber and its uses', *TSAS*, 10, pp. 189–98.
MacKenzie, G. S. (1810), *General Survey of the Counties of Ross and Cromarty*, London.
MacKenzie, G., Earl of Cromartie (1710–12), 'An account of the mosses in Scotland', *Philosophical Transactions of the Royal Society*, 27, pp. 296–301.
MacKenzie, M. (1776), *A Maritim Survey of Ireland and the West of Great Britain*, London.
MacKenzie, N. A. (2002), 'The native woodlands: history, decline and present status', in C. Gimingham (ed.), *The Ecology, Land Use and Conservation of the Cairngorms*, Chichester, pp. 107–19.
McVean, D. N. and Ratcliffe, D. A. (1962), *Plant Communities in the Scottish Highlands*, Monograph of the Nature Conservancy, 1, London (HMSO).
Maidment, J. (ed.) (1836), *Letters and State Papers during the Reign of King James VI, chiefly from the MS Collection of Sir James Balfour of Denmyln*, Abbotsford Club, Edinburgh.
Maitland, J., eighth Earl of Lauderdale (1819), *An Inquiry into the Nature and Origin of Public Wealth and into the Means and Causes of its Increase*, Edinburgh.

Marshall, W. (1794), *General View of the Agriculture of the Central Highlands of Scotland*, London.
Martin, M. (edn 1884), *A Description of the Western Islands of Scotland, c. 1695*, Glasgow.
Mather, A. S. (1990), *Global Forest Resources*, London.
[Maxwell, J. S.], (1913 and 1929), *Loch Ossian Plantations: an Essay in Afforesting High Moorland*, n.p.
Mercer, R. and Tipping, R. (1994), 'The prehistory of soil erosion in the northern and eastern Cheviot hills, Anglo-Scottish borders', in S. Foster and T. C. Smout (eds), *The History of Soils and Field Systems*, Aberdeen, pp. 1–25.
Michie, J. G. (ed.) (1901), *Records of Invercauld*, New Spalding Club, Aberdeen.
Miles, H. and Jackman, B. (1991), *The Great Wood of Caledon*, Lanark.
Mitchell, A. (ed.) (1906), *Geographical Collections Relating to Scotland made by Walter Macfarlane*, 3 vols, Scottish History Society, Edinburgh.
Mitchell, A. and House, S. (1999), *David Douglas, Explorer and Botanist*, London.
Mitchell, J. (2000), *The Shielings and Drove Ways of Loch Lomondside*, Stirling.
Mitchell, L. and House, S. (1999), *David Douglas, Explorer and Botanist*, London.
Monteath, R. (1827), *Miscellaneous Reports on Woods and Plantations*, Dundee.
Monteath, R. (1829), *A New and Easy System of Draining and Reclaiming the Bogs of Ireland*.
Monteath, R. (edn 1836), *Forester's Guide and Profitable Planter*, London.
Moore, P. D. (1993), 'The origin of blanket mire, revisited', in F. M. Chambers (ed.), *Climate Change and Human Impact on the Landscape*, London, pp. 217–25.
Moss, C. E., Rankin, W. M. and Tansley, A. G. (1910), 'The woodlands of England', *New Phytologist*, 9, pp. 113–49.
Munro, J. (1988), 'The golden groves of Abernethy: the cutting and extraction of timber before the Union', in G. Cruickshank (ed.), *A Sense of Place: Studies in Scottish Local History*, Edinburgh, pp. 152–62.
Munro, J. (ed.) (1992), *Inventory of Chisholm Writs 1456–1810*, Scottish Record Society, Edinburgh.
Munro, J. (ed.) (2000), *The Lochiel Inventory, 1472–1744*, Scottish Record Society, Edinburgh.
Munro, J. and Munro, R. W. (eds) (1986), *The Acts of the Lords of the Isles, 1336–1493*, Scottish History Society, Edinburgh.
Murray, D. (edn. 1973), *The York Buildings Company: a Chapter in Scotch History*, Edinburgh.
Murray, W. H. (1962), *Highland Landscape: a Survey*, Edinburgh.
Nairne, D. (1892), 'Notes on Highland woods, ancient and modern', *Transactions of the Gaelic Society of Inverness*, 17, pp. 170–221.
Neeson, E. (1991), *A History of Irish Forestry*, Dublin.
Nicholls, K. (2001), 'Woodland cover in pre-modern Ireland', in P. J. Duffy, D. Edwards and E. FitzPatrick (eds), *Gaelic Ireland c.1250 – c.1650: Land, Lordship and Settlement*, Dublin, pp. 181–206.
Nicholson, A. (1930), *A History of Skye*, Glasgow.

Nicol, W. (1799), *The Practical Planter, or a Treatise on Forest Planting*, Edinburgh.
Nicol, W. (1812), *Planter's Kalendar*, Edinburgh.
Noble, R. (1997), 'Changes in native woodland in Assynt, Sutherland, since 1774', in T. C. Smout (ed.), *Scottish Woodland History*, Edinburgh, pp. 126–34.
Noble, R. [2000], *Woods of Assynt*, project report for the Assynt Crofters Trust, n.p., n.d.
O'Dell, A. C. (1953), 'A view of Scotland in the middle of the eighteenth century', *Scottish Geographical Magazine*, 69, pp. 58–63.
O'Sullivan, P. E. (1973), 'Land-use changes in the Forest of Abernethy, Inverness-shire (1750–1900 AD)', *Scottish Geographical Magazine*, 89, pp. 95–106.
O'Sullivan, P. E. (1977), 'Vegetation history and the native pinewoods, in R. G. H. Bunce and J. N. R. Jeffers (eds), *Native Pinewoods of Scotland*, Cambridge, pp. 60–9.
Oram, R. (1989), 'The Lordship of Galloway, ca. 1000 – ca. 1250', unpublished University of St Andrews Ph.D. thesis.
Parker, A. W. (1997), *Scottish Highlanders in Colonial Georgia: the Recruitment, Emigration and Settlement at Darien, 1735–1748*, Athens, GA.
Parry, M. L. and Slater, T. R. (eds) (1980), *The Making of the Scottish Countryside*, London.
Paton, V. N. (ed.) (1893), 'Masterton Papers, 1660–1719', in *Miscellany of the Scottish History Society*, 1, Scottish History Society, Edinburgh.
Pellew, R. A. (1972), 'An ecological appraisal of the proposed wildlife reserve, Ostaig Estate, Sleat' (unpublished report in possession of Clan Donald Lands Trust).
Pennant, T. (edn 1774), *A Tour in Scotland, 1769*, Warrington.
Pennant, T. (edn 1776), *A Tour in Scotland and Voyage to the Hebrides, 1772*, 2 vols, London.
Photos-Jones, E. and Atkinson, J. A. (1998), 'Iron-making in medieval Perth: a case of town and country?', *Proceedings of the Society of Antiquaries of Scotland*, 128, pp. 887–904.
Photos-Jones, E., Atkinson, J. A., Hall, A. J. and Banks, I. (1998), 'The bloomery mounds of the Scottish Highlands, part 1: the archaeological background', *Journal of the Historical Metallurgy Society*, 32, pp. 15–32.
Pilcher, J. R. (1996), 'A palaeoecologist's view of the Irish landscape', in F. H. A. Aalen (ed.), *Landscape Study and Management*, Dublin, pp. 72–80.
Pilcher, J. R. and Mac an tSaoir, S. (1995), *Woods, Trees and Forests in Ireland*, Dublin.
Pryor, S. N. and Smith, S. (2002), *The Area and Composition of Plantation on Ancient Woodland Sites*, Woodland Trust.
Quelch, P. R. (2000–1), 'Upland pasture woodlands in Scotland', part 1, *Scottish Forestry*, 54, pp. 209–14; part 2, *ibid.*, 55, pp. 85–91.
Rackham, O. (1980), *Ancient Woodland*, Cambridge; new revised edition, Colvend, Dalbeattie, 2003.
Rackham, O. (1986), *The History of the Countryside*, London.

Rackham, O. (edn 1995), *Trees and Woodland in the British Landscape*, London.

Rackham, O. (1996), 'Forest history of countries without much forest: questions of conservation and savanna', in S. Cavaciocchi (ed.), *L'uomo e la foresta secc. XIII–XVIII*, Prato, pp. 297–326.

Redclift, M. (1987), *Sustainable Development: Exploring the Contradictions*, London.

Richards, E. and Clough, M. (1989), *Cromartie: Highland Life 1650–1914*, Aberdeen.

Ritchie, J. (1920), *The Influence of Man on Animal Life in Scotland*, Cambridge.

Roberts, A. J., Russell, C., Walker, G. J. and Kirby, K. J. (1992), 'Regional variation in the origin, extent and composition of Scottish woodland', *Botanical Journal of Scotland*, 46, pp. 167–89.

Roberts, G. (2000), 'The relocation of ancient woodland', *Quarterly Journal of Forestry*, 24, pp. 305–12.

Robertson, J. (1794), *General View of the Agriculture in the Southern Districts of the County of Perth*, London.

Robertson, J. (1808), *General View of the Agriculture of the County of Inverness*, London.

Robson, J. (1794), *General View of the Agriculture of the County of Argyll*, London.

Rogers, C. (ed.) (1879–80), *Rental Book of the Cistercian Abbey of Coupar-Angus*, 2 vols, Grampian Club, London.

Ross, I. (1995), 'A historical appraisal of the silviculture, management and economics of the Deeside forests', in J. R. Aldhous (ed.), *Our Pinewood Heritage*, Farnham, pp. 136–44.

Ross, R. (1881), 'On the peeling and harvesting of native barks', *TSAS*, 9, pp. 58–62.

Royal Commission on the Ancient and Historical Monuments of Scotland (1994), *South-East Perth: an Archaeological Landscape*, Edinburgh.

Royal Commission on the Ancient and Historical Monuments of Scotland (1995), *Mar Lodge Estate, Grampian: an Archaeological Survey*, Edinburgh.

Royal Commission on the Ancient and Historical Monuments of Scotland (1997), *Eastern Dumfriesshire: an Archaeological Landscape*, Edinburgh.

Royal Scottish Geographical Society (1973), *The Early Maps of Scotland*, Edinburgh.

Royal Society for the Protection of Birds [1993], *Time for Pine: a Future for the Caledonian Pinewoods*, n.p., n.d.

Ryle, G. (1969), *Forest Service: the First Forty-five Years of the Forestry Commission of Great Britain*, Newton Abbot.

Sanderson, N. (1998), 'Veteran trees in Highland wood pasture', *Scottish Woodland History Discussion Group Notes*, 3, pp. 4–9.

Schenk, W. (1996), 'Forest development types in Central Germany in pre-industrial times', in S. Cavaciocchi (ed.), *L'uomo e la foresta secc. XIII–XVIII*, Prato, pp. 201–23.

Schwappach, A. (1898), 'Report of a visit to the forests of Scotland in August 1896', *TRSAS*, 15, pp. 11–21.

Scott, A. (2002), *A Pleasure in Scottish Trees*, Edinburgh.

Scottish Green Party (1989), *A Rural Manifesto for the Highlands: Creating the Second Great Wood of Caledon*, Inverness.

Selby, P. J. (1842), *A History of British Forest Trees, Indigenous and Introduced*, London.

Sellar, W. D. H. (1993), *Moray: Province and People*, Scottish Society for Northern Studies.
Shaw, J. (1984), *Water Power in Scotland 1550–1870*, Edinburgh.
Shaw, L. (edn 1882), *History of the Province of Moray*, Glasgow.
Sheail, J. (1976), *Nature in Trust: the History of Nature Conservation in Britain*, Glasgow.
Sheail, J. (1987), *Seventy-five Years in Ecology: the British Ecological Society*, Oxford.
Sibbald, R. (1684), *Scotia Illustrata*, Edinburgh.
Sibbald, R. (edn 1803), *The History Ancient and Modern of the Sheriffdoms of Fife and Kinross*, London.
Simmons, I. G. (1993), 'Vegetation change during the Mesolithic in the British Isles: some amplifications', in F. M. Chambers (ed.), *Climate Change and Human Impact on the Landscape*, London, pp. 109–18.
Simmons, I. G. and Tooley, M. J. (eds) (1981), *The Environment in British Prehistory*, London.
Simpson, G. G. (ed.) (1990), *Scotland and Scandinavia, 800–1800*, Edinburgh.
Sinclair, J. (1795), *General View of the Agriculture of the Northern Counties and Islands of Scotland*, London.
Sinclair, J. (ed.) (1814), *General Report of the Agricultural and Political Circumstances of Scotland*, 5 vols, Edinburgh.
Singer, W. (1829), 'An essay on converting to economical uses trees usually treated as brushwood', *TRHAS*, 7, pp. 137–46.
Skelton, J. (edn 1995), *Speybuilt: the Story of a Forgotten Industry*, Garmouth.
Skene, W. F. (ed.) (1872), *John of Fordun's Chronicle of the Scottish Nation*, Edinburgh.
Skene, W. F. (1876), *Celtic Scotland*, Edinburgh.
Smith, A. (1874), 'On Aberdeenshire woods, forests and forestry', *TRHAS*, 4th series, 6, pp. 264–303.
Smith, A. (edn 1910), *The Wealth of Nations*, 2 vols, London.
Smith, J. (1798), *General View of the Agriculture of the County of Argyll*, Edinburgh.
Smith, W. G. (1911), 'The vegetation of woodlands', *TRSAS*, 24, pp. 6–13, 131–8.
Smout, T. C. (1960), 'Some problems of timber supply in later seventeenth-century Scotland', *Scottish Forestry*, 15, pp. 3–13.
Smout, T. C. (1963), *Scottish Trade on the Eve of Union, 1660–1707*, Edinburgh.
Smout, T. C. (ed.) (1978), 'Journal of Henry Kalmeter's travels in Scotland 1719–20', in *Scottish Industrial History, a Miscellany*, Scottish History Society, Edinburgh, pp. 1–52.
Smout, T. C. (ed.) (1986), *Scotland and Europe, 1200–1850*, Edinburgh.
Smout, T. C. (1989), 'Landowners in Scotland, Ireland and Denmark in the Age of Improvement', *Scandinavian Journal of History*, 12, pp. 79–97.
Smout, T. C. (1993), 'Woodland history before 1850', in T. C. Smout (ed.), *Scotland since Prehistory: Natural Change and Human Impact*, Aberdeen, pp. 40–9.
Smout, T. C. (ed.) (1993), *Scotland since Prehistory: Natural Change and Human Impact*, Aberdeen.
Smout, T. C. (1994), 'Trees as historic landscapes', *Scottish Forestry*, 48, pp. 244–52.

Smout, T. C. (1995), 'The history of the Rothiemurchus woods in the eighteenth century', *Northern Scotland*, 15, pp. 19–32.
Smout, T. C. (1997), 'Cutting into the pine: Loch Arkaig and Rothiemurchus in the eighteenth century', in T. C. Smout (ed.), *Scottish Woodland History*, Edinburgh, pp. 115–25.
Smout, T. C. (ed.) (1997), *Scottish Woodland History*, Edinburgh.
Smout, T. C. (1999), 'The Grant lairds of Rothiemurchus', in T. C. Smout and R. A. Lambert (eds), *Rothiemurchus: Nature and People on a Highland Estate, 1500–2000*, Dalkeith, pp. 14–21.
Smout, T. C. (1999), 'The Norwegian timber trade before 1707 from the Scottish perspective', in [S. I. Langhelle] (ed.), *Timber and Trade*, Localhistorisk Stiftelse, Tysvær, Norway, Fagrapport nr. 1, pp. 24–36.
Smout, T. C. (2000), *Nature Contested: Environmental History in Scotland and Northern England since 1600*, Edinburgh.
Smout, T. C. (2001), 'Woodland in the maps of Pont', in I. C. Cunningham (ed.), *The Nation Survey'd: Timothy Pont's Maps of Scotland*, East Linton, pp. 77–92. The second edition, forthcoming, carries a significant postscript.
Smout, T. C. (ed.) (2003), *People and Woods in Scotland: a History*, Edinburgh.
Smout, T. C. (2003), 'Energy rich, energy poor: Scotland, Ireland and Iceland 1600–1800', in D. Dickson and C. Ó Gráda (eds), *Refiguring Ireland: Essays in honour of L. M. Cullen*, Dublin, pp. 19–36.
Smout, T. C. and Lambert, R. A. (eds) (1999), *Rothiemurchus: Nature and People on a Highland Estate, 1500–2000*, Dalkeith.
Somers, R. (1848), *Letters from the Highlands, or The Famine of 1847*, London.
Somerville, W. (1909), 'Forestry in some of its economic aspects', *Journal of the Royal Statistical Society*, 72, pp. 40–63.
Spiers, A. (1906), 'Some notes of the home timber-trade in the east of Scotland', *TRSAS*, 19, pp. 66–72.
Stebbing, E. P. (1916), *British Forestry: Its Present Position and Outlook after the War*, London.
Stell, G. (1972), 'Two cruck-framed buildings in Dumfriesshire', *Transactions of the Dumfriesshire and Galloway Natural History and Archaeological Society*, 49, pp. 35–48.
Steven, H. M. (1954), 'Changes in silvicultural practice in Scotland, 1854–1954', *Scottish Forestry*, 8, p. 53.
Steven, H. M. and Carlisle, A. (1959), *The Native Pinewoods of Scotland*, Edinburgh.
Stewart, K. (2000), *Abriachan: the Story of an Upland Community*, Abriachan.
Stewart, M. (1997), 'The utilisation and management of the semi-natural woodlands of Lochtayside, 1650–1850', unpublished University of St Andrews M.Phil. thesis.
Stewart, M. (2000), *Loch Tay, its Woods and its People*, Scottish Native Woods, n.p.
Stone, J. C. (1989), *The Pont Manuscript Maps of Scotland: Sixteenth Century Origins of a Blaeu Atlas*, Tring.
Stuart, C. E. and Stuart, J. S. (alias Allan) (1848), *Lays of the Deer Forest*, Edinburgh.

Summers, R. (1999), *The History and Ecology of Abernethy Forest, Strathspey*, RSPB Internal Report, Inverness.
Sunart Oakwoods Research Group (2001), *The Sunart Oakwoods: a Report on their History and Archaeology*, n.p.
Svenning, J. C. (2002), 'A review of natural vegetation openness in north-western Europe', *Biological Conservation*, 104, pp. 133–48.
Symon, J. A. (1959), *Scottish Farming Past and Present*, Edinburgh.
Tait, J. (1884), 'The agriculture of the County of Stirling', *TRHAS*, 4th series, 16, pp. 177–8.
Thomson, A. (1990), 'The Scottish timber trade, 1680 to 1800', unpublished St Andrews University Ph.D. thesis.
Thomson, D. S. and Grimble, I. (eds) (1968), *The Future of the Highlands*, London.
Tinsley, H. and Grigson, C. (1981), 'The Bronze Age', in I. G. Simmons and M. Tooley (eds), *The Environment in British Prehistory*, London, pp. 211–16.
Tipping, R. (1993), 'A "History of the Scottish Forests" revisited', *Reforesting Scotland*, 8, pp. 16–21; 9, pp. 18–22.
Tipping, R. (1994), 'The form and fate of Scottish woodlands', *Proceedings of the Society of Antiquaries of Scotland*, 124, pp. 1–54.
Tipping, R. (2003), 'Living in the past: woods and people in prehistory to 1000 BC', in T. C. Smout (ed.), *People and Woods in Scotland: a History*, Edinburgh, pp. 14–39.
Tittensor, R. M. (1970), 'History of the Loch Lomond oakwoods', *Scottish Forestry*, 24, pp. 103–4.
Tsouvalis-Gerber, J. (1998), 'Making the invisible visible: ancient woodlands, British forest policy and the social construction of reality', in C. Watkins (ed.), *European Woods and Forests: Studies in Cultural History*, Wallingford, pp. 220–42.
Turnbull, J. (2001), *The Scottish Glass Industry, 1610–1750*, Edinburgh.
Tytler, P. F. (1829), *History of Scotland*, Edinburgh.
Ure, D. (1794), *General View of the Agriculture of the County of Dumbarton*, London.
Vera, F. W. M. (2000), *Grazing Ecology and Forest History*, Wallingford, Oxon.
Walker, D. M. (ed.) (1981), *The Institutions of the Law of Scotland by James, Viscount Stair, 1695*, Edinburgh.
Walker, G. J. and Kirby, K. J. (1989), *Inventories of Ancient, Long-established and Semi-natural Woodland for Scotland*, Research and Survey in Nature Conservation, no. 22, Nature Conservancy Council, Peterborough.
Walker, J. (1808), *An Economical History of the Hebrides and Highlands of Scotland*, 2 vols, Edinburgh.
Walker, J. (1808), *Essays on Natural History and Rural Economy*, Edinburgh.
Watkins, C. (ed.) (1998), *European Woods and Forests: Studies in Cultural History*, Wallingford.
Watson, A. (1983), 'Eighteenth century deer numbers and pine regeneration near Braemar, Scotland', *Biological Conservation*, 25, pp. 289–305.
Watson, F. (1997), 'Rights and responsibilities: wood management as seen through baron court records', in T. C. Smout (ed.), *Scottish Woodland History*, Edinburgh, pp. 101–14.

Watson, G. (ed.) (1890), *Bell's Dictionary and Digest of the Law of Scotland*, 7th edn, Edinburgh.
Watson, W. J. (1904), *Place-names of Ross and Cromarty*, Inverness.
Watt, H. B. (1900), 'Scottish forests and woodlands in early historic times', *Annals of the Andersonian Institute*, 2, pp. 89–107.
Westermann, E. (1996), 'Central European forestry and mining industries in the early modern period', in S. Cavaciocchi (ed.), *L'uomo e la foresta secc. XIII–XVIII*, Prato, pp. 927–41.
Whittington, G. (ed.) (1996), *Fragile Environments: the Use and Management of Tentsmuir NNR, Fife*, Edinburgh.
Whittington, G. and Edwards, K. J. (1993), '*Ubi solitudinem faciunt pacem appellant*: the Romans in Scotland, a palaeoenvironmental contribution', *Britannia*, 24, pp. 13–25.
Whittington, G. and Edwards, K. J. (1994), 'Palynology as a predictive tool in archaeology', *Proceedings of the Society of Antiquaries in Scotland*, 124, pp. 55–65.
Whittington, G. and Gibson, A. J. S. (1986), *The Military Survey of Scotland, 1747–1755: a Critique*, Historical Geography Research Series, no. 18, Aberdeen.
Whittington, G. and Smout, C. (1998), 'Landscape and history', in G. B. Corbet (ed.), *The Nature of Fife*, Edinburgh, pp. 24–39.
Whyte, I. D. (1995), *Scotland before the Industrial Revolution: an Economic and Social History, c.1050 – c.1750*, London.
Whyte, I. D. and Winchester, A. (2004), *Upland Landscapes*, Society for Landscape Studies.
Wickham-Jones, C. R. (1994), *Scotland's First Settlers*, London: Historic Scotland.
Wight, A. (1778–84), *Present State of Husbandry in Scotland*, 4 vols, Edinburgh.
Wilkie, T. (1881), 'On the system of oak coppice management recently adopted', *TSAS*, 9, pp. 270–2.
Wilkie, T. (1889), 'Report upon the rearing of underwood for game coverts in high forest', *TRSAS*, 12, pp. 371–3.
Williams, J. (1784), 'Plans for a Royal forest of oak in the Highlands of Scotland', *Archaeologica Scotica*, 1, p. 29.
Williams, J. (1967), 'A mediaeval iron smelting site at Millhill, New Abbey', *Transactions of the Dumfriesshire and Galloway Natural History and Archaeological Society*, 44, pp. 126–32.
Williamson, A. T. (1893), 'The utilisation of smallwood for turnery and other purposes', and 'The manufacture of home-grown timber', *TRSAS*, 13, pp. 145–61.
Winkleman, P. H. (1971), *Nederlandse Rekeningen in de Tolregisters van Köningsbergen 1588–1602*, The Hague.
Woodell, S. R. J. (ed.) (1985), *The English Landscape, Past, Present and Future*, Oxford.
Woodland Trust (2000), 'Ancient woods and translocation', *Position Statement 19*.
Wordsworth, J., 'Report on the investigation of various iron-working mounds in the Ben Wyvis area in November, 1992', report to Scottish Natural Heritage, northern area (Inverness).

World Commission on Environment and Development (1987), *Our Common Heritage*, Oxford.
Wormell, P. (2003), *Pinewoods of the Black Mount*, Skipton.
Worrell, R. and MacKenzie, N., 'The ecological impact of using the woods', in T. C. Smout (ed.), *People and Woods in Scotland: a History* (Edinburgh, 2003), pp. 195–213.
Worster, D. (1977), *Nature's Economy: a History of Ecological Ideas*, Cambridge.
Yalden, D. (1999), *The History of British Mammals*, London.

Index

col.pl. denotes colour plates

Aberdeen, 127, 271
Aberdeen, Earl of, 129
Aberdeenshire, 40, 46, 65
Abergledie, Deeside, 73
Abernethy, 48, 151, 152
Abernethy Forest, 60, 71, 75, 90, 183, 185, 196, 201, 286
    exploitation, 127, 131, 194, 211–16, 270
    extent, 214–15, 216, 290
    prosecutions, 190, 191
    sporting use, 276
    timber from, 129, 131, 148
Abernethy Forest Lodge, 86
Aberuchill, 175
Aboyne, Earl of, 150, 186, 188, 221
Achindennan, 257
Achnacarry, 72, 85–6, 181, 236, 342
Achnashellach, 287
Achray, 238
Acland Committee, 280
Acts of Parliament
    Middle Ages, 38
    16th–17th centuries, 45
    Forestry Act (1919), 280, 281
    protection of woodland, 229
    taking of timber, 135
Adair, John, 52
agriculture *see* farming
alder, 2, 3, 66, 73, 74, 114, 151, 284; *col.pl.7a&b*
    cutting, 174
    in folklore, 79
    uses, 35, 92, 266, 268, 272
Aldscaidrow, Lochaber, 183, 198
Alloa, 127
Alloa Tower, 81
Alloa wood, 154, 165, 172, 174, 175, 178
Allt Broighleachan, 348
Allt Coire Bhiscair, 348
Allt Ghiusachan, 361
Allt Mheuranby, 361
Allt nan Giubhas, 361
Altnahatnich, 301–2, 315
Amat, 336, 337, 338
Amotnatua, 185
An Sgaothach, 337–8
Ancient Woodland Inventory, 61, 68
ancient woods, 1–2, 283, 286, 289
Angus, 26, 65
Angus, Earls of, heraldic seal, 169
animals

fodder for, 99–100
forest, 42–3, 102–3
    see also grazing animals; *names of individual species*
Annan, 250
Annandale, 39
Annexed Estates Commissioners, 140, 146, 149, 155–6, 186, 187, 203–4, 366
Antony, Lambert, 321
apple in folklore, 78–9
arboreta, 277
Ardgour, 104, 105, 194, 200–1, 201
Ardincaple, Seil Island, 176
Ardkinglas, 261, 277
Ardmaddy, 343
Ardmore, 329, 332
Ardmucknish, 343
Ardnamurchan, 93, 117, 256
Ardrostan wood, 180
Ardshiel, 113
Ardtornish, 117–18
Argyll, 113, 114, 115
    charcoal and tanbark, 227, 240–1
    enclosure, 167, 169–70
    oakwoods, 2, 73, 177, 240–4
    woodland cover, 64, 65, 66, 261
Argyll, Dukes of, 83, 117, 118, 129, 192, 253–4
    iron works, 170, 247–8
    tenants' removal of wood, 136–8
Argyll, Earl of, 162, 200
Argyll Furnace Company, 240–1, 247, 248
Arisaig, 176, 239
Ariundle, 244
Armadale, 367, 374, 376, 377, 381, 383, 387
armies, use of wood for fuel, 98–9
Armstrong, Charles, 345, 353
Arneill, Francis, 201
Arnprior estate, 74, 155
Arran, 30, 72
ash, 2, 3, 15, 66, 67, 73, 74, 75, 114, 284
    cutting, 174
    in folklore, 78, 79
    uses, 83, 90, 266, 268, 269, 272
ashwoods, 2, 90
Aslaig, Skye, 367, 371, 373, 374, 375
aspen, 2, 15, 53, 66, 72, 73, 74, 75; *co.pl.5a*
    in folklore, 79
assarting, 39
Assynt, 69, 105, 109; *col.pl.5a&b*
Atholl, Dukes of, 70, 138–9, 149, 162, 178, 276, 277

Index

Atholl, Earls of, 106
    great hunting lodge, 91–2
Auchmore, 142
aurochs, 11, 12, 33, 43, 102–3, 122
'authenticity' of a wood, 7–8, 9–10, 15, 394
Avery, Joseph, 74, 207, 208, 336
Ayrshire, 65, 153, 163

Badenoch, 23, 43, 143
Bains, James, wright, 332–4
Balblair Wood, 72
Balbridie, Deeside, 29–30
Baldoon, 107–8
Balfour, Sir James, of Denmilne, 47
Balfron parish, 159
Ballachulish, 60
Ballindalloch, 152
Ballochbuie, 194, 221, 222, 224, 286
Ballochearn, 159
Balmaha, 263, 268
Balnagown, 69
    exploitation, 207, 332–3, 338; naval timber, 197, 323–4, 325–9, 331–2
Balnagown, Bog of, 139
Baltic, imports from, 124, 126, 129
Banff, 46, 65
Bannerman, John, 283
Bardowie Castle, 81
Bargany, Ayrshire, 176, 181
bark, 172, 179, 181, 225, 375
    transport, 180
    see also tanbark
bark-cutters, 181
barrels, 90, 91, 92, 272
barren timber, 90, 121, 156, 173–4
Barrisdale, 187, 194, 204
Barrmolach, Carradale, 116–17
Barrs, 361
Baxter, John, architect, 129
Baxter, John, surgeon, 305
Baxter, John, timber merchant, 217
Bayne, Patrick, 190
bear, 42
beaver, 42
Bedford, Duchess of, 310
beech
    plantation, 17, 70–1, 72, 74
    uses, 265, 266, 269, 272
Bell, Patrick, 199
Ben Alder, 52, 54–5
Ben Lawers, 75
Ben Lomond, 66
Benderloch, 60
Bennie, river, 301
Beresford-Peirse, Sir Henry, 284, 287–8
Berwickshire, woodland cover, 65
Białowieza forest, 15, 33
biodiversity, 6, 8, 9, 10, 394–7
    coppicing and, 16–17, 394–5
    loss in Highlands, 43
birch, 2, 3, 15, 34, 53, 66, 67, 73, 74, 75, 114, 259, 284, 337; *col.pl.5a*
    cutting, 174

downy, 2, 3
dwarf, 2
in folklore, 78–9
in prehistory, 25, 26, 27
silver, 2; *col.pl.3*
trade in, 151, 152–3
uses, 83, 90–1, 92, 96, 264, 266, 267–8
birchwoods, 2–3, 46, 56, 264, 286, 393; *col.pl.14b–15a*
    loss of, 43, 119, 284
Birkhill, Clackmannan, 188
Birnam oak, 78
Birse, 46, 266
Black Loch, Fife, 34
Black Wood of Rannoch, 61, 88, 115, 131, 287; *col.pl.15a*
    destruction of, 186, 187
    exploitation, 194
    sporting use, 276
    timber from, 149, 150
blackthorn (sloe), 2, 73, 114
    in folklore, 79
Blaeu, John, 20, 21, 47
    Atlas Novus, 48, 49, 51, 52, 55, 234, 336
Blair, 162
blast furnaces, 227, 228, 231, 235, 238, 241
bloomeries, 225–7; *col.pl.10b*
boar, wild, 42, 43, 102–3, 122
bobbins, 77, 91, 264–6, 269, 272, 382, 384
Boece, Hector, 20
bogs (woods for common use), 139
Bonar Bridge, 152
Bonawe, 240, 241–3, 244–8, 261, 354, 356–7
Bo'ness, 127
boring mills, Rothiemurchus, 151, 298, 299
Borthwick, Patrick, 307
Bothwell, 181
Boturich, 174, 175
Bowmont valley, 13–14
Braco, Lord, 143–7
Braemar, 66, 146, 151, 183
bramble, 12
Brandis, Sir Dietrich, 285
Breadalbane, 70, 99, 117, 135, 160
Breadalbane, Earls of, 106, 110, 117, 142, 167, 179, 246, 275
    contracts with Irish, 203, 341, 342, 343, 346, 348–53, 356–9, 360–1
Brechin, 148
British Vegetation Committee, 283
broadleaf woods, 2, 15–16
Brodick, 72
Bronze Age, 13, 29, 31, 388, 398
broom, 97, 98
Bruce, David, of Clackmannan, 178
Bruce, Henry, forester, 188–9
Brundtland Report, sustainable development, 6
Brunstane Park, 181
Buccleuch, Duchess of, 118, 237
Buchanan, 238, 252
Buchanan, Herbert, of Arden, 257
building, timber
    before 1500, 32–5

423

materials for, 80–1, 91–2, 95–7; imported timber, 126, 129–33
Middle Ages, 37–8
Buiston, Ayrshire, 35
Bunawe ironworks, 166–7, 174, 179, 181–2
burning
  by Vikings, 36
  hill grazing, 69, 123
  in prehistory, 12, 13, 28
  see also muirburn

Cabrach, Aberdeenshire, 104
Cadboll, 69
Caddell, Lord, 342, 343
Cadzow, 107, 390
Caenlochan glens, 30
Cairngorm forests, 26
Cairngorms National Nature Reserve, 293, 318
Cairngorms National Park, 318
Caithness, 23, 26, 29, 65
  Flow Country, 12–13, 28, 289
Caithness, Earl of, 152, 331
Calder, West Lothian, 166, 175
Caledonian pine, 3, 23
Caledonian Wood, 13, 20–4, 43–4
Callander, 21, 80, 140
Callanish, Lewis, 30
Callart, 295
Callert, Moor of, 314
Cambusmore, 313, 315
Camden, William, 20
Cameron, Alexander, of Glen Nevis, 95–6
Cameron, Donald, of Locheil, 342, 343
Cameron, Ewen, of Ardtornish, 342, 343, 345
Cameron, Ewen, of Glenevis, 204
Cameron, Sir Ewen, of Locheil, 342
Cameron, John, of Locheil, 342
Camerons of Fassifern, 142
Camerons of Locheil, 139, 142, 177, 324
  commercial enterprises, 181, 192, 198, 203, 295;
  with Irish, 236, 250–1, 342, 345
Campbell, Archibald, timber merchant, 384
Campbell, Colin, of Glenfalloch, 351
Campbell, Colin, of Inveresregan, 342, 343, 345
Campbell, Daniel, of Shawfield, 345
Campbell, Donald, of Ardnamurchan, 117
Campbell, Donald, of Losset, 342, 343, 345
Campbell, Black Duncan, of Glenorchy, 160
Campbell, Captain Duncan, of Inverawe, 167, 179, 247
Campbell, Sir Duncan, of Glenorchy, 81, 110
Campbell, Sir Duncan, of Lochnell, 167, 174, 179, 181, 342, 343
Campbell, General, of Monzie, 206
Campbell, Hugh, of Stronmilchan, 346, 349, 354–5
Campbell, James, of Ardkinglas, 347
Campbell, John, of Cawdor, 148
Campbell, John, of Macorne, 342, 343
Campbell, Patrick, of Barcaldine, 342, 343
Campbell, Patrick, merchant in Inveraray, 342, 343, 347
Campbell, Robert, of Kentrae, 362
Campbell, William, of Glenfalloch, 350, 351

Campbells of Barcaldine, 170, 187
Campbells of Glenorchy, 107, 176, 200
Campbells of Glenure, 164
Campbells of Lochnell, 167, 174, 179, 181, 244, 246
Campsie, Perthshire, 47, 157–8, 162
Camusallach, 118
candle fir, 88, 149
Cannich, River, 209
Canonbie, Dumfriesshire, 118, 174, 237–8
Caolisnacon, 200
Capon Tree, 78
Cardenie, Fife, 40
Carlisle, A., 286, 288
Carmichael, John, 300
Carnhouran, Rannoch, 115
Carr, Ralph, 213
Carriber, West Lothian, 175
Carrick, Ayrshire, 49, 106
Carron Iron Works, 241
Carstaires, George, 154
Cashel, 180
Cassie, Andrew, slater, 332
Cassilis, Earl of, 97, 98, 158–9, 162, 188, 190
Castle Grant, 84–5
Castlecary, Stirlingshire, 166
Cathcart, Earls of, 238
Cathcart, John, 356
cattle, 37, 99, 103, 105, 107, 108–9, 111, 115, 120, 391; *col.pl.9a*
  black, 114, 122
  wild, 43, 122; *see also* auroch
Cawdor, 148
cedar, western red, 277
Celestine of the Isles, 364, 377
Channel Tunnel Rail Link, 7
charcoal, 77, 84, 89, 91, 92, 93, 179, 212, 261
  Irish and, 203
  iron industry, 225, 235, 241, 242, 244, 245, 246–8; *col.pl.10a*
  production of, 181–2
Charles II, king, 322, 323, 329
cherry
  bird, 2, 66, 75, 79, 114, 366
  wild (gean), 2, 73, 74, 79, 92
Cheviots, 31, 35
Chisholm family, 207–9
Chisholm, Roderick, of Strathglass, 295
Chisholm, William, 88
Cistercians, 40, 157
Clackmannan, 65, 80, 130, 188
clear-felling
  native oak and Scots pine, 5, 6
  pinewoods, 184, 187
  plantation after, 17
  rotational, 16
  semi-natural woodland, 16, 17
clearance of woodland
  by military, 40–1
  Middle Ages, 39–40
  in prehistory, 12–13, 29, 31–2
Cleghorn, Hugh, 278
Clerk, Dugald, of Bralecken, 73, 136, 244
Clerk, Sir John, of Penicuik, 181

climate change, effects, 58–9, 339, 388, 389–90, 392
    in prehistory, 12, 25, 28, 29, 30, 34
Clunie, 160, 343, 395
Cockburne, Alexander, 154
Coe, River, 53
Coigach, 113, 183, 184, 205–6
Coill' a Ghasgain, 367
Coille Kynachan, 46
Coille Mhór, Strath Nairn (the Great Wood), 45
Coille na Glas Leitir, 287–8
Coille na Shi, 46
Coinletter, 206
Collins, Samuel, 201, 294
Comer Wood, 115
Compstone, wood of, 154
Comrie, 72, 252
Conaglen, 104, 200, 343
Conecherechan, 343
conifers, non-native, 5, 6, 8, 17, 71, 271, 281–2, 283, 284–5, 288
Convention on Biological Diversity, Rio de Janeiro, 6
Cooper, J. and R., timber merchants, 381, 383, 384
coppice/coppicing
    decline, 260–4
    industrial uses, 252–4
    management, 161, 162, 164, 171, 172–3, 176–8, 393, 394–5
    rotation, 158, 178
    semi-natural woodland, 16–17
    value, 254–5, 257
coppice-with-standards, 16, 161, 176, 177, 243, 254, 264
Coronelli, Vincenzo, 20, 21
Corpach, 181
Corrienamor Firwoods, 53
Corrimony, cruck-framed building, 86, 87
Corrour, 278, 285
Corrycherry, 203
Corryriggar, 203
Coulnakyle, 149, 151
Coupar Angus, 135, 157, 165
Crag Wood, Murthly, 175
Craig, William Gibson, 312
Craigiefrisch, 104
Craignee Wood, 97
Craigroyston, 98, 115
Craleckan, 240–1, 247, 248, 254
Crandirth, 104
Crannach, 347, 348, 358
Crannich, 107, 160
crannogs, 29, 35
Cranston, Midlothian, 174
Crarae, Argyll, 177
Crathes, 183, 184
Crathie, 146, 222
Creag Fhiaclach, 314
Creag Ghiunhais, 222
creel houses, 93–5, 121
Creetown, 153
Creich, woods of, 226
Cromartie, Earl of, 58, 205–6
Cromarty, 36

Cromwell, Oliver, 322, 329
Cronich, 203
Crowe, Sylvia, 281
cruck-framed buildings, 86, 87
Cruixton, 74, 108, 159
cultural associations of trees, 78–9
Cumbernauld, 180
Cumming, Alexander, 213–14
Cumming, David, 184, 186, 188
cup-turning, 181
cutting *see* felling
cypress, Lawson's, 277

Dalavil, Skye, 367–9, 370, 371, 373–4, 379–81, 382, 384
Dalbog, Wood of, 236
Dalchreenacht, 343
Dalhousie, 49
Dalhousie, Earl of, 126
Dalkeith, 49, 171, 261, 283, 390
Dalrymple wood, 97, 99, 158, 188
Darien Scheme, 239–40
Darling, Frank Fraser, 23, 283, 284
Darnaway Castle, 37, 38, 80, 81
Darnaway Forest, 42, 80, 81, 257, 262
David I, king of Scots, 37
David II, king of Scots, 39
Davidson, John, 236–7
Davidson, John, forester, 188
Dawyk, 277
Dee, River, 110
deer, 12, 75, 102, 103, 105, 122, 123, 165, 270–1, 276
deer forests, 276
deer hunt, 106
deer park, 106, 390
Deeside, 143–7
    exploitation, 131, 220–2, 270
    pinewoods, 183, 185
Deirnafuor, 343
Denmark, 118, 398–9, 400–1
Derry, 183
Diebidale, 152, 336
ditches, boundary, 165, 169
Dodsworth, Ralph, 218–19, 220
Doire Darach, 347, 348, 360
Doon, River, 49, 153
Dorret, Jan, military map, 53
Douglas, David, 70, 277
Douglas, George, 130
Douglas, Sir James, of Kelhead, 250
Douglas fir, 1, 17, 70, 277
Dower Report, 283
Drielochan, 343
Drumlanrig Castle, 97
Drummond, James, of Croftnappoch, 252
Drummond, William, 277
Drumnadrochit, 264
Drymen, 180
Duddon Company, 240, 247
Dulnain, 113, 185, 194, 276
Dumbar, Robert, of Newtoun, 198
Dumfriesshire, woodland cover, 65
Dunbar, Sir David, of Baldoon, 107

Dunbar, Patrick, of Sidderay, 335
Dunbartonshire, 65, 66, 73, 163, 227, 261
Dundee, 125, 126, 129, 180, 271
Dunkeld, 160, 163, 171, 173, 263, 276
Dunkeld, Little, 111
Duntreath, 180
Durness, 99
Duthil, 71, 151, 270, 293
dykes, boundary, 165–70, 313, 317, 373–4, 379; *col.pl.9b&c*

East Lothian, 49, 65
Easter Balcroy, 148
Easter Ross, 35–6, 69
ecosystems, 8, 10, 395
Edinbellie, 159
Edinburgh, 132–3, 227, 271
Edinburgh Castle, 80
Edward I, king of England, 40–1
Eglinton Wood, 176
elder, 2, 79
    water, 114
Elderslie, 78
Eliock Wood, Dumfriesshire, 103, 105
elk, 42
elm, 2, 3, 15, 67, 73, 74, 75, 114
    in folklore, 79
    in prehistory, 25, 26, 28–9
    uses, 92
    wych, 2
enclosure, 103, 105, 108, 117–18, 122, 390, 392
    15th–16th centuries, 39, 158, 159, 160
    17th–18th centuries, 160, 161, 162, 163, 169–90, 187
    boundary structures, 165–70
    Rothiemurchus, 313–14, 317
    Skye, 373–4, 379
English interests
    iron industry, 231, 232, 233–4, 237, 238–9, 240–1
    timber, 201, 203–4, 207–8, 211–14, 216–21, 321, 400; Rothiemurchus, 293–8
    see also naval timber
Enzie parish, 46
Eriska, 343
erosion, 14, 31
Erskine, John, 47
Eskdale, 174
Ettrick Forest, 39, 110
export of wood, 89, 192–3
    prevention, 193, 195

fairs *see* markets
Falconar, Hugh, merchant, 209
Falkirk, 180
Falkland Wood, 106
farm implements, wooden, 88, 90, 91, 269
farming
    effects of, 13–14, 19, 31, 46–7, 260
    in prehistory, 13–14, 31
    see also sheepfarming
Farquhar, Alexander, merchant, 329, 330
Farquharson of Invercauld, 111, 143–4, 221
Fasagh, 234

Faskally, 175, 181
Fearn, Abbey of, 319
felling, 172–9
    exemptions, 174, 176
    pinewoods, 182, 184, 186–7
    regulation, 160–1, 162
    seasons, 173, 175
fences, 165–7, 169, 170
Ferrier, James, 138
Field, David, 296–8
Fife, Earls of, 111, 143, 146–7, 221
Fife, woodland cover, 65
Fife Adventurers, 229
Finlanrig, 107
Finzean, 221
    bucket mill, 267
fir
    Douglas, 1, 17, 70, 277
    grand, 70
    noble, 70
    silver, 259
fir candles, 88, 149
Firbush, Lochtayside, 168
fires *see* burning
Fisher, James, of Inveraray, 173, 342, 343, 346, 347, 349, 350
Fisher, John, merchant, 252
Fisher, Robert, merchant, 347
Fleming, John, 345, 354
Fleming, Captain John, 201
Fleming, Lord, 159
Flisk, Fife, 72
floating of logs, 209, 222, 293, 317
    Glenorchy, 356–7
    Speyside, 152, 294, 298, 300, 304, 307
Flow Country, 12–13, 28, 289
folklore, 78–9, 374–5
food from woods, 99–100
Ford, Richard, and Company, 166–7, 179, 181–2
forest laws, 37, 135
    Danish, 401
Forest Parks, 281
foresta, 105
foresters (fosters), 157, 158, 188–90
forestry, 5, 67, 70–3, 277–89, 395, 398
Forestry Act (1919), 280, 281
Forestry Commission, 280–1, 283–4, 287–9
    restoration of native woods, 18
    Rothiemurchus, 318
Forfar, 180
forfeited estates, 139–40, 155–6, 203, 366
Forres, 152
Fortingall, great yew, 78
Francis, William, of Oumby, 217
Fraser, Thomas, of Struy, 207
fuel, wood as, 97–9

Gadgirth, Ayrshire, 108
Gage, John, merchant, 294
Gairloch, 129
Galbraith, Captain Arthur, 251, 345, 346, 351, 353, 354–5, 356
Galloway

INDEX

oakwoods, 2, 153
parks, 107–8
in prehistory, 29
woodland cover, 65
Galloway, James, 232–3
Gardner, Alexander, 190–1
Garioch, 46
Garmouth, 129, 151, 152, 201, 212, 213, 217, 218, 307
   shipbuilding, 89, 218, 220
Gartary Wood, 47
Gateside bobbin mill, 265, 266
gean (wild cherry), 2, 73, 74
General Report of the Agricultural State and Political Circumstances of Scotland (Sinclair), 21, 64–6
Ghiusachan, 287
Gilbert, John, 154
Girvan, Water of, 49, 153
glaciation, 25
Glacknadrochad, 343
Glamis, 70
Glaschoille, 337
Glasdeir, 343
Glasgow, 180, 227, 271
Glass, River, 209
Gleann na Guiserein, 205
Glen Affric, 8, 74, 75, 185, 194, 207, 208, 209, 287, 289; *col.pl.14a*
Glen Alladale, 337–8
Glen Aray, 137
Glen Arkaig, 342
Glen Avon, 66, 109
Glen Barrisdale, 204
Glen Cannich, 74, 115, 194, 207, 209
Glen Dee, 290
Glen Dochart, 67
Glen Einich, 290
Glen Einig, 325, 337
Glen Etive, 185, 194, 200, 202, 207, 352
Glen Falloch, 194
Glen Feshie, 194, 220
Glen Finglas, 99, 100, 106, 123, 390, 391–2; *col.pl.7a*
Glen Fuar, 347, 348
Glen Ghiubhsachain, 53
Glen Ghiubhsachan, 290
Glen Kinglass, 228, 244, 246, 342, 362
   ironworks, 202, 238, 345, 354
Glen Livet, 60, 143
Glen Lochay, 110–11, 342
Glen Loy, 139, 196, 198, 287
Glen Loyne, 209, 210
Glen Lui, 59, 151, 286, 342
Glen Lyon, 187
Glen Mallie, 285, 342, 343
Glen Moriston, 66, 74, 86, 93, 104, 139, 140, 148, 194
Glen Scaddle, 201
Glen Shee, 110
Glen Strae, 347, 348, 351, 361
Glen Tilt, 110
Glen Urquhart, 196
Glencairn, 196

Glencalvie, 152, 183, 325, 327, 335, 336, 337, 339
Glencarrie, 183
Glenchairnich, 150
Glencharnick, 185
Glencoe, 53, 56, 183, 194, 200, 202
Glenelg, 251
Glenesk, 53, 56, 57
Glenfalloch, 207
Glenfua, 203
Glengarry, 18, 60, 180, 194, 198, 201, 209–10, 287, 289
Glenisla, 104
Glenmore, 18, 184, 185, 191, 286, 287, 289, 290, 338
   exploitation, 131, 194, 217–20, 270, 294, 400
   sporting use, 276
Glenmore (naval brig), 89, 218
Glenorchy, 21, 59, 183, 189, 194
   exploitation, 194, 200, 202, 361–3; Irish, 187, 202–3, 246–63, 341, 342, 343
Glenorchy, Lord, 173
Glenorchy Firwood Company, 202–3, 345, 359
Glenshiel, 56, 58
Glenstrathfarrar, 75
Glentanar, 149, 152, 184, 186, 188, 189
   exploitation, 194, 221–2
   sporting use, 276
Glentrool, 284
Glenure, 342, 343
goats, 37, 103, 108–9, 113, 120, 127, 370, 371, 391
Gordon Castle, 129
Gordon, Duke of, 142–3, 184, 217, 219, 294, 298, 300
Gordon, George, constable of Ruthven, 292
Gordon, James, 48
Gordon, Sir Robert, of Straloch, 48, 49, 52, 54, 127, 152, 205, 234, 319, 321, 336
Gordon, William, merchant, 335
gorse, 97
Gortanbeg, 118
Graham, Sir John, 180
Grant, Elizabeth, of Rothiemurchus, 292, 302, 304–5, 311, 312, 314, 316
Grant, James, of Burnhall, 305, 307, 309
Grant, Sir James, of Grant, 127, 211–12, 213, 296, 300
Grant, James, of Rothiemurchus, 295–6, 297
Grant, Sir John, of Freuchie, 184, 196, 198, 321
Grant, Sir John, of Grant, 236
Grant, Sir John Peter, of Rothiemurchus (9th laird), 217, 302–5, 307, 308–9, 316
Grant, Sir Ludovic, of Grant, 112–13, 213
Grant, Patrick, of Muckrack, 292
Grant, Patrick, of Rothiemurchus (6th laird), 295, 297–8
Grant, Patrick, of Rothiemurchus (8th laird), 298, 300–1, 304
Grant, William, of Rothiemurchus, 298–9, 304
Grant, William Patrick, of Rothiemurchus, 305–7, 308, 309, 312–15, 316–17
Grants of Grant, 131, 149, 190, 211–13
Grants of Seafield, 70, 71
Grantown, 71
Grantully, 175

Graphis alboscripta, col.pl.4b&c
grazing animals
　effect on woodland, 18–19, 75, 102–3, 105, 110–15, 120–3, 270–1, 391–2, 393, 395; *col.pl.8a*
　exclusion, 158, 179, 256, 313–14, 315, 390
　Mesolithic Age, 11
　Middle Ages, 37, 40
　semi-natural woodland, 13, 14, 16
　summer grazing, 108, 110, 115, 116
　see also overgrazing; pasture
Great Glen, 74, 201
　survey, 279, 282
Great Michael, 81, 97
Great Wood of Caledon, 13, 20–4, 43–4
Greenock, 127
Grigor, John, 71
Gruinard Island, 53, 59
Gruinard River, 53, 59
Gualinliguish, wood of, 347, 348
Guislich, 315
gunpowder manufacture, 91, 93, 261, 266, 267–8, 269, 272

Haddo House, 129
hagberry *see* cherry, bird
haggs, 16, 171, 172, 352, 392
　pinewood, 186–7
haining, 16, 158, 171
Hamilton, Duke of, 165, 331
Hamilton, Sir William, of Preston, 166
Hardy, Marcel, 283
hawthorn, 2, 73, 79, 92
Hay, Sir Alexander, 193, 195
Hay, Sir Charles, 166, 176
Hay, George, of Kirkland, 199
Hay, Sir George, of Netherliff, 195, 199
Hay, Sir George (Viscount Dupplin; Lord Hay of Kinfauns; Earl of Kinnoull), 229–32, 236
Hay, John, of Carriber, 166
Hay of Yester, 159, 233
hazel, 2, 3, 11, 12, 34, 66, 67, 73, 74, 114, 284
　building, 35
　in folklore, 79
　in prehistory, 25, 26, 27, 28
　uses, 93–5, 99, 266
hazelwoods, Atlantic, 3, 14; *col.pl.4a*
Hebrides, 26
hemlock, western, 277
Hill, Aaron, 211
holly, 2, 3, 15, 67, 74, 75, 114; *col.pl.4b*
　in folklore, 79
　uses, 92, 98, 99
Holyrood Palace, 332–3
horses, 37, 103, 109, 179–80, 391
　draught, 179, 180, 184, 307
Horsey, Colonel, 211
Hugh, count of St Paul and Blois, 42
Humbie wood, 49, 154, 155
Hunter, Robert, 345
hunter-gatherers, 10–13, 28
hunting, 106
hunting parks, 106, 107

hunting reserves, 105–6
Hutton, Robert, timber merchant, 383–4

importation of wood, 78, 84, 124–34, 257, 260, 266, 269, 274–5, 281, 400
Inbhir nan-giubhas, 361
Inchcailliach, 154, 250
Inchketan, 343
Inchnadamph, National Nature Reserve, 75; *col.pl.8a*
Inchtuthil, legionary fortress, 32
industrialisation, 77, 259, 263–9, 271–4, 392
Innercheagich, 343
Innes, John, lawyer, 205
Innes, William, 205, 206
Innis Chonain, 350
Inshstomach, 190
introduction of species, 10, 17–18, 70–2, 271, 277, 394, 397
Inveraray, 129, 130, 162, 261–2
Inverdruie, 306, 307, 309
Invergarry, 238, 239, 342, 343
Invergaunan, 347
Inverghiusachan, 361
Inverguseran, 205
Inverie, 86
Inverinate, 119
Inverkinglass, 343
Inverkip parish, 166
Inverlochy, 324
Invermay, 172, 174, 175
Inverness, 133, 148, 152, 329
　Cromwellian citadel, 327, 328
Inverness-shire, 65, 66, 113, 121
Inverscaddell, 343
Iona, 35
Ireland
　Brehon Law, classification of trees, 79
　exports to, 200, 201
　ironworks, 399
　woodland history, 33, 398, 399
Irish, 341, 342–7, 400
　Glenorchy, 202–3, 341, 342, 343, 346–63
　and iron industry, 236–7, 238
　and tanning industry, 249–51
Iron Age, 13–14, 29, 31, 34, 389
iron-making, 179, 225–7, 228–48, 345, 354, 386, 392
　Abernethy, 212
　banning of ironworks, 195
　Ireland, 399
Islay, 74, 244

James IV, king of Scots, 81, 106
James VI, king of Scots, 160, 193, 195
Jedforest, 40
Jee, Nathaniel, 335
Jeffrey, John, 277
John, Lord of the Isles, 41–2
John of Fordun, 41
John of Lorn, 41
Johnston, William, wright, 154
juniper, 2, 25, 98
　in folklore, 79

428

Index

Keir, David, 385
Keith, Aberdeenshire, 39
Keith, East Lothian, 49
Kendall, Henry, 170
Kennacraig, 175, 181, 343
Kennedy, Alexander, woodkeeper, 376
Ker, Robert, portioner, of Gilmerton, 154
Kettleswell, William, 345, 354, 355
Kilbride, Glenaray, 351
Kilbuco manse, 92
Kilfinan, Argyll, 175
Killin, 107
Kilmahog, 66
Kilmallie, ancient ash, 78
Kilmartin, Argyll, 30
Kilmun, 343
Kilmure, 319
Kincardine, 65, 172, 179, 183, 196, 332
Kingairloch, 141–2
Kingston Port, 218
Kinloch, Skye, 367, 371, 374, 375, 376, 379, 380, 381, 382
Kinlochbeg, 251
Kinlochewe, 199, 229, 234
Kinlochleven, 53
Kinlochmoidart, 113
Kinloss, 304
Kinneil wood, 165
Kinross, 65
Kintail, 56
Kintyre, 116–17
Kinveachy, 276
Kippen, 74, 155
Kirkdale, Kirkcudbrightshire, 175
Kirriemuir, 148
Knapdale, 3
Knoydart, 204, 206, 342, 378
Kyleakin, 374, 375, 381
Kylerhea, 367, 375

Lanarkshire, 65, 163
Lancaster, John, Duke of, 21–2, 40
Langholm, 173
larch, 1, 259
    plantation, 17, 71, 72, 73, 271, 276, 277, 285, 393, 397; Rothiemurchus, 313, 314, 315
    uses, 83, 272
Lasswade, 172
Lauder, George, of Bass, 159
Lauderdale, eighth Earl, 9
laurel, 397
law, taking of timber, 134–5, 136, 141–2, 143; *see also* Acts
Lays of the Deer Forest (Stuart), 22
legislation *see* Acts of Parliament
Leith, 127, 134, 271
Leitir Fura, Skye, 364, 367, 368–9, 371, 372, 373, 374, 375, 380
Lennox, Duke of, 135, 159
Lenzie, 159
Leslie, John, 61
Letterbeg Wood, 83, 97
Letterewe, 234, 343, 345
Letters, wood of, 117, 343
Lettervallin, 343
Leuquhatt, 172
Leven, Earl of, 162
Lewis, Isle of, 29, 110
Liddesdale, 39
Lilburne, Colonel Robert, 323
Limekilns, 236
Lindsay, Sir David, 236
Linlithgow, 180, 227
Linnaea borealis, 397; *col.pl.2c*
Little Park, Kirkcudbrightshire, 165, 179
Liubeg, 183
Loch Achall, 205
Loch Alsh, 199, 229
Loch an Eilein, 10, 304, 314, 315
Loch an Nid, 53
Loch an Niel, 53
Loch Arkaig, 72, 86, 88, 113, 129, 139, 140, 187, 198, 286, 343
    exploitation of forest, 194, 196, 203–4
    iron industry, 236–7
Loch Awe, 60, 252, 342, 347
Loch Ba, Mull, 100
Loch Broom, 58, 59, 60, 113
Loch Broom, Little, 53, 58, 211, 390
Loch Carron, 199, 229, 232, 233
Loch Creran, 60
Loch Duich, 119, 199, 229
Loch Eil, 113
Loch Einich, 304
Loch Etive, 176, 347–8, 352, 361–2
Loch Ewe *see* Loch Maree
Loch Fleet, Sutherland, 72
Loch Fyne, 72, 73, 244, 253
Loch Fyne iron company, 179
Loch Gamhna, 304
Loch Garry, 74
Loch Garten, 29
Loch Goil, 283
Loch Insh, 152
Loch Lee, 53
Loch Leven, Highland, 53, 113, 199, 200
Loch Lochy, 113
Loch Lomond, 115, 116, 154
Loch Lomond Stadial, 25
Loch Long, 283
Loch Loyal, 49
Loch Maree, 14–15, 75, 194, 195, 199, 236
    iron industry, 229, 231–6
Loch Morar, 343
Loch Morlich, 60
Loch na Sealga, 53, 206, 211, 390
Loch Ness, 66, 74, 148
Loch Pityoulish, 29
Loch Rannoch, 48, 60
Loch Sunart, 138
Loch Tarbert, 253
Loch Tay, 48, 49, 275
Loch Tulla, 358, 359–60
Lochaber, 36, 80, 113, 198, 402
Lochans, 304
Lochawe, 173, 175, 343

Lochbroom, 129
Lochbuie, 247
Locheil estates, 75; *see also* Camerons of Locheil
Lochnevis, 343
Lochwood, Dumfriesshire, 283
Logie Almond, 265
Lorn Furnace Company, 240, 241–3, 244–8, 261, 361
Louch Hourn, 204
Lovat, Simon, Lords Lovat, 139, 208, 278, 279, 280, 285
Lovat estate, 276
Lummis, John, 209, 216, 295–6
Luncarty, 265
lynx, 42

McCorquodale, William, forester, 72, 262
McDonald, Alexander, of Aikbrechlan, 251
MacDonald, Alexander, factor, 384
Macdonald, Alexander, of Glengarry, 180
MacDonald, Alexander, Portree, 377
Macdonald, Allan, of Morar, 342, 343
MacDonald, Angus, woodkeeper, 371
Macdonald, Cicely, 79
Macdonald, Donald, of Kinlochmoidart, 239
MacDonald, Donald, of Sleat, 377
MacDonald, Sir James, 372–3
Macdonald, John, of Invergarry, 342, 343
Macdonald, John, of Morar, 342, 343
Macdonald, Ranald, of Clanranald, 239
MacDonald estate, Skye, 155
MacDonalds of Sleat, 141, 364, 366, 377–8, 381
Macdonell, John, 238
Macdonell of Glengarry, houses, 85, 86
MacGregor, James, 300
Machan, James, 181
Mackenzie, John, 129
McKenzie, John, of Gairloch, 229
Mackenzie, John, of Applecross, 239
Mackenzie, Kenneth, of Dalmore, 151
Mackenzie, Thomas, 149
Mackenzies of Kintail, 195, 199, 229, 232–4, 235–6
Mackie, Andrew, 154
Mackie, John, portioner, of Larbert, 154
MacKinnon, Christopher, 375, 376
Mackintosh, Sir Aeneas, of Mackintosh, 220
Mackintosh, Sir William, of Torchastell, 198
McLauchlane, John, 250
MacLean, Allan, of Inverscaddell, 342, 343, 345
MacLean, John, of Ardgour, 342, 343
Maclean, Neil, 311, 378
MacLeod, James, of Raasay, 372
McNeill of Barra, 250–1
MacQueen, Archibald, 369
Maine, Robert, 345, 354
Mains of Park, Glenluce, 166, 176
Maitland, John, of Eccles, 134
Mamlorn, Forest of, 106, 110
Mamore, 53, 56
maps
    16th–17th centuries, 47–64
    woods' portrayal on, 48–53, 336
    see also individual mapmakers

Mar, Earls of, 106, 178
Mar, Forest of, 111, 112, 143–6, 221, 270
    exploitation, 194, 222
    sporting use, 276
markets, burgh, 147–9, 153
Mary, Queen of Scots, 45, 147–8
Mason, Captain John, 184, 196, 321
Masterton, Francis, 171, 178
Mauchline, Ayrshire, 166
Meaghlaich, 337
Mechanach, 118
Meggernie, 131, 186, 194
Melville, Fife, 70
Menteith, 238, 250, 251
Menzies of Weems, 189–90
Mesolithic Age, 11–13, 28
Methven Wood, 135
Middle Ages, 34–44, 388, 389, 398
Midlothian, 49, 65
Miles, Hugh, 24
military depredations, 40–1
Millerghead, 86
Moidart, 93, 176, 239
Monaltry, 146
monastic lands, 40, 93, 157
Moneses uniflora, 72, 397; *col.pl.2b*
monkey-puzzle, 70
Monro, George, of Culraine, 173
Monteath, Robert, 73, 114–15, 171, 254, 255, 257
Montrose, 66, 126, 263
Montrose, Duke of, 108, 172, 238, 250
Montrose, Marquis of, 180
Monymusk, 265
Monzievaird, 175, 180
moon
    effect on regrowth, 173, 173n
    and felling, 374–5, 400
Morar, 196, 198, 342
Moray, woodland cover, 65
Moray, Earls of, 70, 106, 123, 257, 391
Morsaig, Skye, 367, 371, 380, 382
Morvern, 117–18, 136, 138
Muckairn parish, 62–3, 244, 256
Mudalach, Skye, 366, 367, 375, 381, 386
Mugdock, 108, 172, 175
muirburn, 35, 75, 115, 160, 187, 392
    causing fire damage, 190, 191
Mull, 74, 100, 136, 138, 244, 247
Murchison, Donald, 343, 345
Murphey, Roger, 181, 202, 251, 342–6, 347, 348–52, 353–4
Murray, Sir Alexander, of Stanhope, 170, 345
Murray, Sir Patrick, of Ochtertyre, 180
Murray, William, 277
Murthly, 277

Nairn, 65, 152
Nairne, David, 22–3, 276, 285
Napier, John, of Merchiston, 159
Napier Commission, 316
Napoleonic Wars, 129, 221, 257, 274, 304
Narrachan, 343

Index

National Parks, 283, 284
National Trust for Scotland, 286–7
native trees, 1
native woodlands
    before 1500, 20–44
    1500–1920, 45–76
    changing character, 3
    definition, 1
    destruction, 5, 6, 7; measurement, 10
    extent, 1
    restoration, 1
    types, 2–3; in prehistory, 26, 27
Nature Conservancy, 283, 287–8, 289, 318
nature reserves, 283, 284
naval timber, 196–7, 211, 321, 322–3, 330–1
    Strathcarron, 197, 207, 323–32
Neolithic Age, 13, 29–31
Netherlorn, 362
Nethybridge, 191
New Forest, 33
Newhall, East Lothian, 74
Newland Company, 241
Nicholson, Alexander, 141, 372
Nicholson, Sir John, 172
Nicolson, John, 369
Nixon, Edward, 345–6, 351, 354, 356
Noble, William, wright, 154
Norway, 47, 99, 100
    imports from, 124–6, 127, 129, 131

oak, 2, 3, 15, 34, 66, 80, 114, 259, 284; *col.pl.11a,13c*
    coppice, 72, 73, 84, 108, 242, *col.pl.11a–b*;
        decline, 260–4
    cutting, 174–5, 177
    in folklore, 78, 79
    importation, 124
    pedunculate, 2, 71
    plantation, 71, 72–3
    in prehistory, 25, 26, 27
    sessile, 2, 71
    trade in, 152, 153
    uses, 261, 266, 269, 272; building, 35, 37, 38, 80–1, 97; shipbuilding, 42, 81–3, 97; *see also* tanbark
    Wallace's, 49, 78, 174
oakwoods, 22, 72–4, 393, 394; *col.pl.12a–b*
    animal fodder from, 99
    exploitation, 235–6
    loss of, 5, 6, 46
    management, 242–4, 256–7
    upland, 2, 3
Old Doveran, 104
Ollach, Skye, 366, 367, 369, 370, 376
Ord, Skye, 367, 368, 371, 374, 378, 380, 381, 382, 383, 384, 385
Ordie Mill, 265
Ordnance Survey, 48, 57, 61, 63–4, 68, 337–8
Orkney, 26, 30, 128, 331
Ormiston, 49, 172, 174, 175
Osborne, William, 218–19, 220
Ostaig, 376, 383, 387
overgrazing, 66, 67, 123, 392, 393
oxen, 179, 184

Paisley, 382
pan-wood, 39
Panmure House, Angus, 126
pannage, 37, 40
Pannanich wood, 150, 222
parkland, 105, 106–7, 165
Parsons, Benjamin, 201, 293–4
pasture
    conversion of wood to, 19
    summer, 108, 110, 115, 116
    woodland as, 102–23, 390–1; *see also* wood pasture
PAWS, 1, 17–18, 61
peat bogs/formation, 12, 13, 28, 30, 36, 58, 389
Peebleshire, woodland cover, 65
Pencaitland wood, 49, 154, 155
Penn, Sir William, 327
Pepys, Samuel, 327
Perrott, Andrew, 296–8
Perth, 133, 180, 227
Perth, Duke of, 139, 155–6
Perthshire, 163, 172, 227, 261
    hill grazing, 115
    oakwoods, 2, 73, 80
    in prehistory, 26
    woodland cover, 65, 66
Peterken, George, 289
Pett, Phineas, 197, 325–9, 330, 331
Pictish period, 34–6
pigs, 40, 99
Pinchot, Gifford, 5
pine, 15, 66, 67, 259, 397
    Austrian, 271
    Corsican, 71, 314
    in folklore, 79
    imports, 124
    lodgepole, 5, 17, 70, 277, 398
    in prehistory, 26, 27, 29
    trade in, 149–51, 152
    uses, 84–90; building, 35, 84–8; shipbuilding, 88–9, 193, 196–7, 201–2
    see also Caledonian pine; Scots pine
pine marten, *col.pl.15a*
pinewoods, 3, 4, 72, 74–5, 285–6; *col.pl.14a*
    conservation, 286–8
    exploitation, 193–224, 270–1
    loss of, 5, 6, 43, 53, 58–9, 339
    management, 182–8, 393, 394
    regeneration, 145, 182, 186, 187, 216, 217, 220, 223–4, 270, 271
    restoration, 18
pit props, 89, 91, 92, 271
Pitlochry, 180, 264
place-names, 43, 46, 205, 319, 389–90
plane *see* sycamore
plantation, 17–18, 70
    18th–19th centuries, 67, 68, 70–3, 277–8, 394
    20th century, 282
    contemporary, reasons for, 34
    non-native conifers, 5, 6, 8, 17, 71, 271, 281–2, 283, 284–5, 288
    see also forestry
Plantations on ancient woodland sites *see* PAWS

Plater, Forfarshire, 39–40
Plora wood, 166
podsolisation, 30
pollarding, 3, 165; *col.pl.7a,8b,13b–c*
    for fodder, 99–100
Pont, Timothy, 14–15, 47, 75, 195, 319
    maps, 47–8; Highlands, 50, 52–3, 56, 57;
        Lowlands, 49
poplar, 66, 74, 272
Portree, 371, 376, 378, 379
Poullich, 343
Pressmennan Wood, 49, 90, 97, 154, 159, 283
Prestonpans, 126
Primrose, Archibald, 236
prosecutions for woodland offences, 140–1, 189–91
    Rothiemurchus, 301
    Skye, 140–1, 371–2
Ptolemy, 20
pyroligneous acid works, 77, 91, 263, 268

Quoich, 150

Rackham, Oliver, 289
railways, use of wood, 77, 89, 91, 272, 314–15
Rannoch, *col.pl.3*
    Black Wood *see* Black Wood of Rannoch
    pinewoods, 155
Rassal Wood, 90
Rawlinson, Thomas, 239
Reay, Lord, 173
Red Smiddy, 235
regeneration, pinewoods, 182, 186, 187, 216, 217,
    220, 223–4, 270, 271, 393, 394
    Rothiemurchus, 302, 305–6, 310, 311–12, 314,
        318
Renfrewshire, 39, 65, 163
resin, 89
Rhidorroch, 194, 205, 206, 319
rhododendron, 397
Rhododendron ponticum, 72
Riddell, James, 117
Rio de Janeiro environment conference, 6
Ritchie, James, 284
Robert I, king of Scots, 41
Robertson, Alexander, of Faskally, 180
Robertson, Alexander, wright, 378
Robertson, James, 66–7, 74, 88, 93, 109, 296
Robertson, Rev. James, 140
Robertson, William, merchant, 200
Robertson of Struan, 149, 186
Robinson, Roy, 283
Roman period, 20, 21, 22–4, 32, 34, 388
Romisdale, 378
Roosevelt, Theodore, 5
ropes, 88, 91, 92–3
rose
    dog, 2
    guelder, 2
Rose of Kilravock, 331
Roslin wood, 49, 154, 155
Ross, Alexander, of Invercarron, 321
Ross, General Charles, of Balnagown, 205, 207, 336
Ross, David, of Balnagown, 325–9, 331–5, 338

Ross, Colonel David, of Balnagown, 197, 323, 329
Ross, Farquhar, Earl of, 319
Ross, John, Earl (Lord of the Isles), 364
Ross, John, of Little Tarrell, 324
Ross, Robert, forester, 262
Ross, Walter, of Invercarron, 327
Ross, William, of Invercarron, 335
Ross, William, Lord, 335
Ross-shire, 65, 68, 69
Rossdhu, 174, 175
Rosslyn chapel, 78
rotation, 108, 158, 178–9, 246, 254, 392
Rothiemurchus, Doune of, 86, 88, 290, 302, 304,
    307, 310, 316
Rothiemurchus Forest, 8, 10, 74, 201, 286, 288,
    290–2
    conservation, 293
    dykes, 313, 317
    entail, 301, 302–4
    exploitation, 131, 194, 216–17, 270, 292–301,
        302–15
    extent, 290–1, 295–6
    management: 18th century, 301–2; 19th
        century, 305–7, 312–17
    sawmills, 151, 152, 293, 300, 306, 307, 309,
        311, 316
    sporting use, 276, 317
    20th century, 317–18
    value, 307–8, 311–12
roup agreements, 164
rowan, 2, 3, 67, 73, 75, 114; *col.pl.13b*
    in folklore, 78, 79, 80
    in prehistory, 25
    uses, 92, 99, 266
Rowardennan, 72
Roxburghshire, 39, 65
royal forests, 37, 40, 42, 105, 123
Royal Scottish Forestry Society, 286
Roy's Military Survey of Scotland, 53, 57, 59–64,
    244, 290, 336, 337
Rum, 29
Ryfylke, 125, 126, 127

Sage, Alexander, 120
St Andrews Cathedral, 97
sale deeds, 158–9, 160–1, 163–4, 171
Salen, bobbin mill, 264
Saltoun Wood, 49, 97, 181
saltpans, 39
Sandelands, Andrew, 322
Sandeman, George, joiner, 186
saugh *see* willow
sawmills, 149, 150–1, 152, 184, 201, 209, 332
    Rothiemurchus, 151, 152, 293, 300, 306, 307,
        309, 311, 316
    Skye, 379
Scalpay, Skye, 367, 371, 372, 379
Schiehallion, 46
Schwappach, Adam, 263, 276
Scone, 262, 277
Scot, Sir John, of Scotstarvet, 47
Scots pine, 2, 3, 74–5, 103, 259, 270, 284–6;
    col.pl.2a,6a

Index

clear-felling, 5, 6
   plantation, 71, 72
   in prehistory, 25–6, 29
   uses, 84–6
Scottish Arboricultural Society, 277–80, 283, 286
Seaforth, Earl of, 119, 232–4, 235–6, 326
Seaforth estate, 343, 345
Selkirkshire, woodland cover, 65
Sellar, Patrick, 83, 96
semi-natural woods, 1–2, 8, 259, 283, 391, 394, 398; *col.p.1*
   emergence, 14–15
   loss, 388, 394, 395
   management, 13–17
   plantation in, 17–18
Sempill, John, of Aitkinbar, 199–200
servitudes, 141–7, 149
Severus, 21, 22, 41
Sgianadin, Skye, 367, 371, 372
sheep, 37, 40, 75, 99, 103, 108–9, 114–15, 123
sheep farming, 40, 67, 69–70, 187, 260, 393
shelter, woodland as, 102, 164–5, 377
Shetland, 26, 128
Shiel, River, 56
shielings, 108, 109–10, 111, 112
Shin, River, 152
shipbuilding
   Middle Ages, 41–2
   17th century, 193, 196–7, 201–2
   18th century, 218, 220
   20th century, 272, 273
   Vikings, 35–6
   woods for, 81–3, 88–9, 92, 97
   see also naval timber
shuttles, 265–6, 272
Sibbald, Sir Robert, 20–1, 293
Sighthill Tree, 174
Sinclair, Sir John, 21, 64–6
Sinclair, Sir John, of Ulbster, 370
Sinklers, Alexander and Hugh, 154
Sites of Special Scientific Interest, 284
Skail, wood of, 83, 226
Skye, 29
   MacDonald woods, 364–87; exploitation, 374–85
Sleat, woods of, 364–85
Slisgarrow, 198
sloe *see* blackthorn
Small, Ensign James, 186
Smith, David, 188–9
Smith, John, of Doune, 354
Smith, John, shipwright, 217
Smith, Patrick, of Braco, 324
Smith, Robert, ecologist, 283
Smith, William, ecologist, 283
snowberry, 72, 397
Somerville, Dr William, 279, 280
souming, 108–9, 391
Speed, John, 196
Spey, River, 52
Speyside, 113, 151, 212
   Middle Ages, 43
   pinewoods, 183, 185, 201; exploitation, 196, 211–20, 270
Spinningdale, 152
spokewood, 254, 261, 269
sporting estates, 275–6
Spott, 154
Spreull, John, 341
spruce
   imports, 124
   Norway, 1, 8, 17, 71, 72, 259, 271, 285, 393, 397
   Sitka, 1, 5, 8, 17, 70, 71, 277, 278, 281, 285, 398
squirrel, red, *col.pl.15b*
Staffing, John, 154
Steven, H. M., 286, 288
Stewart, Duncan, of Appin, 199–200
Stewart, John, of Grantully, 163–4
Stewart of Appin, 200
Stinchar, River, 49, 153
Stirling, 180, 227
Stirling, Captain, of Keir, 282
Stirling Castle, 80
Stirling Maxwell, Sir John, 278, 279, 280, 285
Stirlingshire, 65, 73, 163, 227, 261
Stobie, Matthew, 372, 377
Strath, Skye, 366, 386
Strath Cuileannach, 337, 338
Strath na Sealga, 53
Strath Nairn, 45
Strath Oykell, 36, 152, 325, 336, 337
Strathaven, 143
Strathbogie, 46
Strathcarron, 36, 152
   exploitation, 185, 194, 319, 321–36, 338–9
   extent of woodland, 60, 336–8
   naval timber, 89, 197, 207, 323–32
Strathconon, 36, 93
Strathdon, 98
Strathgartney, 252
Strathglass, 48, 74, 207, 210
Strathmore, 121
Strathnaver, 49, 50–1, 69, 75, 97, 152–3, 402
Strathnaver, Lord, 331, 335
Strathspey, 23, 29, 43, 48, 131, 152
Stronshira, 169, 170
Strontian, 345
Struan estate, 149
Stuart, C. E. and J. S. (Sobieski Stuarts), 22
Stuart, John, of Blackhall, 166
Sunart, 167, 170, 250, 267, 289; *col.pl.10–12*
sustainability, 5–9, 390, 392–3, 394, 398
   loss of, 18–19
Sutherland, 69, 105, 119–20
   hazelwoods, 3
   in prehistory, 23, 26
   woodland cover, 65
Sutherland, Earl of, 331
Sutherland, William, merchant, 335
Suttie, George, merchant, 334
Swan, John, tanner, 380
Sweden, imports from, 124, 126, 129
sycamore, 70, 72, 74, 266, 268; *col.pl.13a*
Sylvius, Aeneas, 39, 41
Symson, Robert, 199

Tain, 324

tanbark, 77, 84, 91, 92, 203, 227, 248–53, 260
tanning, 93, 154, 155, 225, 227, 248–53, 375, 392
Tansley, Arthur, 283
tar, 89, 197, 335
Tarbat, George, Lord, 205–6
Tarbet, Easter Ross, 69
Tarfside, 53, 56
Taynish, 244
tenants, 401–2
    removal of timber by, 134–47, 155–6
    Skye, 140, 371–2, 374–5, 376–7
    transport provision by, 180, 181
    wood management, 157–8, 160
Tentsmuir, 35
Terrioch, Ayrshire, 238
textile industry, 91, 92, 264–6, 271–2
thorn, 3, 74
Tillydown, Banffshire, 176
timber, 179
Tiree, 138
Tokavaig, Skye, 141, 367, 368, 369, 371, 372, 373, 374
Tomnagrew, Dunkeld, 160
Torwood, 21, 49, 52, 78, 80
    felling, 174, 175
Torwoodhead, 154
Tota-thaoig, Skye, 366
trade
    globalisation, 271, 274–5
    industrialisation, 259–60, 271
    wood, 78, 147–56; external market, 192–257; *see also* export; import
    wool, 40
transport of wood products, 180–1
Traquair, Earl of, 166
tree-line, 389–90
Trossachs, oakwoods, 2
Trotternish, Skye, 30, 83, 366, 377
Tucker, Thomas, 324
Tulliallan, 47
Tullibardine, Earl of, 321, 322, 323
turnery ware, 91, 92, 93, 266
turpentine, 89
Tweeddale, Earl of, 166
Tweeddale, Marquess of, 162, 164
Twenty Shilling Wood, Comrie, 72
Tyninghame, 70, 275

Udwart, Nathaniel, 232–3
Uilder, Skye, 366
Uist, North, 83, 378, 384
uses of wood, 77–101, 264–9, 271–3
Uweth, Fife, 40

value of trees, 97, 155, 347
Vera, Franciscus, 11, 33
Victoria, Queen, 224, 278
Vikings, 35–6, 319

Waldseemüller, Martin, 20, 21
Walker, John, 366–7

Wallace's oak, 49, 78, 174
wattle construction, 35, 93–5
wattle fence, 165–7, 169, 170
Wellingtonia, 70
Wentworth, Alexander, Lord MacDonald, 370
West Lothian, woodland cover, 65
Wester Ross, 53, 58, 113, 227
wet woodlands, 3
whin, 97
whitebeam, 2, 114
    rock, 2
whitethorn, 114
Wigtown, Earl of, 166, 180
wildwood, 11–13, 14–15, 394
Wilkie, Thomas, forester, 261–2, 263
willow, 3, 66, 73, 74
    apple-leaved, 114
    arctic, 75
    bay-leaved, 114
    crack, 114
    dark-leaved, 2
    downy, 2
    eared, 2
    in folklore, 78, 79
    goat, 2, 3, 114
    grey, 2, 74, 114
    mountain, 2
    net-leaved, 2
    in prehistory, 25
    tea-leaved, 2
    uses, 92–3, 266, 272
    whortle-leaved, 2
    woolly, 2; *col.pl.6b*
wintergreen, one-flowered (*Moneses uniflora*), 72, 397; *col.pl.2b*
Winton, 49
wolf, 122
wood ants, 395–7
wood-leave, 138, 141
wood pasture, 3, 11–12, 16, 100, 102, 106, 123, 164–5, 171–2, 390; *col.pl.7a&b*
woodland cover, statistics, 64–5, 258–9, 281, 388
World Commission on Environment and Development, 6
World War I, 279–80, 281
wych elm, 2
Wylie and Company of Ardersier, 381–3

Yester, 49, 70, 74, 159, 162, 166
    Castle Wood, 74
    felling, 174, 176
Yetholm Loch, 14
yew, 2, 114
    in folklore, 79
    great yew at Fortingall, 78
Yggdrasil, 78
York Buildings Company, 89–90, 127, 129, 207–8, 211–12, 296, 345

Zuckerman Committee, 281